Friedrich G. Barth Axel Schmid (Eds.)

Ecology of Sensing

Springer
Berlin
Heidelberg
New York
Barcelona
Hong Kong
London
Milan
Paris
Singapore
Tokyo

Friedrich G. Barth · Axel Schmid (Eds.)

Ecology of Sensing

With 100 Figures

Springer

Professor Dr. FRIEDRICH G. BARTH
Professor Dr. AXEL SCHMID
University of Vienna
Biocenter
Institute for Zoology
Althanstr. 14
1090 Vienna
Austria

ISBN 3-540-66901-9 Springer-Verlag Berlin Heidelberg New York

Library of Congress Cataloging-in-Publication Data
Ecology of sensing / Friedrich G. Barth, Axel Schmid (eds.)
p. cm.
Includes bibliographical references (p.).
ISBN 3-540-66901-9
1. Senses and sensation. 2. Animal ecophysiology. I. Barth, Friedrich G., 1940- II. Schmid, Axel, 1953-
QP435.E26 2001
573.8'7-dc21

Springer-Verlag Berlin Heidelberg New York
a member of BertelsmannSpringer Science+Business Media GmbH

Printed in Germany

Cover illustration by H. Bleckmann ; color-coded water velocity behind a swimming fish
Cover design: Erich Kirchner, Heidelberg
SPIN 10723236 31/3130xz - 5 4 3 2 1 0 - Printed on acid-free paper

Introductory Remarks

For good reasons we think of the brain as the master organ shaping the behavior of an animal. Yet both the central nervous system and behavior are at the mercy of sufficient and adequate sensory information. This information is provided by sensory systems responding to different forms of energy and representing the biological interfaces between the environment and the behaving organism.

As has been known for a long time, there is a considerable degree of filtering, biasing, and distortion of the theoretically available information, starting way out in the sensory periphery and continued at the various stages of central nervous processing. Many kinds of foveae and distorted cortical or other sensory maps (to name just two examples) reflect the sophistication in the evaluation of the biologically significant aspects of complex stimulus patterns. Much of this filtering must have evolved under the constraints of the specific physical and chemical conditions prevailing in the natural species-specific habitat of an animal, in the marine environment of a planktonic crustacean in the same way as in the desert environment of a scorpion. As a consequence of evolution, in many cases the functional properties of sensory systems nicely match at least some of the properties of the stimulus patterns encountered under behaviorally relevant conditions. The detailed spatial and temporal characteristics of these stimulus patterns may often be of frightening complexity. However, the information actually used by the animal for the guidance of its behavior has often proved to be much simpler.

Sensory ecology in its widest sense deals with the acquisition of information and the ways in which an organism responds to sensory information to organize its interaction with its environment. The information addressed here informs about actual environments and habitats as experienced by an animal in its own individual life. It differs from genetically stored information, which reflects the environments experienced by past generations.

The ideas behind *Ecology of Sensing* are not new, and the reader is particularly referred to the seminal book by Dusenbery (1993). However, with all the knowledge and understanding now available of the basic principles at work in many sensory organs, and considering the much-increased appreciation of the need to experiment and observe under natural conditions, the ecology of sensing is more likely than ever before to have a great future. Among the first advocates of sensory ecology was Jakob von Uexküll who published his *Umweltlehre* and his *Umwelt und Innenwelt* der Tiere as early as 1909. Von Uexküll always stressed the uniqueness of the sensory worlds in which different animal species find themselves living and, as a consequence, underlined the importance of knowing the particular features in the environment relevant for them.

The present book is an outcome of a symposium organized by us in Vienna in 1999. Its 16 chapters exemplify the diversity of the problems to be dealt with when we consider the different constraints and opportunities which are associated with the uptake and use of stimuli representing different forms of energy. The book brings together experts for the different sensory modalities. It stresses the events taking place way out in the sensory periphery where the organism is exposed to and comes into contact with its environment. The approach taken is more adequately described by *Ecology of Sensing* than by sensory ecology because the primary concern is sensory biology in relation to ecology, instead of the other way around.

Sections II to VII of the book are devoted to different stimulus modalities, whereas Section **I** covers aspects of general relevance for our topic. Thus, in the first chapter David Dusenbery develops a big picture of sensory ecology, leaving no doubt about the value of a rigorous quantitative understanding of the underlying physics. He demonstrates how much can be learned by applying basic physical principles to phenomena as complex as the motility of bacteria, the mechanism of gradient detection, and the employment of pheromones for mate attraction. Similarly rigorous physical approaches are found in the chapters contributed by Henry Bennet-Clark (II.1) and Joseph Humphrey, Friedrich Barth, and Karl Voss (III.1). The second chapter by Lars Chittka and Adriana Briscoe takes insect color vision to draw our attention to another general issue relevant to the ecology of sensing. This issue is the widespread habit of sensory ecologists (and many others) to rely on adaptive explanations without demonstrating the impact of an "adaptive" trait on fitness and without considering alternative explanations offered by evolutionary biology. In short, a good match between the functional properties of a sense organ or a sensory system and behaviorally relevant natural stimulus patterns should not be called "adaptation" without proper analysis.

The following sections of the book are devoted to sensory systems serving different stimulus modalities. They start with three chapters under the heading *Sound and Hearing* (Section **II**). Henry Bennet-Clark elaborates on impedance matching. This is a topic of fundamental importance in both acoustic signaling and hearing. It stresses the physics at the interfaces between the animal and its environment, which is crucial for a quantitative understanding of the chain of processes in acoustic communication. Heiner Römer focuses on the question of how the active space of acoustic communication can be increased, emphasizing the role played by the transmission channel in animals differing greatly in size (from grasshoppers to elephants), living in different natural environments, and using air or water for sound transmission. Hans Winkler concludes this section with a chapter on birds, the animal group likely to rely most extensively on acoustic communication. The reader will be given answers to questions asked about the specific consequences of ecology for signal transmission, signal properties, broadcasting tactics, and natural habitat use.

The section on *Medium Flow and Vibrations* (Section **III**) is devoted to arthropod motion-sensing hair sensilla, vertebrate vibration communication, and the fish lateral line. In a joint effort of engineering, biology, and mathematics, Joseph Humphrey, Friedrich Barth, and Karl Voss try to better understand the "design" principles of filiform hairs by asking which effects ecology may have had on the adaptive evolution of these extremely sensitive mechanosensors in terrestrial and aquatic habitats, respectively. The question is: what is physically possible, and which biological solutions do we actually find? Peter Narins then devotes his contribution to the ways in which vertebrates detect and utilize vibrational or seismic surface waves to communicate. Finally, taking the lateral line system as an example, Horst Bleckmann, Joachim Mogdans, and Guido Dehnhardt emphasize the importance of applying natural and ecologically meaningful stimulus patterns and noise conditions when striving to fully understand the specializations of a particular sensory system.

Next is *Light and Vision* (Section **IV**), that is the specializations and adaptations used by animals living in different photic environments. Simon Laughlin points to the relationship between sensory ecology and the energy budget of a sensory system. He examines how efficiency can be improved in insect compound eyes by matching the reception and the processing to the properties of the visual stimuli. Eric Warrant also concentrates on arthropod compound eyes, which show extreme adaptability. The author reviews the ways in which this most abundant of all eye types deals with the intensity and direction of natural illumination and how compound eye design differs in different types of habitat. In his chapter on the visual ecology of fish, Ron Douglas concentrates on the adaptation of visual pigments to different photic environments, ranging from the deep sea, with almost no daylight at all, to the variable colored waters of freshwater lakes and rivers. As it turns out variations in spectral sensitivity among species sharing the same habitat are still only little understood.

Problems in the sensory ecology of *Odors and Chemoreception* (Section **V**) are illustrated by Stefan Schulz, who discusses the structural diversity of arthropod pheromones and its relation to signaling against a background of chemical noise. He also emphasizes that it is still hard to extract general rules from the enormous variety of chemical communication systems. Joop van Loon and Marcel Dicke draw our attention to the fascinating complexity of the interactions between herbivorous and carnivorous arthropods and their selected plant species. The chemical world of the plants is a dominant factor in the lives of these arthropods (both the herbivores and their predators).The intriguing relationship of specific phytochemical profiles to chemoreception constitutes one of the major challenges in the ecology of sensing.

In *Hygro- and Thermoreception* (Section **VI**) the interaction between environmental constraints and sensory specialization are shown to be only little understood. Most of our knowledge on thermo- and hygroreception comes from studies on a few insects and a spider. Harald Tichy and Ewald Gingl relate data on the structural properties of thermosensory sensilla to their response characteristics and discuss different models that may explain hygroreceptive transduction mechanisms.

Among the senses alien to humans *Magnetic Field and Electroreception* (Section **VII**) are classical examples. Wolfgang and Roswitha Wiltschko tell us how the geomagnetic field, an omnipresent environmental factor, serves different animal groups as a reliable indicator of direction in oriented behavior. Unfortunately, the primary processes of magnetoreception are still rather obscure. This is different in electroreception, which is the topic of the contribution of Gerhard von der Emde. He introduces us to the electrosensory world of the weakly electric fish and the remarkable adaptations of the electrosensory system to the conditions met by the fish in its freshwater habitat.

Clearly, this book can provide only examples of a topic in sensory biology which is as rich and varied as are both the sensory systems used by different animals and the habitats in which they are living.

Many thanks go to the Austrian Academy of Sciences for making the Symposium from which this book originated both possible and pleasant. Its generous support is gratefully acknowledged. The cooperation of the authors and the help and patience of Dr.D.Czeschlik and the editorial staff of Springer-Verlag are much appreciated.

Wien and Lofer, May 2000 *Friedrich G. Barth*

Axel Schmid

Contributors

BARTH FRIEDRICH G.
Biocenter, Institute of Zoology, Vienna University, Althanstr. 14, A-1090 Vienna, Austria

BENNET-CLARK HENRY C.
Department of Zoology, University of Oxford, South Parks Road, Oxford, OX1 3PS, U.K.

BLECKMANN HORST
Institut für Zoologie, Universität Bonn, Poppelsdorfer Schloß, D-53115 Bonn, Germany

BRISCOE ADRIANA
Department of Molecular and Cellular Biology, Life Sciences South 444/1007 E. Lowell, Tucson, AZ 85721, USA

CHITTKA LARS
Zoologie II, Biozentrum, Am Hubland, D-97074 Würzburg, Germany

DEHNHARDT GUIDO
Institut für Zoologie, Universität Bonn, Poppelsdorfer Schloß, D-53115 Bonn, Germany

DICKE MARCEL
Laboratory of Entomology, Wageningen University, P.O. Box 8031, 6700 EH Wageningen, The Netherlands

DOUGLAS RON
Applied Vision Research Centre, Department of Optometry & Visual Science, City University, 311-321 Goswell Rd, London EC1V 7DD, U.K.

DUSENBERY DAVID B.
School of Biology, Georgia Institute of Technology, Atlanta, GA 30332, USA

GINGL EWALD
Biocenter, Institute of Zoology, Vienna University, Althanstr. 14, A-1090 Vienna, Austria

HUMPHREY JOSEPH A.C.
Department of Mechanical and Aerospace Engineering, School of Engineering and Applied Science, University of Virginia, Charlottesville, VA 22904-4746, USA

LAUGHLIN SIMON
Department of Zoology, University of Cambridge, Downing Street, Cambridge CB2 3EK, U.K.

MOGDANS JOACHIM
Institut für Zoologie, Universität Bonn, Poppelsdorfer Schloß, D-53115 Bonn, Germany

NARINS PETER M.
Department of Physiological Science, UCLA, P.O. Box 160606, Los Angeles, CA 90095-1606, USA

RÖMER HEINER
Institut für Zoologie, Karl-Franzens-Universität, A-8010 Graz, Austria

SCHMID AXEL
Biocenter, Institute of Zoology, Vienna University, Althanstr. 14, A-1090 Vienna, Austria

SCHULZ STEFAN
Institut für Organische Chemie, Technische Universität Braunschweig, Hagenring 30, D-38106 Braunschweig, Germany

TICHY HARALD
Biocenter, Institute of Zoology, Vienna University, Althanstr. 14, A-1090 Vienna, Austria

VAN LOON JOOP J.A.
Laboratory of Entomology, Wageningen University, P.O. Box 8031, 6700 EH Wageningen, The Netherlands

VON DER EMDE GERHARD
Institut für Zoologie, Universität Bonn, Poppelsdorfer Schloss, D-53115 Bonn, Germany

VOSS KARL
Department of Mathematics, College of Arts and Sciences, Bucknell University, Lewisburg, PA 17837, USA

WARRANT ERIC
Department of Zoology, University of Lund, Helgonavägen 3, S-22362 Lund, Sweden

WILTSCHKO ROSWITHA
Fachbereich Biologie der J.W. Goethe-Universität, Zoologie, Siesmayerstrasse 70, D-60054 Frankfurt a.M., Germany

WILTSCHKO WOLFGANG
Fachbereich Biologie der J.W. Goethe-Universität, Zoologie, Siesmayerstrasse 70, D-60054 Frankfurt a.M., Germany

WINKLER HANS
Konrad Lorenz-Institute for Comparative Ethology, Austrian Academy of Sciences, Savoyenstr. 1A, A-1160 Vienna, Austria

Contents

IV. Light and Vision

V. Odors and Chemoreception

VI. Hygro- and Thermoreception

VII. Magnetic Field and Electroreception

Index

I.1 Physical Constraints in Sensory Ecology

David B. Dusenbery

School of Biology, Georgia Institute of Technology, Atlanta, Georgia 30332 USA

Abstract
A "big picture" of sensory ecology is developed. The logic of biological investigation and the meaning of behavior and information are briefly addressed along with the importance of information to organisms. Ties between sensory ecology and physiology and behavior are stressed. The value of understanding physical constraints on what is possible is emphasized. The simplest physical environment is that of planktonic organisms living free of any surface, and consequences of physical constraints and biological generalities are described for such organisms, as examples. This provides new insights into why small bacteria are not motile, why most chemotactic bacteria are rods employing temporal mechanisms of comparison for gradient detection, why small planktonic animals do not employ pheromones for mate attraction, and explaining why male and female mating types are common among sexually reproducing species.

Key words Noise, signal, information, threshold, plankton

1 Introduction

Organisms are extremely complex mechanisms, and it is a challenge to gain even glimpses of understanding how they work. Jumars (1993, Preface) has pointed out that biologists generally tend to focus on the particular and ignore general relationships. This is unfortunate, because science should always strive for the most general understanding possible. This chapter attempts to place sensory ecology in the "big picture", to describe how it relates to other fields of biology, and to discuss some of the general relationships important to the field.

A fundamental question that must be addressed for any biological feature studied is whether organisms possess the feature because it is of adaptive value or whether it is accidental in the sense that it is a sideeffect of some other adaptation, an adaptation to a previous condition that no longer exists, or a pure accident of genetic drift and fixation during evolutionary history.

An important approach in answering this question is to consider what alternatives to the existence of the feature are possible. In other words, how is the evolution of organisms constrained? There are two basic types of constraints: (1) Since organisms evolve from preexisting organisms, the specific history of previous evolution provides one set of constraints. (2) Organisms must live in a particular environment governed by certain physical relationships, and the laws of physics and chemistry provide another set of constraints.

Even the understanding of evolutionary history requires sorting out adaptive changes in the past from the effects of genetic drift, and physical constraints are a major source of adaptive pressure - in the past as well as the present. Since all evidence indicates that the laws of physics and chemistry have remained unchanged since the beginning of life on earth, determining how these laws constrain evolutionary adaptations will help answer questions about both past and present adaptations.

The constraints of evolutionary history are difficult to determine and limited by the fact that they apply to only one lineage. In contrast, some physical constraints are universal, and understanding their implications provides powerful insights, applicable to all organisms. For example, the limited rate of diffusion in liquid water provides a strong constraint on the maximum size of organisms without special transport systems (Dusenbery 1996, pp.6-9; Withers 1992, pp. 571-572). Similarly, recognizing the conservation of atoms and energy and the second law of thermodynamics has provided powerful insights to understanding cellular metabolism, organismal physiology, and ecosystem processes. In sensory biology, understanding the laws of optics has similarly provided powerful insights into understanding many features of the eyes of vertebrates and insects and the differences between them (Land 1981).

What can be done in sensory ecology to exploit physical constraints to reveal general relationships that will improve understanding? In previous work, I illustrated how understanding physical constraints on how stimuli move through the environment (Dusenbery 1992, pp. 49-87) provides insights into the potential range (Dusenbery 1992, p. 239) and relative costs of communication by different modalities (Dusenbery 1992, p. 327).

In this chapter, I review some more recent analyses of physical constraints on small organisms dispersed in water and hopefully provide inspiration for others to exploit this approach further. First, however, let us consider what the field of sensory ecology encompasses.

2 What Is Sensory Ecology?

2.1 Information and Causation

All individuals interact with two kinds of "spaces" - a pool or flow of genetic information (contained in a population of similar organisms) and a material environment (consisting of both animate and inanimate components, with which materials, energy, and sensory information are exchanged). Ecology has generally focused on the exchanges of matter and energy, while sensory interactions have generally been studied as influences on behavior and functions of certain physiological systems (sense organs). The relatively new area of sensory ecology has emerged as more researchers focus on questions concerning information in the environment. What kinds of useful information are available?

This new focus emphasizes a distinction between matter/energy that can make things happen (like applying forces to move objects or doing work in the sense of thermodynamics) and matter/energy that cause changes only after their effects are amplified (e.g., by electronic amplifiers, sensory systems, or physiological effectors). The former influences constitute causal inputs from the environment to the organism, while the latter constitute informational or sensory inputs to the organism.

For example, light may play a causal role by providing energy to heat an organism, energy for photosynthesis, or (especially in the case of ultraviolet light) causing damage to tissues. In addition, organisms with appropriate sensory systems can respond to light in many other ways, employing it as a source of information. Often sensory systems are so proficient that organisms respond to much lower intensities than are required for significant causal effects. Confirmingly, marine biologists have determined that sufficient light for photosynthesis is limited to about the top 200 m of the ocean, while sufficient light for vision can extend down to depths of 1000 m, and both these limits define boundaries between distinct biological communities (Lythgoe 1979, p. 100; Jumars 1993, p. 136).

This kind of distinction (between information and causation) was at the heart of transforming biochemistry (with its focus on material and energy flows) into molecular biology (with its emphasis on information flows) and is currently transforming psychology, with growing emphasis on cognitive processes. (Traditionally, psychology lumped sensory and causal inputs together in both stimuli and reinforcers.) The development of sensory ecology is thus part of a more general trend.

2.2 Behavior

Another important concept is that of behavior. Precisely what is behavior? How should it be defined in biology?

Although organisms might evolve many different ways of storing genetic information, all known organisms employ DNA. From this and similar observations, we conclude that all known organisms share a common ancestry. Consequently, the distinction between organisms and nonliving systems is much clearer than any distinction among organisms, and the basic concepts of biology should be applicable to all organisms. However, on the rare occasions when students of behavior have bothered to define the word, "behavior" has usually been defined so as to require muscles or glands and thus exclude the vast majority of organisms from having behavior (McKechnie 1978; Zuriff 1985).

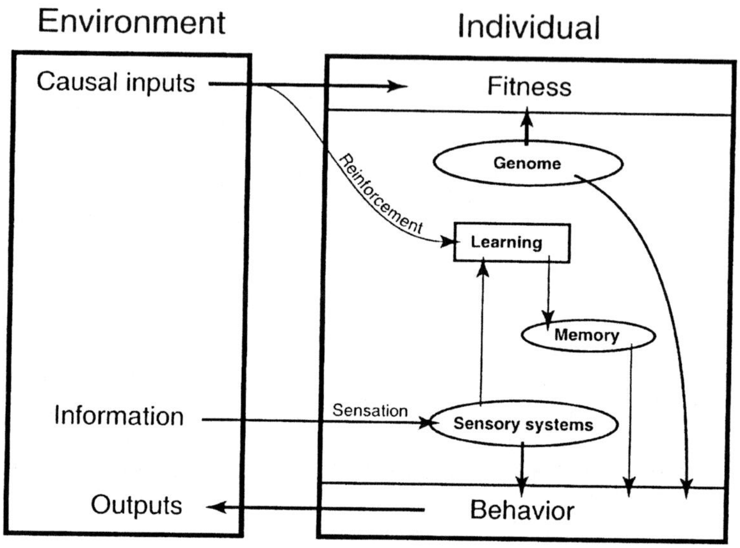

Fig. 1. Input-output relationships of the individual organism. Every individual interacts with two distinct "spaces": its material environment and the gene pool or flow to which it belongs. Behavior of the individual creates its outputs to the environment and its fitness in the environment controls its output (reproduction) to its gene pool. This figure illustrates the influences (*arrows*) between the environment and the individual and major pathways controlling behavior within the individual

An attempt to generalize the concept of behavior from its use in animal behavior studies to all organisms (Dusenbery 1996) has led to the conclusion that behavior constitutes those activities that generate outputs from the organism to its environment. Behavior includes those activities that change the environment either by moving the organism (or part of it) to a new location or modifying the present location. With this view, all the interactions between an organism and its environment are included within one of the three categories: causal and sensory inputs and behavioral outputs. These concepts are summarized in Fig. 1.

All three types of interaction involve states of the environment, and are thus interdependent because behavior indirectly changes both types of input (indeed that is behavior's function). By themselves, sensory inputs have no value to the organism; they gain value only when they are associated with important causal inputs in ways that allow predictions about the occurrence of particular causal inputs at another place or later time. The organism usually exploits these predictions by using them to control some aspect of its behavior, which changes its relationship with its environment, and this, in turn, leads to changes in both kinds of inputs. Thus, organisms are often engaged in chains of causation that cycle between these categories of inputs and outputs:

sensory → behavior → sensory → → → behavior → causal → → → fitness

Understanding *why* an organism responds to a particular sensation with a particular behavioral response (i.e. the *function* of the sensory system and the behavior) requires knowing the causal input later in the chain; and understanding *how* the sensory input is amplified and how it controls behavior (i.e. its *mechanism*) requires knowledge of physiology. Thus sensory ecology must be intimately involved with ecology, behavior, and physiology, as well as evolution.

2.3 Information Is Important

The laws of thermodynamics require that all organisms must obtain causal inputs, including energy and nutrients, but there is no such requirement for obtaining information. A fundamental characteristic of information is that it is easily ignored; it causes nothing to happen unless special mechanisms are present to amplify its effects. Consequently, sensory ecology might have covered only minor, idiosyncratic aspects of biology, and this probably was the case during early phases of evolution. However, we observe among extant organisms: (1) large investments in sensory systems in many bacteria and most animals, (2) frequent modification of external features (camouflage) to alter visual information, (3) daily vertical migration among many pelagic animals probably driven by visual predation (Mangel and Clark 1988), (4) bacteria that transmit signals (by bioluminescence) to vertebrates, (5) bacteria that communicate with one another by chemicals (quorum sensing) (Surette et al. 1999), and (6) flowering plants that make significant investments in sending signals to their pollinators. These observations demonstrate that during evolution, many (most? all?) organisms have found it advantageous to obtain information about their environment. In fact, one might conjecture that all extant organisms have some kind of sensory system and obtain some kind of information about their environment.

Whether or not this conjecture is true, it is clear that information about the external environment is an important resource to many organisms and especially to the large animals most familiar to us. There is also an accumulating body of work suggesting that plants obtain much more information about their environment than previously realized (Baldwin and Schultz 1983; Rhoades 1985; Kasperbauer 1987; Bradburne et al. 1989; Ballaré et al. 1990). Thus the study of the processes by which information is obtained (sensory ecology) is an important part of biology, and this is increasingly recognized as the development of electronic technology leads us to think about information as we design devices that respond to their environment.

3 Time Limits Sensitivity

Sensory ecology is often concerned with the magnitude of the stimulus that is necessary for reliable detection. This sensitivity sets limits on what information can be obtained, the distance over which information is available, and the costs of transmitting information. In some cases, basic physical principles allow us to set some limits on how low such thresholds can be. There are always limits, because all real systems have at least some noise associated with any process due to thermal agitation of the molecules they contain.

When the stimulus arrives in discrete packets occurring randomly in time, such as photons or diffusing molecules, fluctuations in the number of packets received also imposes a limit on sensitivity. This kind of limit is often analyzed by calculating a signal-to-noise ratio (S/N). Reliable detection requires that the ratio be well above 1, and a ratio well below 1 indicates that reliable detection is not possible (Dusenbery 1992, pp. 90-95). The signal is often represented by the increase in the average number (n) of packets received in some time interval, and the noise is often taken as the standard deviation (σ) of the number of packets expected in the same interval. When the packets arrive randomly in time, the number received in a given time interval has a Poisson distribution, and the standard deviation is simply the square root of the mean number.

If the average number of packets received is n_1 in the presence of the signal and n_2 in its absence, the signal-to-noise ratio is limited to

$$\frac{S}{N} \leq \frac{n_1 - n_2}{\sqrt{\sigma_1^2 + \sigma_2^2}} = \frac{n_1 - n_2}{\sqrt{n_1 + n_2}} \qquad \textit{Eqn. } 1$$

For estimating threshold, the reference state is $n_2 = 0$, $\sigma_2 = 0$, and

$$\frac{S}{N} \leq \frac{n}{\sigma} = \frac{n}{\sqrt{n}} = \sqrt{n} \qquad \textit{Eqn. } 2$$

The appropriate value of the signal-to-noise ratio will depend on what response criterion is used for defining the threshold, but any reasonable criterion will correspond to a signal-to-noise ratio near unity. Taking $S/N = 1$ as our criterion, $n = 1$. At threshold, one packet is received on average in the time interval.

Applying these concepts to the problem of estimating the threshold intensity (I_{Th}) for detecting the presence of light against no light, the best any receptor that fits within a sphere of radius r can do is to count the number of photons (n)

absorbed. For a parallel beam of light (such as direct sunlight), this can be expressed as

$$n = \pi r^2 f I t \qquad \textit{Eqn. 3}$$

where f is the fraction of light absorbed in passing through the area of the receptor (πr^2), I is light intensity, and t is the time taken to make the measurement. In this case,

$$\frac{S}{N} \leq \sqrt{\pi r^2 f I t} \qquad I_{Th} \geq \frac{1}{\pi r^2 f t} \qquad \textit{Eqn. 4}$$

using $S/N = 1$. Land (Land 1981, equation 14) gives a formula for estimating the rate at which individual receptor cells absorb photons, and could also be used.

For estimating the threshold concentration (C_{Th}) of a chemical, the best any receptor of a given size can do is to count the new molecules that diffuse to it from all directions. This can be done by measuring the fraction of receptor molecules that have ligands bound and disposing of the molecules by degradation or sequestration in order to maintain the concentration gradient and avoid confusing new molecules with those previously detected (Berg and Purcell 1977). Considering cases where fluid flow is not important, if the receptor fits within a sphere of radius r, the flux to the receptor is at best

$$n = 4\pi r D C \qquad \textit{Eqn. 5}$$

where D is the diffusion coefficient of the chemical and C is the number concentration of the chemical far from the receptor. Thus,

$$\frac{S}{N} \leq \sqrt{4\pi r D C t} \qquad C_{Th} \geq \frac{1}{4\pi r D t} \qquad \textit{Eqn. 6}$$

In both these examples, the signal-to-noise ratio improves with the square root of the time taken to make the measurement, and the threshold is inversely proportional to this time. Excluding other sources of noise, it is possible to detect an arbitrarily small signal if enough time or size is available. Thus, an estimate of the available time and size of the receiving surface is essential to an estimate of sensitivity.

Predictions of the optimal integration time for many sensory systems might be taken from the fact that the swimming speed of organisms from bacteria to whales is on the order of magnitude of 10 body lengths per second (Mann and Lazier 1991; Dusenbery 1996, p.45). Thus, an organism swimming through a stimulus gradient would experience a new sample roughly ten times a second and a sensory integration time near 0.1 s would seem appropriate. Likewise, if a visual or auditory system is designed to detect such movements, it should have a similar integration time. Observations indicate that, in fact, many sensory systems have integration times in this range.

Table 1. Constraints on size.

Stimulus	Mechanism	Constraint formulas	$2r$, µm
None	Diffusion	$\frac{D_m}{D_0} = \left(\frac{4\pi\eta}{kT}\right)^2 \left(\frac{1}{2}\right) u^2 \; r^6$	0.64
Chemical	Spatial	$\frac{S}{N} \leq \left(\frac{4\pi\eta}{kT}\right)^{1/2} (2\pi DC)^{1/2} \left(\frac{3}{L}\right) r^3$	0.58
Chemical	Temporal	$\frac{S}{N} \leq \left(\frac{4\pi\eta}{kT}\right)^{3/2} (2\pi DC)^{1/2} \left(\frac{u}{L}\right) r^6$	0.65
Light	Spatial	$\frac{S}{N} \leq \left(\frac{4\pi\eta}{kT}\right)^{1/2} (2\pi If)^{1/2} \left(\frac{2}{L}\right) r^{7/2}$	1.77
Light	Temporal	$\frac{S}{N} \leq \left(\frac{4\pi\eta}{kT}\right)^{3/2} (4\pi If)^{1/2} \left(\frac{u}{L}\right) r^{13/2}$	1.05
Light	Direction	$\frac{S}{N} \leq \left(\frac{4\pi\eta}{kT}\right)^{1/2} (\pi If)^{1/2} \; \alpha \; r^{7/2}$	1.24
Temp.	Spatial	$\frac{S}{N} \leq \left(\frac{4\pi\eta}{kT}\right)^{3/4} \left(\frac{(4\pi H_T)^3}{k^2 H_c}\right)^{1/4} \left(\frac{2}{L}\right) r^{13/4}$	0.74
Temp.	Temporal	$\frac{S}{N} \leq \left(\frac{4\pi\eta}{kT}\right)^{7/4} \left(\frac{(4\pi H_T)^3}{k^2 H_c}\right)^{1/4} \left(\frac{u}{L}\right) r^{25/4}$	0.69

In the constraint formulas, the parameters are grouped so that the term on the left is a performance factor (signal-to-noise ratio or ratio of diffusion coefficients with and without motility), the first term to the right of the equal sign involves hydrodynamics, the second term involves stimulus intensity, the third term involves gradient decay length (L) and/or relative swimming speed (u), and the right-most term involves the size of the organism (radius r). See Dusenbery (1997) for definitions of other symbols. The right-hand column gives the minimum diameter an organism can have and usefully employ the indicated mechanism, assuming best estimates of parameter values in Dusenbery (1997, Table 1) and $S/N = (D_m - D_0)/D_0 \geq 1$ for useful function.

4 Consequences For Small Plankton

For microscopic organisms that live in water free of any constraint on movement but viscosity (i.e., not attached to any surface and not containing magnets), Brownian motion causes continual random translation and rotation. The rates at which these occur set limits on the time available for obtaining information useful for guiding locomotion.

Such small organisms live in a world where viscous effects are far stronger than inertial effects. The Reynolds number is well below unity, and flow is laminar, never turbulent. In this situation, hydrodynamic drag can be rigorously calculated for simple shapes, and it is possible to relate speed to energy consumption. Using Stokes law for spherical shapes (Rouse 1961, pp. 212-216) and observing that specific metabolic rates (power per unit volume) are relatively uniform for all kinds of organisms, it can be demonstrated that speed of swimming should be proportional to size (Dusenbery and Snell 1995). With the best estimates of efficiency, the typical speed is estimated to be in the vicinity of 10 diameters per second, consistent with many observations (Dusenbery 1996, p. 45; Mann and Lazier 1991).

4.1 Size and Motility

In addition, the rates of translation and rotation caused by Brownian motion can be rigorously calculated (Berg 1993). Taking advantage of these relations, formulas have been derived for estimating the signal-to-noise ratio for detecting gradients of light, chemicals, and heat (Table 1).

These results lead to the surprising conclusion that there exists a sharp size limit (about half a micrometer diameter) below which free bacteria have no use for motility. A review of published literature (Fig. 2) indicated that the prediction is supported, and about 20% of free, nonmotile genera are below this size limit but the smallest of 97 motile genera was about 0.8 μm in length (Dusenbery 1997).

4.2 Consequences of Shape

Perrin (1934, 1936) derived expressions for the motion of ellipsoids subject to Brownian motion. Unfortunately, they involve definite integrals that have no solution in common functions, but approximations for high axial ratios are available, and have been much used to estimate the shapes of molecules (Tanford 1963; van Holde 1985). With computers it is now easy to evaluate these integrals, and these relations have been used to quantitatively evaluate the effects of constant-volume changes in shape (including small axial ratios) on several possible functions (Dusenbery 1998a).

As seen in Fig. 3, this analysis has led to the surprising conclusion that rod-like shapes with axial ratios greater than two (as have most motile bacteria) *increase* hydrodynamic drag. However, elongated rod-like shapes slow rotation caused by Brownian motion, and this increases the time available to measure concentration gradients. This provides large benefits in detecting gradients, especially by temporal comparison. This explains why rods with axial ratios greater than two are common among motile bacteria.

In addition, it explains why temporal comparison is the common mechanism of gradient detection, which is not necessarily superior to spatial comparison for spherical cells (Dusenbery 1998b).

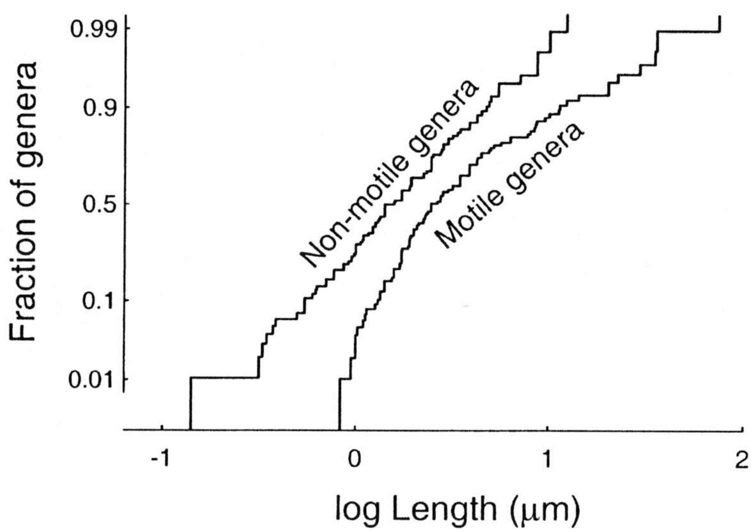

Fig. 2. Cumulative frequency plots of length distributions for motile and nonmotile genera of bacteria. The *stepped curves* represent the fraction of genera for which the length is less than the value on the horizontal axis. The *vertical axis* is a normal probability scale, which causes a cumulative normal distribution to fall on a straight line. Note that the small-size limit of the motile group is cut off more sharply than a normal distribution would be and remains above the predicted size limit of about half a micrometer. In addition, the motile group has more large genera than expected for a normal distribution. (Data from Dusenbery 1997)

Surface area

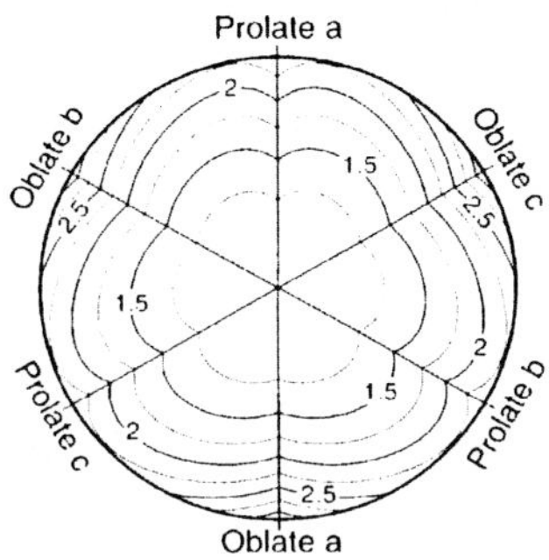

Contours are surface area of the body relative to an equal-volume sphere.

Largest surface area occurs with oblate or disk-like shapes:

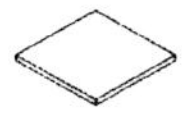

Diffusion

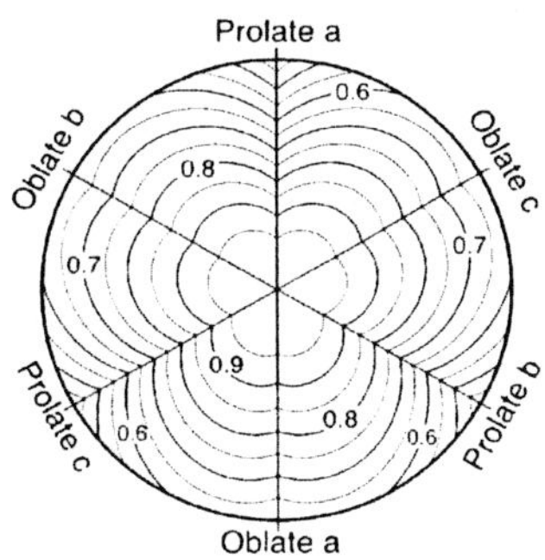

Contours are diffusion coefficient of the body relative to an equal-volume sphere.

Largest rate of diffusion occurs with spheres:

Smallest sinking rate with random orientations occurs with prolate or rod-like shapes:

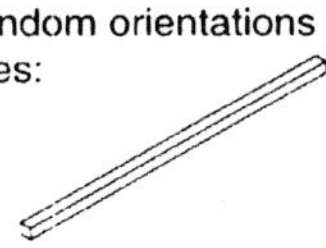

Drag

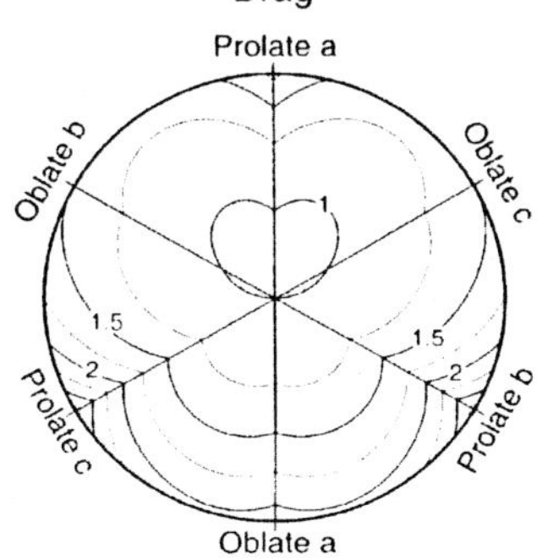

Contours are hydrodynamic drag of the body relative to an equal-volume sphere.

Smallest drag occurs with prolate or rod-like shapes of axial ratio 1.952 moving along the long axis:

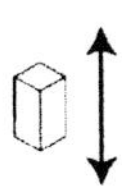

Largest drag occurs with prolate or rod-like shapes moving perpendicular to the long axis:

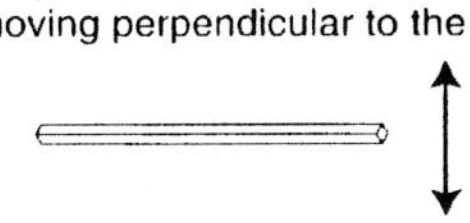

Temporal

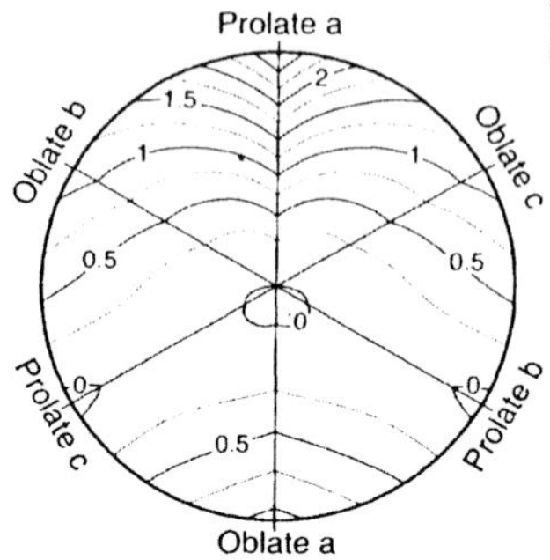

Contours are are proportional to log S/N for orientation to gradients by a temporal mechanism, in which present simulation is compared to past stimulation.

Best sensitivity occurs with prolate or rod-like shapes swimming parallel to their long axis.

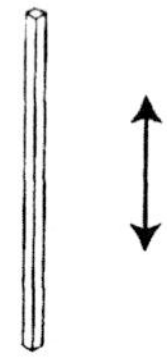

Fore / aft

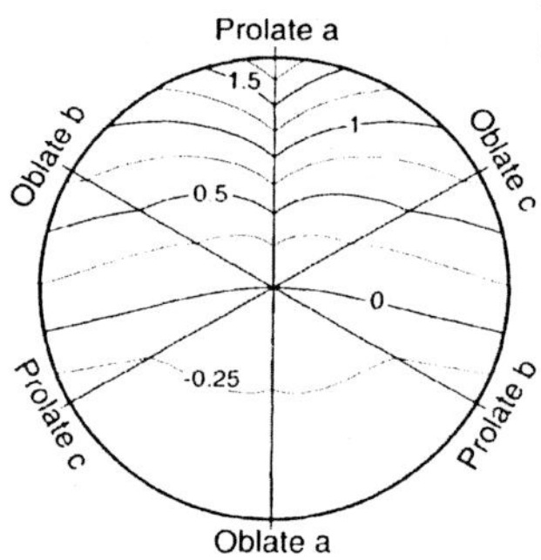

Contours are are proportional to log S/N for orientation to gradients by comparing stimulation of leading and trailing edges.

Best sensitivity occurs with prolate or rod-like shapes swimming parallel to their long axis.

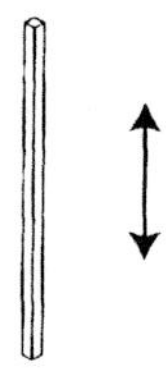

Lateral

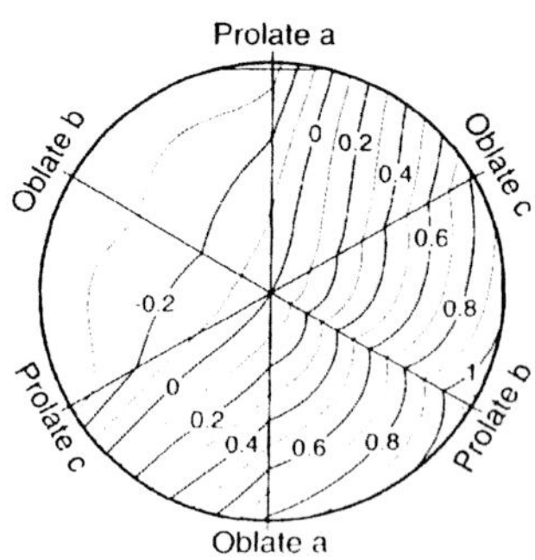

Contours are are proportional to log S/N for orientation to gradients by comparing stimulation left and right of swimming direction (ends of b axis).

Best sensitivity occurs with prolate or rod-like shapes swimming perpendicular to their long axis and comparing stimulation at the ends of the long axis.

Fig. 3 A, B. (On the previous two pages). Contour plots of possible fitness components for equal-volume ellipsoids of all shapes, with axes *a*, *b*, *c*. The values of the contours are relative to an equal-volume sphere. All possible equal-volume ellipsoidal shapes not too distorted from spherical are represented in these plots. The sphere ($a = b = c$) is at the *center*, and the distance from the center is proportional to the negative log of the minimum radius of curvature occurring in each ellipsoid compared to the radius of the equal-volume sphere. At the *outer edge* of the plots, the ellipsoids are most distorted from spherical and have a minimum radius of curvature of 1% that of the sphere. The three axes of the plots encompass shapes in which two axes of the ellipsoid are identical (ellipsoids of revolution). At one end of each axis, prolate ellipsoids, with semiaxes equal to some permutation of (10, $10^{-1/2}$, $10^{-1/2}$), resemble rods, with axial ratios of 32; and at the opposite end oblate ellipsoids, with semiaxes ($10^{-4/5}$, $10^{2/5}$, $10^{2/5}$) resemble disks with axial ratios of 0.063. Where swimming occurs, it is along the *a* axis. For clarity, boxes are illustrated that are just large enough to contain the optimal ellipse. (Data from Dusenbery 1998a)

4.3 Constraints on Pheromone Use

At small size scales, diffusion transports chemicals faster than does flow. This permits a simple calculation of the range at which the pheromone released by a small organism can be detected (Dusenbery and Snell 1995). Making a best estimate of the cost of pheromone production, the benefits of producing pheromone to attract a mate can be compared to the benefits of allocating the same energy to swimming.

Pheromone production can increase the search rate several orders of magnitude for sufficiently large organisms (Fig. 4). But there is again a sharp size limit below which pheromone production is a complete waste. The pheromone diffuses away faster than it can be produced and a threshold concentration never builds up.

The actual value of the size limit depends on the values of several parameters that are poorly known, but the best estimates suggest a size limit on the order of 1mm, and available information is consistent with this (Dusenbery and Snell 1995).

Crustaceans use mate attractant pheromones but are larger than 1 mm. Rotifers are smaller and do *not* employ attractant pheromones. One ciliate is known to employ attractant pheromones, but it usually stays in contact with surfaces. No bacteria are known to employ pheromones for mate *attraction*, although pheromones are used for regulating genes where more time is available (Surette et al. 1999).

4.4 Advantage of Male and Female Mating Types

The size limit on effective pheromone use suggests a possible explanation for why male and female mating types, producing sperm and eggs, have been so successful that almost all familiar organisms have males and females, although they are not necessary for sexual reproduction. Employing an analysis similar to the above (Dusenbery 2000), it has been demonstrated that at low population densities of broadcast spawners, fertilization is more probable if one sex produces a few large gametes (eggs) that release pheromone, while the other sex produces many small motile gametes (sperm), (see Fig. 5).

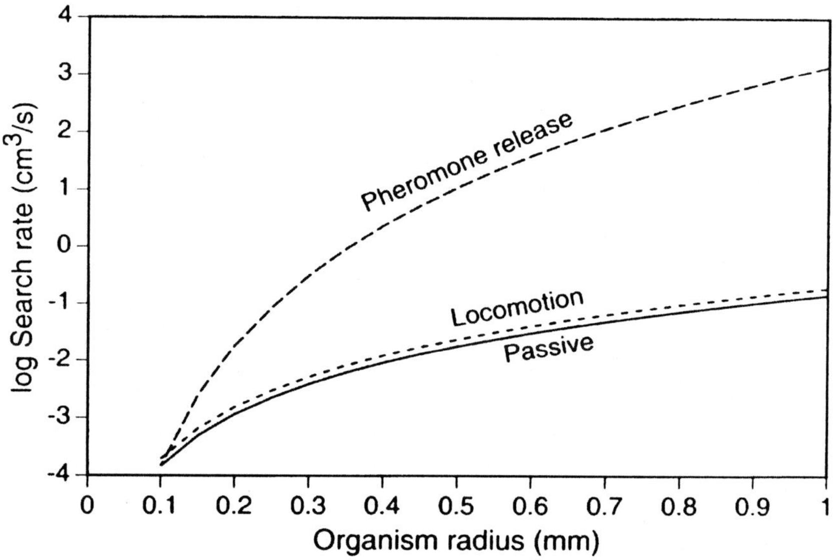

Fig. 4. Search rate dependence on size and target behavior. The target organism can remain passive or devote energy to random locomotion or to pheromone release. Calculations assumed spherical organisms of the indicated radius; power applied to mate location = 6000 $ergs^{-1}cm^{-3}$ of organism; conversion efficiency for pheromone release = 1×10^{-14} mole/erg; threshold sensitivity to pheromone = 1×10^{-7} molar; efficiency of propulsion = 0.1%, which gives a swimming speed of about 10 diameters per second. The critical size with these values is 215 μm diameter. (Data from Dusenbery and Snell 1995)

Thus, lineages with male and female mating types can better survive population crashes and persist longer than lineages with other types of sexual reproduction. This allows more time to evolve other features such as multicellular individuals. Here we have a possible explanation for why male and female mating types are so common among the largest and most complex plants and animals.

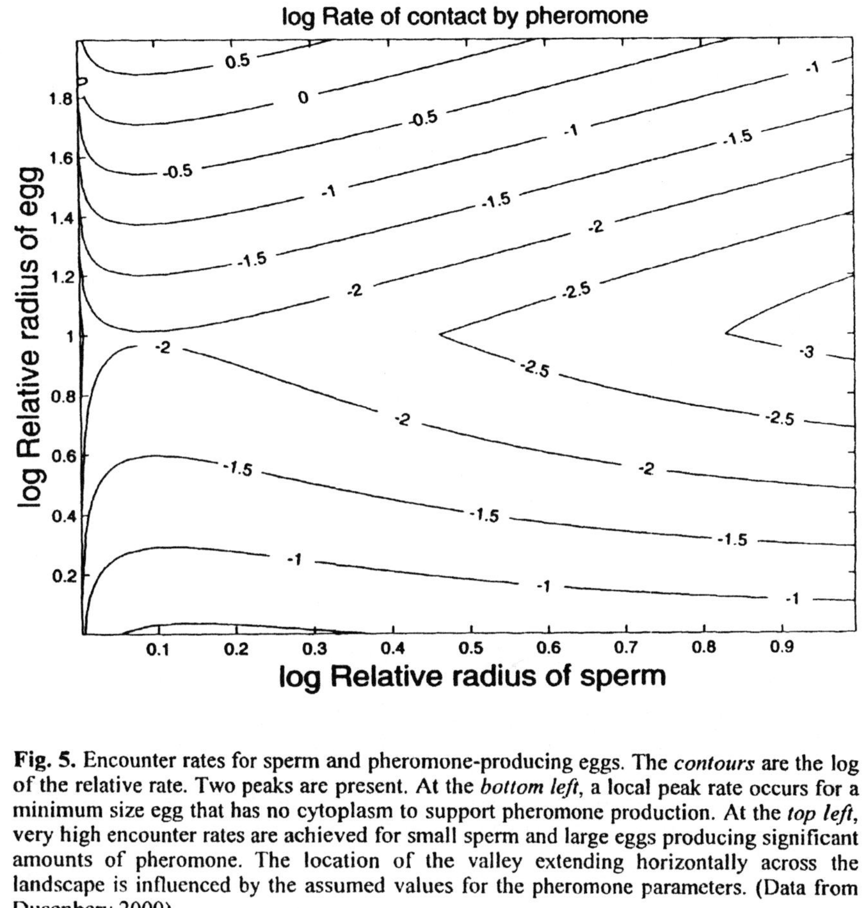

Fig. 5. Encounter rates for sperm and pheromone-producing eggs. The *contours* are the log of the relative rate. Two peaks are present. At the *bottom left*, a local peak rate occurs for a minimum size egg that has no cytoplasm to support pheromone production. At the *top left*, very high encounter rates are achieved for small sperm and large eggs producing significant amounts of pheromone. The location of the valley extending horizontally across the landscape is influenced by the assumed values for the pheromone parameters. (Data from Dusenbery 2000)

5 Conclusion

We have seen that for small organisms living free of contact with surfaces or other constraint on motion, certain processes are relatively easy to evaluate quantitatively from basic physical principles. These organisms live in the simplest possible environment, and we have an unusual opportunity to apply physical principles to aid us in understanding them. This is fortunate, because their world is quite foreign to us, and our ordinary experience is often misleading. I hope the success of considering physical constraints in this situation will encourage others to apply this approach in a wider variety of situations.

References

Baldwin IT, Schultz JC (1983) Rapid changes in tree leaf chemistry induced by damage: evidence for communication between plants. Science 221: 277-279

Ballaré CL, Scopel AL, Sánchez RA (1990) Far-red radiation reflected from adjacent leaves: an early signal of competition in plant canopies. Science 247: 329-332

Berg HC (1993) Random Walks in Biology. Princeton University Press, Princeton

Berg HC, Purcell EM (1977) Physics of chemoreception. Biophys J 20: 193-219

Bradburne JA, Kasperbauer MJ, Mathis JN (1989) Reflected far-red light effects on chlorophyll and light-harvesting chlorophyll protein (LHC-II) contents under field conditions. Plant Physiol 91: 800-803

Dusenbery DB (1992) Sensory Ecology. WH Freeman and Company, New York

Dusenbery DB (1996) Life at Small Scale. Scientific American Library, New York

Dusenbery DB (1997) Minimum size limit for useful locomotion by free-swimming microbes. Proc Natl Acad Sci USA 94: 10949-10954

Dusenbery DB (1998a) Fitness landscapes for effects of shape on chemotaxis and other behaviors of bacteria. J Bact 180: 5978-5983

Dusenbery DB (1998b) Spatial sensing of stimulus gradients can be superior to temporal sensing for free-swimming bacteria. Biophys J 74: 2272-2277

Dusenbery DB (2000) Selection for high gamete encounter rates explains the success of male and female mating types. J Theor Biol 202: 1-10

Dusenbery DB, Snell TW (1995) A critical body size for use of pheromones in mate location. J Chem Ecol 21: 427-438

Jumars PA (1993) Concepts in Biological Oceanography. Oxford University Press, New York

Kasperbauer MJ (1987) Far-red light reflection from green leaves and effects on phytochrome-mediated assimilate partitioning under field conditions. Plant Physiol 85: 350-354

Land MF (1981) Optics and vision in invertebrates. In: Autrum H (ed) Handbook of Sensory Physiology. Springer, Berlin, pp 471-593

Lythgoe JN (1979) The Ecology of Vision. Clarendon Press, Oxford

Mangel M, Clark CW (1988) Dynamic Modeling in Behavioral Ecology. Princeton University Press, Princeton, New Jersey

Mann KH, Lazier JRN (1991) Dynamics of Marine Ecosystems. Blackwell Scientific Publications, Boston

McKechnie JL (1978) Webster's New Twentieth Century Dictionary of the English Language. William Collins Publishers, Inc

Perrin F (1934) Mouvement Brownien d'un ellipsoide (I). Dispersion diélectrique pour des molécules ellipsoidales. Le Journal de Physique et Le Radium V 10:497-511

Perrin F (1936) Mouvement Brownien d'un ellipsoide (II). Rotation libre et dépolarisation de fluorescence. Translation et diffusion de molécules ellipsoidales. Le Journal de Physique et Le Radium VII 1:1-11

Rhoades DF (1985) Pheromonal communication between plants. In: Cooper-Driver GA, Swain T (eds) Chemically Mediated Interactions Between Plants and Other Organisms. Plenum Press, New York, pp 195-218

Rouse H (1961) Fluid Mechanics for Hydraulic Engineers. Dover Publications, New York

Surette MG, Miller MB, Bassler BL (1999) Quorum sensing in *Escherichia coli*, *Salmonella typhimurium*, and *Vibrio harveyi*: A new family of genes responsible for autoinducer production. Proc Natl Acad Sci USA 96: 1639-1644

Tanford C (1963) Physical Chemistry of Macromolecules. Wiley, New York

van Holde KE (1985) Physical Biochemistry. Prentice-Hall, Englewood Cliffs, NJ

Withers PC (1992) Comparative Animal Physiology. Saunders College Publishing, Fort Worth

Zuriff GE (1985) Behaviorism: A Conceptual Reconstruction. Columbia University Press, New York

I.2 Why Sensory Ecology Needs to Become More Evolutionary – Insect Color Vision as a Case in Point

Lars Chittka[1] and Adriana Briscoe[2]

[1] Zoologie II, Biozentrum, Am Hubland, 97074 Würzburg, Germany
[2] Department of Molecular and Cellular Biology, Life Sciences South 444/1007 E. Lowell, Tucson, Arizona 85721, USA

Abstract
Many of the subtleties in the evolutionary tuning of sensory systems still escape us. Insect color vision is a typical case. While we know much about its mechanisms, the abundant adaptive explanations of its components are often ornate speculations. We advocate using an evolutionary approach to understand why different animals see the world in different colors. Such an approach must include (1) phylogenetic analyses, which should help identify patterns of adaptation, constraint and history; (2) molecular studies, to predict how plastic the relevant genes will be in the face of particular selective pressures; (3) assessments of interindividual variance, to see if the raw material for evolution exists; (4) a consideration of pleiotropic effects, where selection on visual pigments may be affected indirectly through correlated characters; (5) biogeography, to explore if populations living in different visual habitats have adapted to those differences; (6) a consideration of random evolutionary processes; (7) selection experiments, to test for heritability of traits and to simulate the influences of strong directional selection; (8) fitness tests: to show that a trait is adaptive, we must show empirically that this trait confers greater fitness to its bearers, compared with conspecifics which lack this trait.

Key words Color vision, evolution, insects, visual ecology, visual pigments

1 Introduction

One major focus of sensory ecology has long been the question why many animals see the world through color receptors so different from our own. Traditionally, the field has concentrated on adaptive explanations. We are used to thinking that if there were differences between species, this must reflect adaptations to different photic environments, whereas if animals share similar color receptors, they must live under similar selective pressures. Fish dwelling in progressively deeper habitats possess rods whose sensitivity is more and more shifted into the blue, thus matching the changing spectral distribution of the ambient light in deep water (Lythgoe 1972; Hunt et al. 1996; Douglas et al. 1998). Other presumed cases of adaptive spectral tuning include humans, whose trichromatic systems might have evolved as a response to frugivory (Mollon 1989), and bees, whose receptors were suggested to be evolutionarily tuned for flower color coding (Chittka and Menzel 1992). Such intuitively appealing stories of adaptation easily find their way into textbooks. Studies which do not find an obvious match between sensory traits and the environment do exist (Crandall and Cronin 1997; Fleishman et al. 1997), but they often achieve much less fame. Even in deep-sea fish, which have long been held as classical examples of adaptive visual pigment tuning, the situation is far from resolved. Fish which live in similar photic environments, but belong to different taxa, often have different visual pigments, whereas closely related fish species sometimes have similar visual pigments, even if they inhabit different light habitats (Douglas et al. 1998). Turning to bees, the notion that pollinator color vision is tuned to floral colors is compromised by the finding that arthropods living under entirely different visual conditions, such as the beach isopod *Ligia* (Hariyama et al. 1993), the freshwater-dwelling bug *Notonecta* (Bruckmoser 1968), nocturnal hawkmoths (White et al. 1994), and the larval ocelli of some Lepidoptera (Ichikawa and Tateda 1982) have similar sets of color receptors. On the basis of such difficulties, Goldsmith (1990) concluded that phylogenetic and molecular constraints might play more important a role in determining the wavelength positioning of color receptors than is good for any pan-adaptionist scenario. We wish to reiterate this warning, and to add several additional ones.

Our most important caveat is that to show that a trait is adaptive, we must demonstrate that it has an impact on fitness (Endler 1986; Reeve and Sherman 1993). What is the evolutionary significance of a model, for example, which shows that for a given visual task, one set of color receptors is 5% better than another? If this really translates into 5% more lifetime reproductive success, the effect of selection will probably be significant over evolutionary time. On the other hand, it is just as possible that 5% improved performance in some criterion will be absolutely irrelevant to fitness. To our knowledge, there is not a single study in sensory ecology that resolves this problem. Such fitness tests are challenging, but should be possible. In what follows, we will lay out a research agenda that include several steps towards such tests. We hope this treatise will stimulate a more evolutionary approach to sensory ecology, and a better understanding of why many animals see the world in colors so differently from ourselves.

2 Uses and Limitations of Model Calculations

Ten years ago, one of us (L.C.), in collaboration with R. Menzel, set out to identify the adaptive significance of bee color vision. The idea was to generate a theoretically optimal color vision system for the task of flower color coding, and to compare this with the system really implemented in bees. We had considerable reasons for optimism. First, Menzel and his coworkers had established a database that included the color receptor sensitivity functions of a large variety of hymenopteran species (Peitsch et al. 1992) and insects in general (Menzel and Backhaus 1991). This database suggested that insects could, in principle, produce pigments with values of maximum sensitivity (λ_{max}) anywhere from 320 to 630 nm, and that the number of color receptor types varied widely between species. This, in combination with the fact that insects occupy a very wide range of visual environments, made studying the sensory ecology of their color vision look promising. Second, Backhaus (1991) had just developed a physiological model of bee color vision, which allowed quantitative predictions of the similarity of flower colors, and of flowers and their background. Such a model is an essential tool to measure the quality of a color vision system and, to date, such models are still not available for any animal species besides bees and humans. Third, unlike many other animals, bees seemed an ideal study subject because the relevant visual tasks are comparatively easily identified: bees obtain their food from flowers, and so selection should favor color vision systems that allow for swift detection of flowers and reliable identification of the most rewarding species (Chittka 1997).

Spectral sensitivity functions of color receptors have roughly Gaussian characteristics, and the exact shape of the curve can be easily predicted if the λ_{max} is known (Stavenga et al. 1993). Our evolutionary model calculations consisted of moving three color receptor sensitivity curves along the wavelength scale. For each theoretical combination of receptors thus generated, the quality of the color vision system for flower color coding was determined. The result was striking: the optimal color receptors generated by the evolutionary model invariably occurred at λ_{max} = 330, 430, and 550 nm, which is very close the most common λ_{max} really found in flower-visiting bees (Chittka and Menzel 1992). This result was independent of whether we varied one, two, or three photoreceptors. It was also independent of the particular set of flowers used (Chittka 1996a).

Since the optimal set of color receptors might also depend on the particular kind of opponent coding in the brain, the mode of this processing, too, was varied – the result remained unchanged (Chittka 1996a). An engineer could hardly design a better receiver for flower colors than the color vision system of bees. Does this, however, mean that flower colors indeed drove the evolution of bee color vision? This is an attractive notion for sensory ecologists. It is joined by other studies in which correlations between the results of model calculations and reality were taken as evidence for adaptation (Lythgoe and Partridge 1989). Thus, many colleagues took our finding to mean that bee color vision indeed adapted to flower colors, although we explicitly stated in the discussion of the original paper that this is not necessarily the case (Chittka and Menzel 1992).

Indeed, there are several complications. Models are useful to generate hypotheses of optimality, but a correlation between a model and biological traits does not resolve how these traits evolved. Using models to reject a hypothesis of evolutionary causality is much more straightforward. Had the optimal color receptors derived from the model calculations been different from the ones found in "real animals", then this would have indicated that evolution has not optimized the photoreceptors according to the same criteria. At the very least it would mean that there are other, more important, criteria, or that evolutionary constraints might have hindered the animal from evolving along the same lines as the model calculations. The fact that the calculations arrive at similar color vision systems as nature is tempting, but it does not necessarily imply that one has found the criterion which has driven the evolution of bee color vision. In fact, sets of color receptors similar to those of bees occur in animals that occupy entirely different ecological niches.

Finally, close inspection of the model results reveals that the color receptors of bees are only nearly optimal. For example, the optimum long wave receptor for coding flowers is at λ_{max} =550 nm, whereas the visual pigments of bee green receptors are maximally sensitive at 540 nm. Since the modeled optima are fairly broad, performance of the real receptors is only 2% below the theoretical optimum for flower coding. This discrepancy is small when considering the large range over which the receptors were varied, but what causes it? If bee visual pigments can be freely tuned, why have they not achieved a perfect match with floral colors? One possibility is that there are tradeoffs with other visual activities. For example, it has been assumed that the bee green receptor is optimally matched to green foliage, and might thus serve as a background detector (Menzel 1979); but leaves also reflect most strongly at 550 nm (Chittka 1996a), so that, again, the theoretical optimum is at longer wavelengths than the peak sensitivity of the bee green receptors.

We are confident that there are other adaptive explanations that might be tried, and eventually, one might be successful: but we also wish to warn that trying a large number of adaptive explanations can lead to speculation. Gould and Lewontin (1979) caricaturized this approach in the following terms: "If one adaptive argument fails, try another" and "In the absence of a good adaptive argument... attribute it to imperfect understanding of where an organism lives or what it does." We are sure that some of our readers will recognize their own thinking in these words. We do not wish to discourage sensory ecologists to continue searching for adaptive explanations where at present they seem hard to find. In what follows, however, we list a number of tools that should make this search less speculative.

3 Phylogenetic Studies

One possible reason why animals in different ecological contexts have conserved traits is phylogenetic constraint. For example, there is little reward in searching for the adaptive significance of why bees have six legs, because leg number is evolutionarily conservative in insects. Mapping traits on an established phylogenetic tree will reveal if the trait is variable within a given taxon, and whether the search for adaptations will be worth our time (Brooks and McLennan 1991; Harvey and Pagel 1991). It is this simple evolutionary reasoning that is absent in many studies of sensory ecology. Often, each species was regarded as an entity that could freely vary all its traits in all directions, and its history seemed to be of minor importance. Some physiologists, in fact, dismissed the possibility of evolutionary constraints entirely: ".... there is also the possibility that the evolutionary history of a species might be important in determining its visual pigments .... but taking a broader view .... the evolutionary explanation is not helpful." (Lythgoe 1972). Quite often, this view has led workers to detect biological adaptations where, in fact, there were none (see Chittka and Dornhaus 1999 for examples).

To test if flower signals drove the evolution of bee color vision, it must be shown that the ancestors of bees possessed different sets of color receptors prior to the advent of the flowering plants. To this end, we have to evaluate members of arthropod taxa whose evolutionary lineages diverged from those of bees before there were flowers. A phylogenetic analysis reveals that the values of peak sensitivity in the Crustacea and Insecta fall into three distinct clusters around 350, 440 and 520 nm (Chittka 1996b). The few insect species in which one of the three types is absent (*Periplaneta* and *Myrmecia*) represent secondary losses. Red receptors show up irregularly in both the Crustacea and Insecta; they have evolved several times convergently. The photoreceptor wavelength positions of UV, blue, and green receptors are surprisingly conserved in the Mandibulata. We conclude that the Cambrian ancestors of extant insects and crustaceans possessed UV, blue, and green receptors. Insects were well preadapted for flower color coding more than 500 million years ago, about 400 million years before the extensive radiation of the flowering plants that started in the mid-Cretaceous (100 million years ago).

To be sure, there *are* differences of 20-30 nm between the measured peak sensitivities within each cluster of insect color receptors. This means that our analysis does leave open the possibility of fine-tuning of receptors to particular visual tasks in each species. For example, microspectrophotometry reveals that the long wave pigment of fireflies differs by 12 nm in λ_{max} between nocturnal and crepuscular species, a difference which can be explained by the specific requirements posed by detecting and identifying conspecific flashes under different visual conditions (Cronin et al. 2000). Unfortunately, however, the data in many other studies were collected by electrophysiological measurements and are therefore noisy, so we do not yet understand whether the differences between many species can be attributed to measurement error, different methods, or actual variation between species.

At first sight, however, considering the large variety of light habitats and feeding habits of insects, it is surprising how little variation there is. If traits are conserved within a taxon, although we have reasonable grounds to predict that they should differ on the basis of optimality arguments, then we must also take the possibility of phylogenetic constraint seriously. Similar phylogenetic studies performed on other sensory traits, or on color vision systems in other animals, may reveal a different pattern; but it is through phylogenetic studies that we can decipher patterns of adaptation and constraint, convergence, and homology.

4 Molecular Tuning, Constraints, and Adaptation

To understand how easily visual pigments can be matched to the colored environment, it is necessary to look at their molecular structure, the mechanisms of tuning, and the genes that encode the pigments. For example, we would expect that color receptors readily adapt if small genetic changes (e.g., single mutations) cause large changes in spectral sensitivity. We predict slower adaptation if most mutations are adaptively neutral. Or if several mutations are necessary in conjunction to alter spectral sensitivity, this may mean that any functional changes require a comparatively improbable sequence of mutation events.

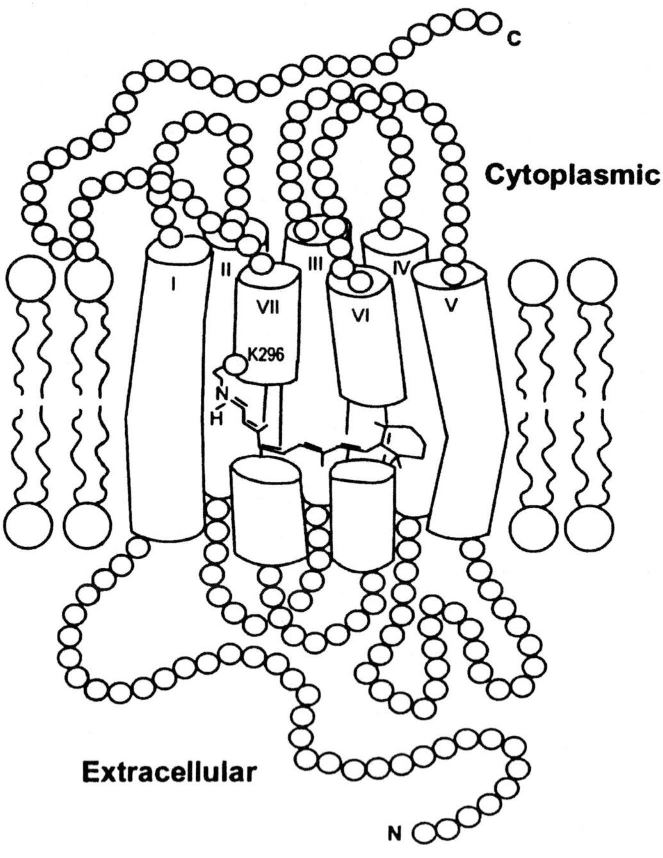

Fig. 1. Diagram of an opsin bound to 11-*cis*-retinal at lysine 296. The seven transmembrane helices are numbered *I-VII*. Specific amino acid side-groups interact with the chromophore and shift its absorption maximum

We would expect particularly strong constraints, if pigments sensitive at two peak wavelength values are separated by adaptive valleys, so that, for example, two combined changes are necessary to alter a receptor's sensitivity, but each single mutation renders the pigment nonfunctional. We would also expect that pigment adaptations might be compromised by molecular spell-checkers that pit protein folding requirements against spectral tuning (Nakayama et al. 1998).

Each class of color receptor contains a distinct visual pigment, which consists of two components. One is the chromophore, retinal (or one of its congeners), which changes its configuration on absorption of a single photon (Seki and Vogt 1998). The other component is a protein moiety, the opsin (Fig.1). Opsins are integrated into the membrane of photoreceptive organelles of the receptors, and contain about 370 amino acids (Deeb and Motulsky 1996). They contain seven transmembrane helices, arranged in a quasicircle so that they form a pocket. This pocket holds the chromophore. Specific amino acids in the transmembrane helices oriented towards the center of the pocket (and, thus, interact electrostatically with the chromophore) are responsible for spectral tuning (Hope et al. 1997).

To understand how visual pigments in insects changed over evolutionary time, it is informative to evaluate the phylogeny of their opsins. To this end, we compared the amino acid sequences of the opsins of 54 species of arthropods, as well as different opsins found within the same animal species. The basis of such an analysis is that one groups together those proteins that are most similar; the nodes of the tree represent (hypothetical) ancestral opsins (Goldsmith 1990). It is immediately apparent that invertebrate opsins fall into distinct functional clades according to spectral sensitivity (Fig.2).

There is one cluster of UV pigments, a distinct group of blue pigments, a third group of long wave pigments, which includes pigments with peak sensitivity from green to red. Most interestingly, chelicerate and crustacean green sensitive pigments are more similar to insect green pigments than they are to UV and blue pigments, which confirms the phylogenetic analysis above: opsin clades diverged from one another before the major groups of arthropods diverged, and it is therefore likely that ancient arthropods already possessed (at least) UV and green visual pigments. Somewhat puzzling are the origins of two clades of blue-green (480 nm) pigments of *Drosophila* and the crab *Hemigrapsus sanguineus*. They might be the result of convergent evolution, or of several independent gene losses. On the basis of electrophysiology, not many insects are expected to have pigments that fall in the 480 nm spectral class, which favors the latter hypothesis. We advocate testing this hypothesis by looking for opsin pseudogenes to members of this clade.

Another important result of this comparative analysis is that rather large portions of opsin amino acid sequence can be exchanged while the spectral function remains surprisingly constant. Bee UV and *D. melanogaster* Rh3, which have nearly identical λ_{max} values (353 and 345 nm) differ by 36.6%, while the two *Limulus* sequences, which are thought to differ by 10 nm, differ by only 1%!

Of course, our argument is dependent on the assumption that differences between opsins do not represent cases of convergent evolution, possibly as a response to similar selective pressures.

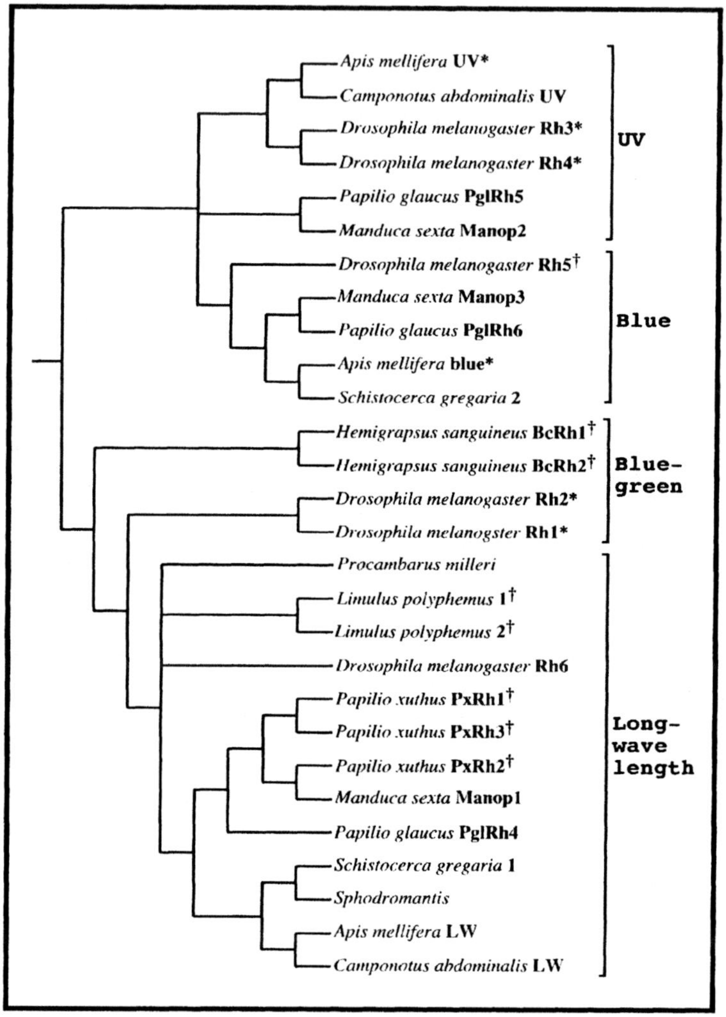

Fig. 2. Phylogeny of insect, chelicerate and crustacean opsins, based upon a maximum parsimony analysis of opsin amino acid sequences. Tree shown is simplified from the analysis of a larger data set of 54 opsin sequences (Briscoe 1998b). Only representative species from available orders or suborders are shown. *Brackets* indicate measured (*single asterix*) or inferred (*cross*) spectral properties of the visual pigments in each clade. Inferred spectral properties are based upon in situ hybridization or immunohistochemistry

in combination with electrophysiological studies. References for measured spectra: *Apis mellifera* UV and blue, *Drosophila melanogaster* Rh1-Rh4 (see Briscoe 1999). References for in situ hybridization: *D. melanogaster* Rh5 (Papatsenko et al. 1997); Rh6 (Huber et al. 1997); *Hemigrapsus sanguineus* (Sakamoto et al. 1996), *Limulus polyphemus* 1 and 2 (Smith et al. 1993); *Papilio xuthus* PxRh1-3 (Kitamoto et al. 1998). Reference for immunohistochemistry: *D. melanogaster* Rh5 (Chou et al. 1996)

This, in turn, is dependent on the ratio of amino acid substitutions that are adaptively neutral to those that cause functional changes. If only a few amino acid substitutions are responsible for spectral tuning, then the overwhelming majority of amino acid changes used to reconstruct the phylogeny are going to reflect the actual history of the gene family.

Although we do not yet know how many amino acid residues are involved in spectral tuning of the insect visual pigments, we infer from research in vertebrates (Hunt et al. 1996; Neitz et al. 1991) that this number is comparatively small relative to the total number of amino acids of opsins. Therefore the pigments which fall into distinct clades are similar because they share a common ancestry, and not because natural selection has erased their history. This view is corroborated by the fact that opsins in the distinct clades have intron splice sites that are shared (Briscoe 1999).

In order to predict how easy it is to change spectral sensitivity as a response to specific selective pressures, it is necessary to see how many amino acid replacements are required to cause such changes. The majority of studies on spectral tuning have focused on the vertebrate visual pigments. These studies, which use mutagenesis to tease apart the amino acids relevant to the spectral properties of visual pigments, have made significant progress in furthering our understanding of the relationship between vertebrate opsin structure and function.

Almost all of the variation in cone pigment absorption spectrum has been accounted for: a mere one to nine specific amino acid substitutions are responsible for the 10-100 nm differences between the vertebrate cone pigments (Asenjo et al. 1994; Sun et al. 1997; Lin et al. 1998). This suggests that adaptive spectral tuning will be achieved relatively easy. On the other hand, previous studies have only looked at the ease of tuning between pigments that already exist in nature. These studies do not necessarily imply that spectral tuning in all spectral domains will happen as readily. For example, the fact that only three amino acid replacements are necessary to turn a human green receptor pigment (λ_{max}=530 nm) into a red receptor pigment (λ_{max}=560 nm) does not mean that generating a hypothetical pigment with λ_{max}=590 nm will require equally few changes.

In contrast to the large body of work that has led to the development of our current model of spectral tuning in the vertebrate pigments, only one study to date examines spectral tuning of insect opsins. Using a *Drosophila* transgenic expression system, Britt et al. (1993) studied the spectral properties of 13 chimeric opsins created by exchanging one or more transmembrane domains of the *Drosophila* Rh1 and Rh2 opsin genes. These chimeric opsin-encoding genes were introduced into a mutant *Drosophila* strain using P element germline transformation, and expressed in the R1-6 photoreceptor cells. The expression of the chimeric opsin genes in the R1-6 cells restored normal spectral sensitivity function to the mutant flies. Using electrophysiology and microspectrophotometry, Britt et al. (1993) measured changes in, respectively, the spectral sensitivities of the R1-6 photoreceptor cells and the absorption spectra of the chimeric visual pigments. This approach allowed them to monitor the effects of exchanging different Rh1 and Rh2 transmembrane domains on the spectral tuning of the opsins.

Britt et al. (1993) found that no single transmembrane domain was responsible for the 60 nm difference between Rh1 (480 nm) and Rh2 (420 nm). Exchange of a single Rh1 transmembrane domain (TM) for the corresponding Rh2 domain resulted in a 4-10 nm shift towards shorter wavelengths for most domains except TM 4, which resulted in an 11 nm shift to longer wavelengths. Replacement of almost all TM domains (TM2-7) was required to convert a Rh1 transgene into a Rh2-like opsin (436 nm). Unlike the three to seven amino acid residues responsible for tuning the primate red and green cone pigments (Neitz et al. 1991), which differ by 15-30 nm, the *Drosophila* Rh1 and Rh2 TM domains do not interact in an additive fashion. For example, replacement of only Rh1 TM6 (with a Rh2 TM6) results in a 12 nm shift to shorter wavelengths, and replacement of Rh1 TM7 alone results in a 4 nm in the same direction. However, replacement of both Rh1 TM6 and 7 simultaneously results in a 20 nm shift to longer wavelengths. The authors propose two mechanisms to account for spectral tuning, one of which is involved in large-scale or coarse spectral tuning, and the other in fine-scale tuning, as exemplified by the human red and green cone pigments. While fine tuning in the vertebrate pigments is apparently nearly additive in effect (Hunt et al. 1996), coarse tuning in the case of the *Drosophila* opsins occurs in a combinatorial manner, involves many more TM domains (Britt et al. 1993), and occurs over a larger evolutionary time scale. Finally, not all mutations in visual pigment genes are neutral or cause functional changes; some are downright deleterious. For example, three different missense mutations in the gene that codes for the human blue sensitive pigment all cause complete loss of the human short-wave receptor function, or even cell death (Deeb and Motulsky 1996).

Besides spectral fine tuning of existing photopigments, color vision systems can also evolve by changing the number of color receptor types. Although clearly the number of such types is conservative in many taxa, there is also some variation. For example, more than 40 species of Hymenoptera have 3 color receptor types most sensitive in the UV, blue, and green, but there are at least three species which have red receptors in addition (Peitsch et al. 1992). Such increases in color receptor types occur through gene duplication and subsequent spectral tuning (Briscoe 1999; Briscoe 1998a). How common are opsin gene duplications? The crab *Hemigrapsus sanguineus* (Sakamoto et al. 1996) and the horseshoe crab *Limulus polyphemus* (Smith et al. 1993) have unique opsin gene duplications, found in no other species so far. The problem is that we are not sure if these duplications are species specific, or whether they are more basal in larger taxonomic groups, such as genera or even orders. The butterflies *Papilio glaucus* and *Papilio xuthus*, for example, share two gene opsin gene duplications (Briscoe 1998a; Kitamoto et al. 1998); as do *Drosophila melanogaster* and *Drosophila pseudoobscura* (Zuker et al. 1985; Carulli and Hartl 1992). All other insects that have been surveyed for opsins have been undersampled, so that we cannot be sure how constrained individual species are in terms of photoreceptor number.

So far we have emphasized changes in the opsin protein as a mechanism for the evolution of color vision. Arthropod chromophores come in five forms that are all derived from retinal. When extracted in ethanol, their absorption maxima differ by several nanometers: RAL1 (383 nm), RAL2 (400 nm), RAL3 (379 nm) and RAL4 (377 nm). Visual pigments reconstituted with the same opsin but different chromophores have slightly different absorption maxima (Seki and Vogt 1998). In nature, most species use only one chromophore but there are exceptions. In the firefly squid, for instance, there are two kinds of chromophore and one kind of opsin giving rise to three kinds of visual pigment (Seidou et al. 1990). Clearly, the particular chromophore used has an effect on spectral tuning, and in some cases, the opsins and chromophores may be coevolving.

5 Interindividual Variance

Variance between individuals is the raw material for evolution. This does not mean that the lack of such variance in extant species indicates the traits are not adaptive - in fact, if a trait is strongly favored by selection, it is likely that it becomes fixed in a population, and all variance might be eliminated (Endler 1986; Reeve and Sherman 1993). The results of many evolutionary "experiments" may no longer exist in our time, but variance is important, both for animals and for scientists studying adaptation. It allows populations to respond to ongoing changes in environmental pressures, and to colonize new habitats. Where there is lack of heritable variation, such changes cannot occur (Chittka 1997; Goldsmith 1990).

For us, interindividual variance offers the possibility to study predictions of adaptation. One phenotype may be favored in one photic environment and another phenotype in a different one. Many physiologists, however, treated such variance as noise, which needed to be eliminated by averaging large numbers of measurements from different animals. Sometimes this may be legitimate. Physiological measurements are often so noisy that extracting any information at all is not possible without averaging, and strong deviations from expected observation will in fact often mean that the measurement is imperfect, for example in electrophysiology: but we may have lost much valuable information through such averaging! Could it be that the reason for much of the conservatism in arthropod color receptors exists because there simply is no variance between individuals of some of the species in question? A large number of scientists have worked on, e.g., the color receptors of honeybees, and the results differed within studies as well as across studies, but the debate about these differences mostly focused on the possible contributions of artifacts or different electrophysiological methods (Menzel et al. 1986). To be sure, both of these may add noise to the measurements, but unfortunately, the possibility that interindividual variance may also contribute has not been considered.

In some vertebrates, conversely, such variance exists and has been well quantified, for example in guppies (Archer et al. 1987) and primates including humans (Deeb and Motulsky 1996; Shyue et al. 1995). In invertebrates, to our knowledge, only a single published study reported intraspecific variance between the sequences of visual pigments. Ayala et al. (1993) sampled five Rh3 alleles from each of four species in the *Drosophila melanogaster* subgroup and three alleles from *D. pseudoobscura,* and found a single amino acid polymorphism in one of the five surveyed species. One of us (A.B.) also found intraspecific amino acid variation in some of the opsin loci from *Papilio glaucus.* We cannot be sure, however, if any of these naturally occurring variants differ in their spectral sensitivities. Clearly, we need more data.

6 Pleiotropy – Selection Through Correlated Characters

A possibility rarely considered by sensory ecologists is that a sensory trait under scrutiny may be favored indirectly, through correlated characters, that do not necessarily have anything to do with the selective pressures associated with the perception of sensory stimuli. For example, Hope et al. (1997) tested the possibility that the pigments of abyssal fish might be adaptations to resist denaturation by the elevated pressure in deep waters, rather than adaptations to the photic environment. They rejected this possibility, but it is an approach well worth considering. Crandall and Hillis (1997) suggested that rhodopsin might have a previously unrecognized function, possibly in the control of circadian rhythms. This conclusion was based on lack of apparent differences in the rate of molecular sequence evolution between the opsin homologues of blind subterranean crayfish and their "sighted" relatives. In another example of possible nonvisual function of opsin, Alvarez et al. (1996) found that *Drosophila* Rh2 opsin is expressed in the testes of male flies, in addition to the ocelli (Pollock and Benzer 1988). If opsins have such dual functions, then spectral sensitivity may sometimes be an effect of pleiotropy, i.e., selection on correlated characters that we have not recognized.

7 Adaptation, Genetic Drift, and Biogeography

It is a widespread misconception of non-evolutionary biologists that beneficial mutations will rapidly spread through a population and eventually wipe out the pre-mutation genotype. However, new mutations are frequently lost immediately upon their introduction because of stochastic processes commonly designated *genetic drift* (Gould and Lewontin 1979). This may cause considerable evolutionary inertia particularly in large populations, or if the adaptive value of the new mutation is relatively small. There are many cases where local populations are kept considerably below their adaptive peaks because of continued gene flow with a large parental population (Stanton and Galen 1997).

Island populations may adapt more readily to local conditions because gene flow with the parental population is disrupted (Emlen 1978). On the other hand, deleterious mutations may also be common on islands, especially where the populations started out from very few animals, or experienced occasional strong reductions in population size (*founder* and *bottleneck effects*) (Endler 1986). It is important to keep these points in mind when considering sensory adaptations.

Of course, a particularly strong case for adaptation could be made if we found biogeographical differences in color vision systems within species, and if we could link these differences to particular environmental pressures exerted on different populations of the species. Unfortunately, no studies in insects are available to date, but a few observations on humans are worth considering. Humans have a single amino acid polymorphism at position 180 in the opsin protein. 62% of white Caucasians have serine at this position, and their red receptor absorbs maximally at 557 nm. The remaining 38% have alanine, and their red receptors are maximally sensitive at 552 nm. This difference has been shown to be relevant in color discrimination, and males with Ser at residue 180 have higher sensitivity to red light (Deeb and Motulsky 1996). This is just the variation we need for selection to act on! Even more interestingly, this polymorphism is different in different human populations: it is 80% Ser; 20% Ala in African Americans and 84/16% in Japanese (Deeb and Motulsky 1996). Are these differences adaptive, however? We favor the notion that these differences are due to genetic drift. Likewise, Ayala et al. (1993), who discovered a single amino acid polymorphism in *Drosophila,* concluded that the alleles are evolving by selectively neutral processes.

Another finding from human color vision is worth consideration. On the tiny South Sea island Pingelap, 75 of the 700 inhabitants are totally colorblind: they have only rods for scotopic vision. This is the result of a classic bottleneck effect. In 1775 this island had almost 1000 inhabitants, when it was struck by a typhoon, which reduced the population of the island to 20 survivors, one of whom was the king. After a few generations, the population was almost back to its pretyphoon level. Unfortunately, the king was carrier of the gene responsible for color blindness, so that today one third of the population carries the recessive gene that is responsible for this defect, and more than 10% of the population is phenotypically color blind (Sacks 1997). In other human populations, the frequency of this defect is about 1 in 30 000.

We do not rule out the possibility that adaptive differences in sensory systems exist between populations. In sticklebacks, for example, McDonald and Hawryshyn (1995) were able to link between-population differences with the light environment; but we wish to caution that not all differences between populations may be adaptive, and most likely random evolutionary processes can explain some of the differences between species as well. The most important message is that the data base is slim, and that we need more studies.

8 Selection Experiments

If we predict that animals will respond to different light habitats (or differently colored objects in their diet) by adjusting their spectral sensitivity, we must test this prediction by means of selection experiments. Such experiments are also necessary to separate phenotypic from genotypic variance. Given this importance, it is surprising that such studies are missing almost entirely in sensory ecology. To be sure, such experiments are demanding, but they are so essential to the reasoning of sensory ecology that we emphatically wish to encourage the taking on of such studies!

We are aware of only a single selection experiment on spectral sensitivity in guppies (Endler et al., subm.) and this experiment showed significant heritability for traits of the visual system. It also showed that animals may respond in multiple ways to the same selective pressure. For example, if animals are selected for higher sensitivity to long wavelengths, they may respond evolutionarily by adjusting their spectral sensitivity, by increasing their overall sensitivity, or by changing the relative strengths of postreceptor neuronal wiring (Endler et al., subm.).

9 Fitness Tests

If we cannot show that a trait confers greater fitness to its bearer, then we cannot claim that it is adaptive (Endler 1986). We must show, for example, that common wild-type animals indeed produce more viable offspring than do those with deviant traits, for example with a different set of color receptors. This is not trivial. Quite often, we will find that performance of one phenotype is better than another at a given task, but this may not have any impact on fitness. To return to the human red receptor polymorphism above, would we predict that someone with a red receptor with λ_{max}=557 nm will be able to *raise more children* than someone with a red receptor with λ_{max}=552 nm, even if it turns out that the person with λ_{max}=557 nm is slightly better at detecting red fruit? Probably not. Even the colorblind on the Island of Pingelap are able to detect and identify ripe fruits (Sacks 1997), so in conditions which are not strongly limiting, even strong deviations from the wild-type phenotype may not be selected against. In another example, Caine and Mundy (2000) were recently able to show that trichromatic marmosets were better at detecting orange fruit against a dappled foliage background than were their dichromatic conspecifics. Nevertheless, dichromats persist in the population.

Whether this occurs because of an unrecognized advantage of dichromats over trichromats at a task not related to frugivory remains to be tested. Without additional data, we also have to consider the possibility that the advantage of trichromats at detecting fruit is so small under natural conditions that it is irrelevant to fitness. The take-home message is that we need fitness tests in sensory ecology.

For such experiments, it is critical that animals be tested in environments which are realistic in terms of food distribution, predator and mate density, and abiotic parameters (Endler 1980). If we test the fitness of bearers of different color vision systems, e.g., bees with different color receptors, in an environment where detectability of flowers is not a limiting factor to search time, or where identification of the most rewarding flowers is not critical, then it is likely that we will find no fitness differences.

In this sense, pollinators such as bumble bees and solitary bees are ideal subjects because we need not worry about "building" an environment: they can forage in natural arrays of flowers, but because they raise all their offspring in one place, we can readily evaluate their fitness. The second prerequisite for fitness tests is that we find variance between individuals of a species in sensory traits. If there is such variation, we can either use naturally occurring phenotypes for fitness tests, or use strongly deviant phenotypes created by selection experiments.

If we do not find differences between individuals, we might consider creating phenotypes by experimental manipulation. This may be difficult in color vision systems (we cannot selectively paint over only one type of color receptor, or alter its spectral sensitivity), but in other sensory systems such manipulations appear feasible. An elegant alternative may be to test organisms in which mutants for sensory traits are available, such as *Drosophila melanogaster*. It is also feasible to create transgenic *Drosophila* with a modified repertoire of visual pigments (Britt et al. 1993), and to measure their fitness. Using these techniques, *Drosophila* has become a model to study the receptor-neuronal control of insect behavior; but these tools have not yet been used to study the role of such mechanisms for survival in nature. The difficulty, in this case, is to create an environment for *Drosophila* flies that is natural enough to allow for realistic tests of fitness. Unfortunately, however, we have not yet succeeded in creating transgenic animals more amenable to realistic fitness tests, such as bees.

10 Conclusion

We do not wish to abandon the notion that sensory systems are adaptive. Instead, we want sensory ecologists to consider constraint, evolutionary inertia, and random processes as possible alternatives to adaptive explanations, not to replace an adaptive scenario entirely. We are also emphasizing that if we want to demonstrate adaptiveness, it is not sufficient to show that sensory traits appear well (or even optimally) matched to the environment. Instead, we must show that animals carrying the sensory characteristics in question are fitter than those that do not. To do this, we must exploit heritable variation in natural populations, create new phenotypes by manipulation of existing traits or selection experiments, or use transgenic animals. Using such methodology, we may even eventually understand why bees have the color receptors they do, why most bee species lack red receptors, and why some other insects with entirely different life-styles have color receptors similar to bees.

References

Alvarez C, Robison K, Gilbert W (1996) Novel Gq alpha isoform is a candidate transducer of rhodopsin signaling in a *Drosophila* testes-autonomous pacemaker. PNAS 93:12278-12282

Archer SN, Endler JA, Lythgoe JN, Partridge JC (1987) Visual pigment polymorphism in the guppy *Poecilia reticulata*. Vision Res 27:1243-1252

Ayala FJ, Chang BSW, Hartl DL (1993) Molecular evolution of the Rh3 gene in *Drosophila*. Genetica 92: 23-32

Asenjo AB, Rim J, Oprian DD (1994) Molecular determinants of human red/green color discrimination. Neuron 12:1131-1138

Backhaus W (1991) Color opponent coding in the visual system of the honeybee. Vision Res 31:1381-1397

Briscoe A (1998a) Molecular diversity of visual pigments in the butterfly *Papilio glaucus*. Naturwissenschaften85:33-35

Briscoe A (1998b) Evolution of the visual pigments in the butterfly *Papilio glaucus*. PhD Thesis, Harvard University

Briscoe A (1999) Intron splice sites of *Papilio glaucus* Pg1Rh1 corroborate insect opsin phylogeny. Gene 230:101-109

Britt SG, Feiler R, Kirschfeld K, Zuker CS (1993) Spectral tuning of rhodopsin and metarhodopsin in vivo. Neuron 11:29-39

Brooks DR, McLennan DH (1991) Phylogeny, Ecology, and Behavior. University of Chicago Press

Bruckmoser P (1968) Die spektrale Empfindlichkeit einzelner Sehzellen des Rückenschwimmers *Notonecta glauca* L. (Heteroptera). Z vergl Physiol 59:187-204

Caine NG, Mundy NI (2000) Demonstration of a foraging advantage for trichromatic marmosets (*Callithrix geoffroyi*) dependent on food colour. Proc R Soc Lond B 267: 439-444

Carulli JP, Hartl DL (1992) Variable rates of evolution among *Drosophila* opsin genes. Genetics 132: 193-204

Chittka L (1996a) Optimal sets of colour receptors and opponent processes for coding of natural objects in insect vision. J Theor Biol 181:179-196

Chittka L (1996b) Does bee colour vision predate the evolution of flower colour? Naturwissenschaften83:136-138

Chittka L (1997) Bee color vision is optimal for coding flower colors, but flower colors are not optimal for being coded - why? Isr J Plant Sci 45:115-127

Chittka L, Dornhaus A (1999) Comparisons in physiology and evolution, and why bees can do the things they do. Ciencia al Dia 2:1-17

Chittka L, Menzel R (1992) The evolutionary adaptation of flower colors and the insect pollinators' color vision systems. J Comp Physiol A 171:171-181

Chou W-H, Hall KJ, Wilson DB, Wideman CL, Townson SM, Chadwell LV, Britt SG (1996) Identification of a novel *Drosophila* opsin reveals specific patterning of the R7 and R8 photoreceptor cells. Neuron 11: 1101-1115

Crandall K, Hillis D (1997) Rhodopsin evolution in the dark. Nature 387:667-668

Crandall K, Cronin TW (1997) The molecular evolution of visual pigments of freshwater crayfishes (*Decapoda: Cambaridae*). J Mol Evol 45:524-534

Cronin TW, Järvilehto M, Weckström M, Lall AB (2000) Tuning of photoreceptor spectral sensitivity in fireflies (Coleoptera: Lampyridae). J Comp Physiol A 186:1-12

Deeb SS, Motulsky AG (1996) Molecular genetics of human color vision. Behav Genet 26:195-207

Douglas RH, Partridge JC, Marshall NJ (1998) The eyes of deep-sea fish I: lens pigmentation, tapeta and visual pigments. Progress in Retinal and Eye Research 17:597-636

Emlen JT (1978) Density anomalies and regulatory mechanisms in land bird populations on the Florida peninsula. American Naturalist 112:265-268

Endler JA, Basolo A, Glowacki S, Zerr J (submitted) Genetic variance and covariance structure of vision in guppies, *Poecilia reticulata*. Am Nat

Endler JA (1980) Natural selection on color patterns in *Poecilia reticulata*. Evolution 34:76-91

Endler JA (1986) Natural Selection in the Wild. Princeton University Press, Princeton

Fleishman LJ, Bowman M, Saunders D, Miller WE, Rury MJ, Loew ER (1997) The visual ecology of Puerto Rican anoline lizards: habitat light and spectral sensitivity. J Comp Physiol A 181:446-460

Goldsmith TH (1990) Optimization, constraint, and history in the evolution of eyes. Q Rev Biol 65:281-322

Gould SJ, Lewontin RC (1979) The spandrels of San Marco and the Panglossian paradigm: a critique of the adaptationist programme. Proc R Soc Lond B 205:581-598

Hariyama T, Tsukahara Y, Meyer-Rochow VB (1993) Spectral responses, including a UV-sensitive cell type, in the eye of the isopod *Ligia exotica*. Naturwissenschaften80:233-235

Harvey PG, Pagel MD (1991) The Comparative Method in Evolutionary Biology. Oxford University Press, Oxford

Hope AJ, Partridge JC, Dulai KS, Hunt DM (1997) Mechanisms of wavelength tuning in the rod opsins of deep-sea fishes. Proc R Soc Lond B 264:155-163

Huber A, Schulz S, Bentrop J, Groell C, Wolfrum U, Paulsen R (1997) Molecular cloning of *Drosophila* Rh6 rhodopsin: the visual pigment of a subset of R8 photoreceptor cells. FEBS Letters 406:6-10

Hunt DM, Fitzgibbon J, Slobodyanyuk SJ, Bowmaker JK (1996) Spectral tuning and molecular evolution of rod visual pigments in the species flock of cottoid fish in Lake Baikal. Vision Res 36:1217-1224

Ichikawa T, Tateda H (1982) Distribution of color receptors in the larval eyes of four species of Lepidoptera. J Comp Physiol 149:317-324

Kitamoto J, Sakamoto K, Ozaki K, Mishina Y, Arikawa K (1998) Two visual pigments in a singe photoreceptor cell: identification and histological localization of three mRNAs encoding visual pigment opsins in the retina of the the butterfly *Papilio xuthus*. J Exp Biol 201:1255-1261

Lin SW, Kochendoerfer GG, Caroll KS, Wang D, Mathies RA, Sakmar TP (1998) Mechanisms of spectral tuning in blue cone pigments. J Biol Chem 273:

24583-24591

Lythgoe JN (1972) The adaptation of visual pigments to the photic environment. In: Dartnall HJA (ed) Photochemistry of Vision (Handbook of Sensory Physiology, Vol.VII/1). Springer, Berlin, pp 566-603

Lythgoe JN, Partridge JC (1989) Visual pigments and the acquisition of visual information. J Exp Biol 146:1-20

McDonald CG, Hawryshyn CW (1995) Intraspecific variation of spectral sensitivity in three-spine stickleback (*Gasterosteus aculeatus*) from different photic regimes. J Comp Physiol A 176:255-260

Menzel R (1979) Spectral sensitivity and colour vision in invertebrates. In: Autrum H (ed) Invertebrate photoreceptors (Handbook of Sensory Physiology, Vol.VII/6A). Springer, Berlin, pp 503-580

Menzel R, Backhaus W (1991) Color vision in insects. In: Gouras P (ed) Vision and Visual Dysfunction. Macmillan, London, pp 123-145

Menzel R, Ventura DF, Hertel H, de Souza JM, Greggers U (1986) Spectral sensitivity of photoreceptors in insect compound eyes: comparison of species and methods. J Comp Physiol A 158:165-177

Mollon JD (1989) "Tho' she kneel'd in that place where they grew..." - the uses and origins of primate colour vision. J Exp Biol 146:21-38

Nakayama T, Zhang W, Cowan A, Kung M (1998) Mutagenesis studies of human red opsin: Trp-281 is essential for proper folding and protein-retinal interactions. Biochemistry 37: 17487-17494

Neitz M, Neitz J, Jacobs GH (1991) Spectral tuning of pigments underlying red-green color vision. Science 252:971-974

Papatsenko D, Sheng G, Desplan C (1997) A new rhodopsin in R8 photoreceptors of *Drosophila* - evidence for coordinate expression with Rh3 in R7 cells. Development 124:1665-1673

Peitsch D, Fietz A, Hertel H, de Souza J, Ventura DF, Menzel R (1992) The spectral input systems of hymenopteran insects and their receptor-based colour vision. J Comp Physiol A 170:23-40

Pollock JA, Benzer S (1988) Transcript localization of four opsin genes in the three visual organs of *Drosophila*; RH2 is ocellus specific. Nature 333:779-782

Reeve HK, Sherman PW (1993) Adaptation and the goals of evolutionary research. American Naturalist 68:1-32

Sacks O (1997) The Island of the Colorblind. Alfred A. Knopf, New York

Sakamoto K, Hisatomi O, Tokunaga F, Eguchi E (1996) Two opsins from the compound eye of the crab *Hemigrapsus sanguineus* . J Exp Biol 199:441-450

Seidou M, Sugahara M, Uchiyama H, Michinomae M, Yshihara K, Kito Y (1990) On the three visual pigments in the retina of the firefly squid, *Watasenia scintillans*. J Comp Physiol A 166: 769-773

Seki T, Vogt K (1998) Evolutionary aspects of the diversity of visual pigment chromophores in the class insecta. Comp Biochem Physiol 119B:53-64

Shyue S-K, Hewett-Emmett D, Sperling HG, Hunt DM, Bowmaker JK, Mollon JD, Li W-H (1995) Adaptive evolution of color vision genes in higher primates. Science 269:1265-1267

Smith WC, Goldsmith TH (1990) Phyletic aspects of the distribution of 3-Hydroxyretinal in the Class Insecta. J Mol Evol 30:72-84

Smith WC, Price DA, Greenberg RM, Battelle B-A (1993) Opsins from the lateral eyes and ocelli of the horseshoe crab, *Limulus polyphemus* . Proc Natl Acad Sci USA 90:6150-6154

Stanton ML, Galen C (1997) Life on the edge: adaptation versus environmentally mediated gene flow in the snow buttercup, *Ranunculus adoneus* . Am Nat 150:143-178

Stavenga DG, Smits RP, Hoenders BJ (1993) Simple exponential functions describing the absorbance bands of visual pigment spectra. Vision Res 33,8:1011-1017

Sun H, Macke JP, Nathans J (1997) Mechanisms of spectral tuning in the mouse green cone pigment PNAS 94: 8860-8865

White RH, Stevenson RD, Bennett RR, Cutler DE (1994) Wavelength discrimination and the role of ultraviolet vision in the feeding behavior of hawkmoths. Biotropica 26:427-435

Zuker CS, Cowman AF, Rubin GM (1985) Isolation and structure of a rhodopsin gene from *D. melanogaster*. Cell 40:851-858

II.1 Impedance Matching in Sound Production and Hearing: a Comparative Study

Henry C. Bennet-Clark

Department of Zoology, University of Oxford, South Parks Road, Oxford, OX1 3PS, UK

Abstract
Sound production involves stages of impedance matching between the higher-density body of the animal and lower-density air. Hearing involves further impedance matching, from air to the higher-density sensory cells. Initial stages of sound production may include a frequency multiplier that converts slow muscle contractions into higher frequency mechanical vibrations. The frequency multiplier may also determine the sound frequency. Larger sound sources allow better impedance amtching with the air so sound radiation often exploits acoustic transformers to increase the effective size of the sound source. In many cases the conversion efficiency of muscle power to sound power is high, giving a large effective range for the signal. Sound detection often uses acoustic transformers to concentrate the sound onto relatively dense vibrating sturctures that are coupled to the sensory cells, providing the inverse of the impedance matching that occurs in sound production. These transformers may be associated with directional mechanisms and may drive arrays of receptors that allow frequency analysis. Sound production and hearing are special cases of the general phenomenon of impedance matching that occurs throughout biomechanics.

Key words Sound production, impedance matching, efficiency, hearing, frequency analysis

1 Introduction: the Stages in Sound Production and Hearing

The chain of processes in acoustic communication in animals involves the production of a signal which is radiated as sound, the propagation of the sound through the medium, its reception, and its effect on the receiving animal. Here, in the context of sensory ecology, I shall confine myself to the physical processes of the interfaces between the sound-producing animal and its environment or between the environment and the receiving animal. In sound production the relevant stages are: (1) conversion of muscle power at low velocity and high force into mechanical vibration; (2) the transformation of the vibration into acoustic vibration; (3) the radiation of sound. In sound reception the stages are: (1) sound power is collected; (2) it is converted into mechanical vibration; (3) the vibration stimulates the sensory cells. For all these stages, efficiency is desirable. I shall endeavour to show that in both sound production and hearing, efficiency depends on impedance matching between the animal and the medium in which the communication takes place.

2 Physics of Transfer of Sound Between Different Media

2.1 Sound Waves and Sound Intensity

Sound is propagated as a wave of compression and rarefaction of the fluid medium which travels outwards from the source at the speed of sound. Because sound depends on the compressibility of the medium, a sound wave has two components: the sound pressure that causes the compression, and the effective velocity of the particles of the medium under compression. In the transfer of power between two media, such as air and water, it is necessary to consider the different acoustic properties of the media. In a sound wave, the sound intensity is expressed, by reference to the pressure (force/area) and fluid particle velocity, as power per unit area: values for sound intensity are usually given in decibels (dB) relative to 10^{-12} Wm^{-2}. The acoustic properties of air and water are shown in Table 1. Sound intensity can be expressed in terms of (1) the sound pressure P, and the particle velocity U, or (2) of either one of these and the specific acoustic resistance of the medium $\rho \cdot c$, where ρ is the density of the medium and c is the velocity of sound in the medium:

$$Sound\ Intensity = P \cdot U = \frac{P}{\rho \cdot c} = U^2 \cdot \rho \cdot c \qquad Eqn.1$$

Table 1. Acoustic properties of air and water

	Medium		
	Water		Air
Density ρ (kgm^3)	1000		1.2
Velocity of sound c (ms^{-1})	1500		340
Specific acoustic resistance ρ/c (kgm^2s^{-1})	$1.5 \cdot 10^6$		410
Ratio of specific acoustic resistances	3650	to	1
Sound pressure ratio	60.4	to	1
Particle velocity ratio	1	to	60.4

2.2 Transfer of Sound Power Between Different Media

The maximum power transfer theorem (Langford-Smith 1953) states that, for optimal source-to-load coupling, the source and load resistances should be equal, or, extended to oscillatory systems, if the source also has a reactive component such as a mass, there should be a matching spring component in the load and vice versa. Thus, for maximal power transfer, the source and load impedances should be equal. At the interface of a sound wave in one medium with another,

$$\textit{the ratio of the transmitted power to the incident power} = \frac{\rho c_1 \cdot \rho c_2}{\left(\rho c_1 + \rho c_2\right)^2} \qquad Eqn.2$$

where ρc_1 and ρc_2 are the specific acoustic resistances of the two media (Olson 1957). At an interface between air and water, which differ in their specific acoustic resistances by a factor of 3650 to 1 (Table 1), only 0.001 of the sound power in either medium is transmitted to the other: 0.999 is reflected at the interface.

From Table 1, it can be seen that maximum power transfer, from tissue to air as in sound production, requires an acoustic transformer that produces a 60-fold decrease in the pressure in the system (e.g. by a relative increase in area between the driver and the radiating surface) while increasing the velocity by 60-fold (e.g., by a mechanical or acoustic "lever"). From air to tissue, the transformer should work the other way: to increase the pressure 60-fold and to reduce the velocity 60-fold.

3 Impedance Matching in Sound Production

3.1 Desiderata for a Sound Signal Used in Communication

From the sender's point of view, it is important to be heard and to be recognized, this implies first that the signal should be loud and second that it should carry such species-specific information as defined frequency and pattern. Loudness implies a high acoustic power which implies good transformation of muscle power into sound power and impedance matching at all stages outlined in Section 1. In the remainder of Section 3, I shall explore the properties of these links.

3.2 Muscle as a Power Source in Sound Production

There is a maximum strain rate at which the muscles may shorten, which is usually less than 25 times the original length per second, at which no external force is produced, and a maximum pressure that can be produced, about $400 \cdot 10^3$ Nm^{-2}, at which no shortening occurs (Weis-Fogh and Alexander 1977). At about one third of the maximum velocity, the muscle produces its maximum mechanical power (the product of its force and shortening velocity). Power outputs of up to 250 Wkg^{-1} have been recorded.

Contraction frequencies of over 40 Hz for the flight muscles of humming birds (Weis-Fogh 1972) and of between 100 and 500 Hz from insect sound-producing muscles (Josephson and Young 1985) have been recorded. At a frequency of contraction of about 100 Hz, the tymbal muscle of the cicada *Cyclochila australasiae* produces forces of $22 \cdot 10^3$ Nm^{-2} (Bennet-Clark and Daws 1999). One atmosphere is $100 \cdot 10^3$ Nm^{-2} , so these tymbal muscles are producing a working pressure of 0.2 atm. An acoustic change of ± 1 atm corresponds to a sound pressure of 194 dB (!). However, the highest sound pressures recorded inside the abdominal air sac of *C. australasiae* are 155 to 159 dB, or ± 0.01 to 0.02 atm (Young and Bennet-Clark 1995). There is a 20-fold stage of reduction of pressure between the muscle and the inside of the insect; this must be a stage of impedance matching because the transduction efficiency between the insect's muscles and the sound that it radiates is at least 18 %. and possibly as high as 46 %. (Bennet-Clark and Daws 1999).

Muscle pressures are appropriate for locomotion or feeding but are far larger than those in acoustic communication; loud music in a concert hall, 94 dB, is 10^{-5} atmospheres and many animals can hear sound pressures of below 10^{-3} Nm^{-2} or 10^{-8} atm.

Frequency multiplication usully occurs between the muscle and the first mechanical stage. The muscle contraction rate in *C. australasiae* is about 120 Hz but the song frequency is 4.3 kHz (Josephson and Young 1981). Each muscle contraction causes the tymbal to buckle and produce a series of clicks at close to the song frequency (Young and Bennet-Clark 1995; Bennet-Clark 1997). The tymbal acts as a frequency multiplier (see Sect. 3.3).

In summary, muscle produces pressures that vastly exceed those encountered in everyday acoustics in air at frequencies that are far lower than those typical of acoustic communication.

3.3 The Role of Oscillators and Other Vibrating Structures

Typically, a single muscle contraction powers one or a series of sound pulses. Where continuous sounds are produced, they are produced by the overlapping alternating contraction of contralateral muscles as in the whistling song of the periodical cicada *Magicicada septendecim* (Young and Josephson 1983). The ubiquity of frequency multiplier mechanisms that convert a long slow muscle contraction into a higher frequency has been highlighted by Michelsen (1983).

Frequency multipliers may be driven by a forced air current, by impacts, by escapements etc. (for a comprehensive early review, see Dumortier 1963a) but, although such structures may set the frequency of the sound, they do not always act as the effective sound source.

Typically, the structure that is driven has resonant properties which control the sound frequency within a narrow band. Examples of air-driven resonators are the mammalian larynx or bird syrinx (for reviews, see Busnel 1963). In many of this type of structure, a reed or membrane is set in vibration by the air current, the velocity of which is modulated by the vibration of the mechanical element but the overall tonal quality may also be affected by accessory structures (see section 3.4).

A wide variety of stridulatory mechanisms is found in insects, usually with a file or row of teeth that makes periodic contact with a ridge whereby, with the passage of one across the other, a series of impacts is produced. The effect of these impacts depends on the shape of the teeth and scraper, the speed of the cycle of catch and release of the teeth and on the nature of the structures that receive the impacts (Dumortier 1963a, b).

Many of this type of structure appear to be resonant (see e.g Nocke 1971 and, for a recent review, Bennet-Clark 1999). The resonant properties depend on the interaction of an elastic element with an inertial one; the resonant frequency, F_O, of the simplest mass-and-spring resonator is given by:

$$F_O = \frac{1}{2\pi} \cdot \sqrt{\left(\frac{1}{m \cdot C_m}\right)} = \frac{1}{2\pi} \cdot \sqrt{\left(\frac{s}{m}\right)} \qquad \textit{Eqn. } 3$$

where m is the mass, C_m is the compliance and s is the stiffness (the compliance of a spring is the reciprocal of its stiffness). Where the other parameter remains constant, the resonant frequency is proportional to $1/\sqrt{\text{mass}}$ or to $\sqrt{\text{stiffness}}$ (Morse 1948; Alexander 1988). Unless either the mass or the stiffness of the structure can be varied, the frequency of vibration is fixed; this appears to be approximately true for the songs of many crickets (Orthoptera-Ensifera) as well as of many cicadas (Hemiptera-Homoptera) (Leroy 1966; Bennet-Clark 1999) although in many cases that have been examined carefully, there is a steady change in the frequency throughout each song pulse (Simmons and Ritchie 1996; Bennet-Clark 1999), which implies that the mass and/or stiffness of the system do not remain constant .

Where long song pulses are produced, sound production may depend on regular and repeated excitation of the vibrating structure at its resonant frequency: mechanisms for sustaining the vibration have been described for the harp region of the forewings of crickets (Elliott and Koch 1985; Koch et al. 1988) and for the tymbals of cicadas (Bennet-Clark 1997). Resonances in both of these determine the song frequency and multiple excitation sustains the resonance over many cycles, producing song pulses of between 12 cycles in some cicadas (see Bennet-Clark 1999) through 100 cycles in many gryllids (Leroy 1966) to over 1000 cycles in the whistling songs of many amphibians, birds or mammals.

The pulse duration correlates with the number of excitations given to the resonator and on the rate of excitation. In the cicada *Cyclochila australasiae*, the 12-cycle song pulse is excited by two or three impulses, the first of which occurs at the start of the pulse with further impulses at intervals of two or three cycles. In crickets such as *Gryllus campestris*, the sound pulse is sustained on a cycle-by-cycle basis by the catch and release of the teeth of file on one wing by the plectrum or scraper on the edge of the other wing (see reviews by Bennet-Clark 1989, 1999); between 35 and 65 %. of the 100 or more teeth on the file are engaged to make the pulse (Leroy 1966).

Whether resonant or not, the muscle-powered vibrating structure appears to have two major roles: to act as a frequency multiplier and as a determinant of the sound frequency. However, this primary vibrator may not be the effective sound radiator: sound radiated from the outside of the tymbal of the cicada *C. australasiae* is 13 dB quieter than that radiated *via* its tympana (or ear drums). For optimum impedance matching between the muscle power source and the vibrating structure, it is mechanically feasible to drive stiff structures with a large mass because the muscle can produce large pressures.

Thus, the primary vibrating structure may be a small thick structure, with large mass and high stiffness, rather than one that is lighter and less stiff with the same resonant frequency. Examples of this are cicada tymbals or vertebrate larynges and syrinxes, none of which act as the primary or sole sound radiating surface. In many animals there is an additional stage of transduction and impedance matching (Sect. 3.4).

3.4 Problems of Sound Radiation

Impedance matching between source and medium is well known from acoustical engineering (Olson 1957). The specific acoustic resistance of a source depends on its configuration: whether it is a freely-suspended piston, a piston in the side of a closed box, a pulsating sphere etc ; sources of all these types are found in animals.

The relationship between the radius of a source and its specific radiation resistance relative to that of the surrounding medium is shown in Fig. 1 for two types of source: a monopole source or a vibrating piston in an infinite baffle (Fig. 1A) and a dipole source or a free-edged piston vibrating normal to its plane (Fig. 1B). Because there is no back-to-front leakage with the monopole source, its specific acoustic resistance decreases less rapidly with decreasing size than that of the free edge piston, where back-to-front leakage becomes worse as the source becomes smaller.

There will be good impedance matching between the source and its surroundings if the radius of a momopole exceeds 0.16 of the sound wavelength λ, or, if a dipole, it exceeds 0.24 λ (Fig.1C). Table 2 gives examples of these types of source found in animals.

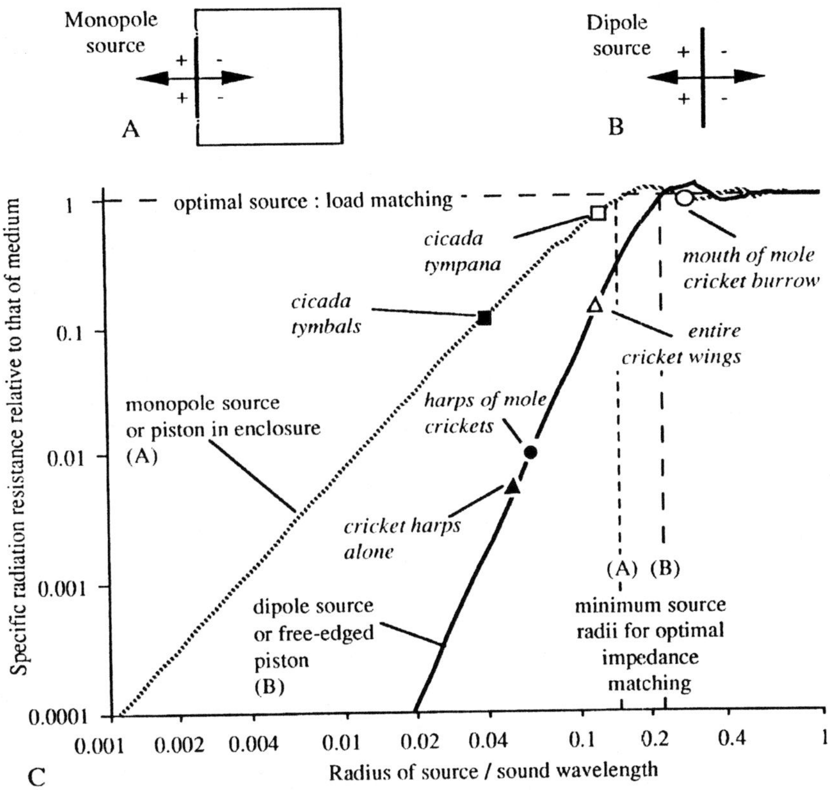

Fig. 1 A-C. Acoustic properties of sound sources. **A** With a monopole source; sound is radiated from the open (*left*) side of the piston. **B** With a dipole source, sound is radiated from both sides of the piston and interacts destructively at the edge of the piston. **C** Plot of the specific radiation resistance of these two types of source. For optimal source to load matching, the source resistance should equal that of the medium (*horizontal dashes*). *Solid symbols* show the specific acoustic resistances of the primary sound sources and *open symbols* show the specific acoustic resistances of the actual sound radiating structures of the same three insects. (Data from Olson 1957; Bennet-Clark 1970, 1987, 1989; Young 1990)

Good impedance matching between the animal's sound-producing structure(s) and the surrounding medium implies a correlation between the size of the sound source and the dominant frequency of the sound that is produced (Bennet-Clark 1998). Two factors may conflict: the source should be large enough to provide good impedance matching, but will be cumbersome if it is too large.

Over a range of sound frequencies between 2 and 100 kHz, there is an approximately linear fit between the source size and the sound frequency produced by many groups of insects (Bennet-Clark 1998) and one species of frog, *Hyla cineraria*, (see Duellman and Trueb 1985; pers. observ.; Fig. 2). This suggests that the sound-producing mechanisms are of appropriate size for good impedance matching. However, in all the cases shown in Fig. 2, the source size is smaller than the minimum for ideal impedance matching (as shown by the diagonal lines on Fig. 2) which suggests that overlarge sound sources may be avoided because they are otherwise disadvantageous.

Table 2. Examples of the types of sound source found in animals

Source type	Animal (popular name)	Group	Location of source (reference)
Monopole	*Cyclochila australiasiae* (cicada)	Insecta Hemiptera Homoptera	Tympana on abdomen (Young 1990)
Dipole	*Gryllus campestris* (field cricket)	Insecta Orthoptera Ensifera	Harp area on fore wings (Nocke 1971)
Pulsating sphere	*Hyla cinerea* (tree frog)	Amphibia Anura	Buccal vocal sac (Duellman and Trueb 1985)

3.5 Increasing the Radiation Resistance of the Source

As has been indicated in Section 3.4, it is advantageous to match the specific radiation resistance of a source to that of the medium. In particular, an immediate advantage is gained if a small source is surrounded by a baffle or plate (Fig.1C). A dipole will benefit most from any increase in its effective area (Fig.1 B, C). One example where this may be seen is the pent wings of singing crickets (e.g. *Gryllus* spp.), in which a pair of 5 x 3 mm vibrating surfaces are set in the middle of wings that, when raised and held together in the singing position, make a source that has an effective radius of c. 12 mm.

As Fig.1C shows, the probable gain in specific acoustic resistance is considerable. A similar trick is used by tree crickets of the genus *Oecanthus*, which sing with their wings extended across the gap between leaves (Prozesky-Schulze et al. 1975) thus raising the source impedance - in one example, the loudness was increased by c. 3.5 x or by 11 dB.

Other insect singers use of more specialized mechanisms in which the small, stiff large-mass primary vibrator (Sect. 3.3) is used to drive a larger secondary structure from which the sound is radiated. In mole crickets (Bennet-Clark 1987), conventional cicadas (Bennet-Clark and Young 1992) and bladder cicadas (Bennet-Clark and Young 1998), the secondary structures are resonant at the same frequency as the primary one. In such a system, the purely resistive impedance of the secondary resonator provides a mechanism for drawing power from the primary resonator without altering its resonant frequency (Bennet-Clark 1987, 1995).

Male mole crickets (Orthoptera-Ensifera) sing from burrows (Bennet-Clark 1970, 1987; Daws 1996) which have two main elements, an outer horn that opens at the soil surface and an inner ovoid bulb (Fig.3). The insect sings at the junction between these with its wings raised across the throat of the horn (Bennet-Clark 1970).

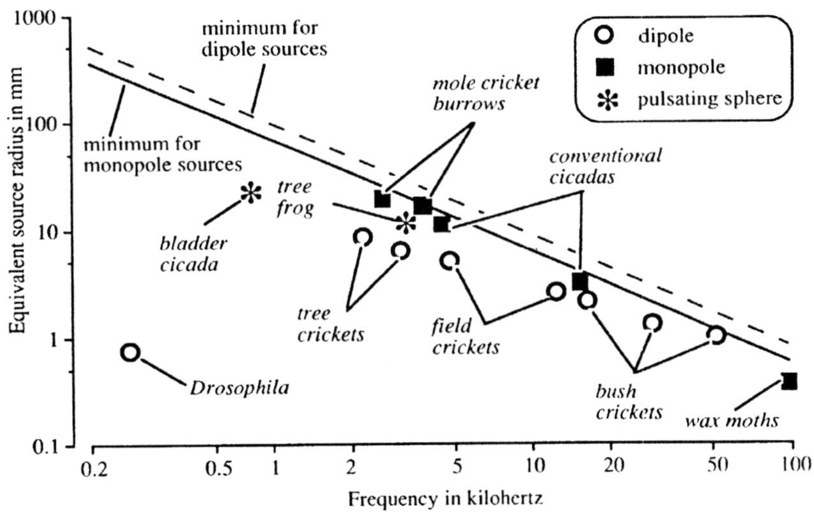

Fig. 2. Plot of the equivalent source radius versus sound frequency for various insects and a tree frog. The *solid line* and *dashes* show the minimum source size for optimal source to load impedance matching (see Fig.1) with monopole (Fig.1A) and dipole (Fig.1B) sources, respectively. The optimal size for a pulsating sphere source (*) is similar to that for a monopole source. The *inset* shows the symbols used for the three types of source. (Data from Bennet-Clark 1970, 1971, 1987, 1998, pers. observ.; Young 1990)

The bulb and horn tune the burrow to the insect's song frequency (Bennet-Clark 1987, Daws et al. 1996); the burrow provides appropriate coupling for the primary sources on the insect's wings to the throat of the horn (Daws et al. 1996). The exponential flare of the horn provides an acoustic transformer that provides impedance matching between the small wings and the large mouth of the horn at the soil surface (Figs. 1C and 3). The gain due to the burrow is 24 dB (which represents a pressure gain of 15.8-fold or a power gain of about 240-fold: Eqn. 1) (Bennet-Clark 1987). The efficiency of conversion of muscle power into sound power may be over 30 % (Bennet-Clark 1970).

Table 3. Comparison of primary and secondary resonators involved in sound production

Insect	Mole cricket *Scapteriscus acletus* (Fig.3)	Typical cicada *Cyclochila australasiae*	Bladder cicada *Cystosoma saundersii*
Song frequency	2.7 kHz	4.3 kHz	830 Hz
Primary resonator (see Sect. 3.3)	Paired harps on fore wings	Paired tymbals on abdomen	Paired tymbals on abdomen
Size of 1^{ary} resonator	5 × 3 mm	6 × 5 mm	4 × 3 mm
Secondary resonator	Horn-shaped burrow	Helmholtz resonator	Thin-walled bladder
Inertial element	Effective mass of horn in burrow	Effective mass of tympanal openings	Mass of flat areas Abdominal cuticle
Compliant element	Volume of bulb in burrow	Volume of abdominal air sac	Volume of abdominal air sac
Acoustic output from secondary resonator	Mouth of horn	Large area of thin tympana	Flat region of thin wall of abdomen
Dimensions of 2^{ary} sound source(s)	30 × 40 mm	10 × 22 mm	Two 10 mm diam + one 15 × 10 mm
Song wavelength	126 mm	79 mm	410 mm
Optimal source diameter	40 mm	25 mm	130 mm

For references, see text

In typical cicadas such as males of *Cyclochila australasiae*, the large abdominal air sac and the thin tympana (or eardrums) act as the elements of a Helmholtz resonator (Bennet-Clark and Young 1992). Although the tymbals are the primary sound sources (Pringle 1954), the sound they emit is 13 dB quieter than that emitted through the tympana which act as the effective sound source (Young 1990). The efficiency of transduction of muscle power into sound power may be as high as 46 % (Bennet-Clark and Daws 1999).

In male bladder cicadas such as *Cystosoma saundersii*, the elements of the secondary resonator are the compliance of the large abdominal air sac and the mass of the flat areas of abdominal cuticle from which the sound is radiated (Table 3). The abdominal bladder resonates sharply at the 830 Hz frequency of the insects' song (Bennet-Clark and Young 1998).

All these secondary resonators share two features: first, the secondary resonator considerably increases the source size and hence provides a closer match between the source and the surrounding medium (Fig.1C); second, the sound is radiated from the mass-like or inertial elements of the resonator.

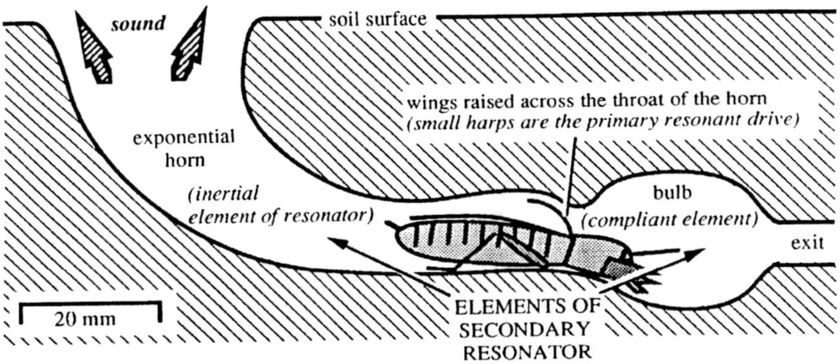

Fig. 3. Diagram of the inertial (mass-like) and compliant (spring-like) elements involved in the tuning of the secondary resonator from which sound is radiated by a mole cricket such as *Scapteriscus acletus*. The specialized horn-shaped burrow is tuned to the frequency of vibration of the harps on the fore-wings. (Bennet-Clark 1995; Daws et al. 1996)

3.6 Impedance Matching in Sound Production: How Effective Is It ?

Many animal sounds are remarkably loud. *Gryllotalpa vineae*, a 3 g mole cricket, produces peak sound levels of 95 dB at 1-m range (Bennet-Clark 1970); cicadas of similar size produce peak levels of around 100 dB at 1 m (e.g., Young 1990). Such intensities occur in music or from machinery, but weight for weight, animals can be very noisy.

4 Impedance Matching in Hearing

4.1 Desiderata for an Auditory System

For information to be received, the recipient must have an auditory system that is sensitive to the relevant sound, is able to evaluate its quality and can provide information about the direction from which it is coming. The neural response requires mechanical excitation of sensory neurons and hence impedance matching between the sound wave and the sensory chain. Directionality will be considered only briefly here.

Hearing involves a chain of elements that couple the sound wave to a compliant structure that is set into vibration which then provides mechanical drive to the sensory cells and finally produces a neural response; the response may be dependent on the intensity and frequency of the sound.

4.2 The Primary Vibrating Structure

The simplest type of sound detector is a vane extended in the sound field which is driven by the particle velocity or fluid movement due to the sound wave and actuates sensilla at the base of the vane: examples are the antennae of mosquitoes (Göpfert et al. 1999) or the sensory hairs of many insects and arachnids (e.g., Michelsen 1983; Fletcher 1978, 1992; Barth et al. 1993).

In pressure-driven ears, a thin diaphragm extends over an air-filled cavity; the pressure in the sound wave causes vibration of the diaphragm which drives the sensory cells. Because acoustic pressures are very small relative to, e.g., atmospheric or ventilatory pressure changes, the cavity is usually vented by a narrow pipe, such as the tracheal system in insects or the eustachian tube in tetrapods. Such ears are found in insects such as locusts, cicadas or moths and throughout the tetrapods (see, e.g. Autrum 1963, Michelsen 1983).

In both types of sound detector, the primary vibrating structure tends to be light in weight but to have a large effective surface area. In antennae, lightness with large surface area may be achieved, as with a feather, by a central rod bearing lateral hairs. At the low velocities at which such structures vibrate, the relatively thick air boundary layer increases the effective area of the antenna. If the vibrating structure is too thin, it will bend along its length, and if it is too thick, it will be too heavy to bend in the sound wave.

In pressure-driven ears, the diaphragm may be thin: thicknesses of 0.5 to 2 μm have been reported for parts of the tympana of locusts (Stephen and Bennet-Clark 1982) and cicadas (Young 1990). However, if the diaphragm is thin, it will readily transmit rather than absorb sound and it may be difficult to extract energy from its vibration; a thick heavy diaphragm may have a small amplitude of vibration but it will be able to drive a larger group of sensory cells.

Many sound sensors have resonant properties. In antennae, the mass or rotational moment of inertia of the vane will interact with the stiffness of the basal articulation (Humphrey et al. 1993). In a pressure-driven ear, the mass of the diaphragm and its stiffness will interact (Fletcher 1992).

In general, smaller ears of either type tend to respond at higher frequencies. Simple resonant properties give rise to a V-shaped threshold response curve, with maximal sensitivity at the resonant frequency (e.g. Göpfert et al. 1999). At lower frequencies, the response will be limited by the stiffness, and at higher frequencies by the mass or inertia of the system. Ears tend to have arrays of sensory cells that allow analysis over a band of frequencies: these have been described in as diverse animals as bush crickets (Oldfield 1982), locusts (Michelsen 1971) and mammals (von Békésy 1960). In many cases, the mechanism of tuning is incompletely understood.

4.3 Coupling the Medium to the Primary Vibrating Structure

Ear drums tend to be thin and fragile and are often protected by an external tube or cover. Such tubes or covers can be shaped as horns or reflectors to increase the sound pressure reaching the ear drum approximately in the ratio of the relative areas of the reflector and the ear drum.

The pinnae of therian mammals such as rabbits provide gains of from 20 to nearly 30 dB (a 100- to 1000-fold increase in sensitivity) between 2 and 5 kHz (Lewis 1983, see also Shaw 1974). These horn-shaped reflectors are frequency sensitive and only start to be effective when their dimensions approach 1/3 λ, as is predicted from their specific acoustic resistance (Fig. 2C) or, with a typical rabbit ear that is 100 mm long, above 1 kHz. Significant further gain may be provided by the tapering external auditory meatus (Fig.4).

A horn-like tracheal sound collector is connected to the ears of tettigoniid bush crickets. Hill and Oldfield (1981) measured the effective gain due to the tracheal horn in two bush cricket species using a probe microphone inserted into the trachea at the cut distal end of the fore femur. The effective gain was +15 to +20 dB between 15 and 20 kHz which was similar to the difference in auditory sensitivity in the intact animal and after removal of the sound input via the acoustic trachea and also correlated with the gain predicted from the dimensions of the acoustic trachea. Diffraction by the side of the head or body also provides a useful increase in sound pressure of ca. 6 dB on the side facing the sound source at frequencies above one half to one third (see Olson 1957).

4.4 Coupling the Primary Vibrating Structure to the Sensory Cells

The major impedance mismatch in an auditory system is likely to be between the light thin primary vibrating structure (e.g., an ear drum) and the sensory cells because the specific acoustic resistances of the two structures differ so greatly (Sect. 2.2, Table 1): ideally a 60-fold transformation of pressure is required for optimal impedance matching. Pressure transformers are common in engineering and typically use levers or hydraulic mechanisms. In the simple ear of noctuid moths, two sensory cells are attached near the centre of a thin tympanic membrane: the attachment site of these cells is only about 1/10 the diameter of the tympanum so their relative areas are about 1 to 100 (measured from figures in Treat and Roeder 1959). The coupling between the tympanum and the sensory cells in cicada ears appears to involve a lever system (Daws 1996, Doolan and Young 1981). Unfortunately, there are no data about the relative areas or lever ratios in these systems.

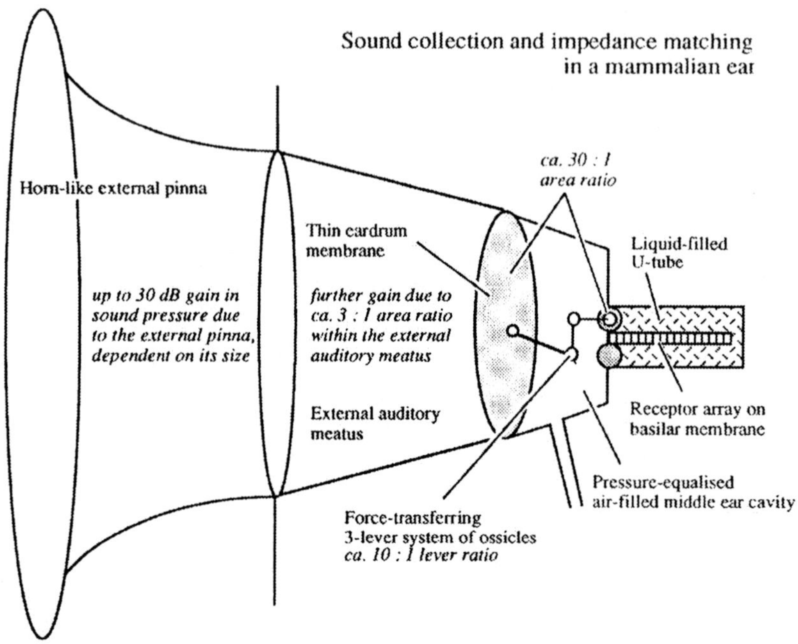

Fig. 4. Diagram of the ear of a therian mammal to show the sound-collecting roles of the external pinna and external auditory meatus with the impedance matching roles (in *italics*) of the chain of ear ossicles and the relative areas of the eardrum and oval window in the cochlea (in *italics*). (From data in Shaw 1974; Møller 1974; Lewis 1983)

In the ears of mammals, the large-area pinna collects sound and leads into the tapering external auditory meatus (Fig.4); these act together to increase the sound pressure at the tympanum. The middle ear ossicles of the mammalian ear act as a lever system that transfers the movements of the tympanum to the oval window of the cochlea (Fig. 4). The lever ratio of the ossicular chain is about 1.3 to 1 and the area ratio between the tympanum and the oval window is about 14 to 1, so the product of these values gives an overall transformer ratio between the tympanum and the oval window of about 18 : 1 (Møller 1974); this gives to an impedance ratio of $(18)^2$ (Møller 1974; Fletcher 1992) or 330 to 1 which is about one tenth of the optimum for ideal impedance matching (cf. Table 1). The threshold sound sensitivity of the ear (20 μPa) corresponds to a vibration amplitude in air of about 10^{-11} m, which, after transduction via the eaxternal auditory meatus, tympanum and ossicular chain, leads to a movement of the basilar membrane and hence of the hair cells of 10^{-11} m (Fletcher 1992). Despite the considerable impedance mismatch, the auditory sensitivity is close to the level of ambient noise in the environment.

Within the inner ear, impedance matching between the liquid medium and the sensory cells is associated with frequency analysis. The sensory cells are attached to the basilar membrane which, near to the oval window is narrow and stiff so it can only vibrate at high frequencies but, towards the distal end of the cochlea, the basilar membrane is both broad and compliant so it vibrates at lower frequencies: this oversimplified account of cochlea mechanics highlights the fact that good impedance matching may be highly frequency-dependent. More detailed accounts of cochlear mechanics are found in von Békésy (1960) or Fletcher (1992).

Other evidence for frequency-dependent impedance matching comes from locust ears (Stephen and Bennet-Clark 1982), which are situated on either side of the abdomen. Each ear consists of a tympanic membrane with a thick flexible anterior region and a thin posterior dorsal region which bears three sclerites (Fig.5A) to which the four different groups of the sensory cells that make up Müller's organ are attached (Gray 1960). Müller's organ is suspended by its sensory cells on the tympanum and connected to the rest of the insect only by the tympanal nerve.

The thickness of the tympanal membrane varies from region to region: around the folded and styliform bodies, it is between 2 and 4 µm thick, but there is a thin compliant arc around the posterior edge of the tympanum that is less than 1 µm thick. The different regions show independence of movement when driven by high pressure sound (Stephen and Bennet-Clark 1982; Breckow and Sippel 1985). Calculations of the resonant frequencies (Eqn. 3) of the different regions (Fig. 5B), using an estimate of the mass of the different regions based on the thickness of the membrane, and its measured compliance (Stephen and Bennet-Clark 1982), gave values close to the measured frequencies of maximal amplitude of vibration of the membrane (Breckow and Sippel 1985). The best frequencies of the C, B and D groups of sensory cells (Michelsen 1971) suggest that the mechanics of the membrane determines the different responses of the different regions (Stephen and Bennet-Clark 1982; confirmed by Breckow and Sippel 1985).

The adequate stimulus requires mechanical strain of the sensory cells. The attachment of Müller's organ to the tympanum is highly compliant and its calculated resonant frequency is lower than that of any region of the tympanum (Fig. 5B; Stephen and Bennet-Clark 1982). When the tympanum vibrates, the relatively heavy Müller's organ vibrates at a lower amplitude and at a different phase to the tympanum, so there is relative movement between the ends of the sensory cells by which Müller's organ is suspended; this thereby provides frequency-dependent adequate stimuli.

The locust ear also provides evidence of close-to-optimal impedance matching between the tympanum and the sensory cells. When the vibration of the folded body region was compared before and after removal of Müller's organ, first, the resonant frequency was unchanged but, second, the amplitude of vibration of the unloaded membrane increased by 4 dB (Stephen and Bennet-Clark 1982). The first observation showed that the loading of the tympanum due to Müller's organ acted as viscous damping on the membrane, so mechanical power was being absorbed from the membrane by the sensory cell regions of Müller's organ. The second observation showed that approximately half the energy of vibration of the system was being transferred from the tympanum (as the source) to Müller's organ (as the load) suggesting that the system closely meets the requirements of the maximum power transfer theorem (Sect. 2.2).

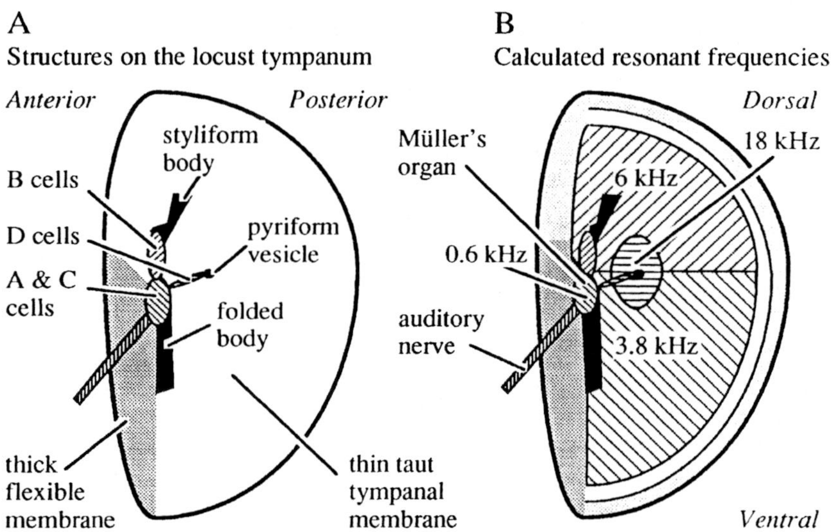

Fig. 5 A, B. Diagram of the right ear of a locust viewed from the inside. **A** shows the sclerites in the tympanum and the attachments to them of the different groups of sensory cells. **B** shows the resonant frequencies that were calculated for the differently *cross-hatched* regions of the tympanum and for Müller's organ (all groups of sensory cell s) based on measurements of the compliance of the different structures and their estimated masses. The *stippled thick membrane* and the arc of thin cuticle around the posterior part ot the tympanum (*left in hatched*) appear to delimit the vibration of the tympanum. (After data in Stephen and Bennet-Clark 1982)

4.5 How Effective is Impedance Matching in Auditory Systems ?

Throughout Section 4, I have endeavoured to show that some form of impedance matching occurs between the different steps in the auditory chain. However, few systems have been studied consistently and so hard numbers are difficult to obtain. Even so, the picture that emerges is that stage-to-stage impedance matching occurs in auditory systems as well as in sound-producing systems. As a consequence, ears typically have thresholds of sensitivity that allow the detection of sounds that are close to the background noise level in the environment.

5 Conclusion

The special case of sound communication is merely one example of the general phenomenon of selection for optimization of function that has occurred in animals (Alexander 1996). Other examples, which have important parallels with sound communication, include the loading of muscles for maximum efficiency in locomotion and similar tasks (Alexander 1973). In the context of sound communication, there is ample evidence that the loudness and species specificity of sound production and also the sensitivity and frequency selectivity of hearing are of great biological significance. My aim in this chapter has been to show how natural selection in different animal groups has led to a variety of solutions to the shared physical problems of impedance matching between the animal and the medium in which it lives.

Acknowledgments. Grateful thanks to the Austrian Academy of Sciences for their generous support of this symposium and to Professor F G Barth and his colleagues for organizing such an interesting programme and inviting me to participate.

References

Alexander RMcN (1973) Muscle performance in locomotion and other strenuous activities. In: Bolis L, Schmidt-Nielsen K, Maddrell S H P (eds) Comparative Physiology: Locomotion, Respiration, Transport and Blood. North-Holland: Amsterdam. pp 1-21

Alexander RMcN (1988) Elastic Mechanisms in Animal Movement. Cambridge University Press, Cambridge

Alexander RMcN (1996) Optima for Animals. 2nd edn Princeton University Press, Princeton, NJ

Autrum H (1963) Anatomy and physiology of sound sensory cells. In: Busnel R-G (ed) Acoustic Behaviour of Animals. Elsevier, Amsterdam. pp 412-433

Barth FG, Wastl U, Humphrey JAC, Devarakonda R. (1993) Dynamics of arthopod filiform hairs. II Mechanical properties of spider trichobothria (*Cupiennius salei* Keys.). Phil Trans R Soc Lond B. 340: 445-461

Békésy G von (1960) Experiments in Hearing. McGraw-Hill, New York

Bennet-Clark HC (1970) The mechanism and efficiency of sound production in mole crickets. J Exp Biol 52: 619-652

Bennet-Clark HC (1971) Acoustics of insect song. Nature 234: 255-259

Bennet-Clark HC (1987) The tuned singing burrow of mole crickets. J Exp Biol 128: 383-409

Bennet-Clark HC (1989) Songs and the physics of sound production. In: Huber F, Moore TE, Loher W (eds) Cricket Behavior and Neurobiology. Cornell University Press, Ithaca, NewYork pp 227 - 261

Bennet-Clark HC (1995) Insect sound production: transduction mechanisms and impedance matching. In: Ellington C P, Pedley T J (eds) Biological Fluid Dynamics. Company of Biologists, Cambridge. pp. 199-218

Bennet-Clark HC (1997) Tymbal mechanics and the control of song frequency in the cicada *Cyclochila australasiae*. J Exp Biol 200: 1681-1694

Bennet-Clark HC (1998) Size and scale effects as constraints in insect sound communication. Phil Trans R Soc Lond B 353: 407-419

Bennet-Clark HC (1999) Resonators in insect sound production: how insects produce loud pure-tone songs. J Exp Biol (in press)

Bennet-Clark HC, Daws AG (1999) Transduction of mechanical energy into sound energy in the cicada *Cyclochila australasiae*. J Exp Biol 202: 1803-1817

Bennet-Clark HC, Young D (1992) A model of the mechanism of sound production in cicadas. J Exp Biol 173: 123-153

Bennet-Clark HC, Young D (1998) Sound radiation by the bladder cicada *Cystosoma saundersii*. J Exp Biol 201: 701-715

Breckow J, Sippel M (1985) Mechanics of the transduction of sound in the tympanal organ of adults and larvae of locusts. J Comp Physiol A 157: 619-629

Busnel R-G (1963) (ed) Acoustic Behaviour of Animals. Elsevier, Amsterdam

Daws AG (1996) Resonance and frequency selectivity in insect sound communication. Ph.D. thesis, University of Melbourne

Daws AG, Bennet-Clark HC, Fletcher NH (1996) The mechanism of tuning of the mole cricket singing burrow. Bioacoustics 7: 81-117

Doolan JM, Young D (1981) The organization of the auditory organ of the bladder cicada, *Cystosoma saundersii*. Phil Trans R Soc Lond B 291: 525-540

Duellman WE, Trueb L (1985) Biology of Amphibians. McGraw-Hill, New York

Dumortier B (1963a) Morphology of sound emission apparatus in Arthropoda. In: Busnel R-G (ed) Acoustic Behaviour of Animals. Elsevier, Amsterdam. pp 277-345

Dumortier B (1963b) The physical characteristics of sound emissions in Arthropoda. In: Busnel R-G (ed) Acoustic Behaviour of Animals. Elsevier, Amsterdam. pp 346-373

Elliott CJH, Koch UT (1985) The clockwork cricket. Naturwissenschaften 72: 150-153

Fletcher NH (1978) Acoustical response of hair sensory cells in insects. J Comp Physiol 127: 185-189

Fletcher NH (1992) Acoustic Systems in Biology. Oxford University Press, Oxford

Göpfert MC, Briegel H, Robert D (1999) Mosquito hearing: Sound-induced antennal vibrations in male and female *Aedes aegypti*. J Exp Biol. 202: 2727-2738

Gray EG (1960) The fine structure of the insect ear. Phil Trans R Soc Lond B 243: 75-94

Hill KG, Oldfield BP (1981) Auditory function in Tettigoniidae (Orthoptera: Ensifera). J Comp Physiol 142: 169-180

Humphrey JAC, Devarakonda R, Iglesias I, Barth FG (1993) Dynamics of arthropod filiform hairs. I Mathematical Modelling of the hair and air motions. Phil Trans R Soc Lond B. 340: 423-444

Josephson RK, Young D (1981) Synchronous and asynchronous muscles in cicadas. J Exp Biol 91: 219-237

Josephson RK, Young D (1985) A synchronous insect muscle with an operating frequency greater than 500Hz. J Exp Biol 118: 185-208

Koch UT, Elliott CJH, Schäffner K, Kleindienst H (1988) The mechanics of stridulation in the cricket *Gryllus campestris*. J Comp Physiol A 162: 213-223

Langford-Smith F (1953) Radio Designer's Handbook. (4th ed) Iliffe and Sons, London

Leroy Y (1966) Signaux acoustiques, comportement et systématique de quelques espèces de Gryllidae (Orthoptères, Ensifères). Bull Biol Fr Belg 100: 1-134

Lewis DB (1983) Directional cues for auditory localisation. In: Lewis D B (ed) Bioacoustics: A Comparative Approach. Academic Press, London. pp. 233-257

Michelsen A (1971) The physiology of the locust ear. I Frequency sensitivity of single cells in the isolated ear. Z vergl Physiol 71: 49-62

Michelsen A (1983) Biophysical basis of sound communication. In: Lewis B (ed) Bioacoustics: A Comparative Approach. Academic Press, London. pp 3-38

Møller AR (1974) Function of the middle ear. In: Keidel WD, Neff WD (eds) Handbook of Sensory Physiology V/1 Springer, Berlin. pp. 491-517

Morse PM (1948). Vibration and Sound. McGraw-Hill, New York

Nocke H (1971) Biophysik der Schallerzeugung durch die Vorderflügel der Grillen. Z vergl Physiol 74: 272-314

Oldfield BP (1982) Tonotopic organisation of auditory sensory cells in Tettigoniidae (Orthoptera: Ensifera). J Comp Physiol 147: 221-241

Olson HF (1957) Acoustical Engineering. Van Nostrand, Princeton

Pringle JWS (1954) A physiological analysis of cicada song. J Exp Biol 32: 525-560

Prozesky-Schulze L, Prozesky OPM, Anderson F, van der Merwe GJJ (1975) Use of a self-made sound baffle by a tree cricket. Nature 255: 142-143

Shaw EAG (1974) The external ear. In: Keidel WD, Neff WD (eds) Handbook of Sensory Physiology V/1 Springer, Berlin. pp. 454-490

Simmons LW, Ritchie MG (1996) Symmetry in the songs of crickets. Proc Roy Soc Lond B 263: 305-311

Stephen RO, Bennet-Clark HC (1982) The anatomical and mechanical basis of stimulation and frequency analysis in the locust ear. J Exp Biol 99: 279-314

Treat AE, Roeder KD (1959) A nervous element of unknown function in the tympanic organs of moths. J Insect Physiol 3: 262-270

Weis-Fogh T (1972) Energetics of hovering flight in hummingbirds and in *Drosophila*. J Exp Biol 56: 79-104

Weis-Fogh T, Alexander RMcN (1977) The sustained power output obtainable from striated muscle. In: Pedley TJ (ed) Scale Effects in Animal Locomotion. Academic Press, London. pp 511-525

Young D (1990) Do cicadas radiate sound through their ear drums? J Exp Biol 151: 41-56

Young D, Bennet-Clark HC (1995) The role of the tymbal in cicada sound production. J Exp Biol 198: 1001-1019

Young D, Josephson RK (1983) Pure-tone songs in cicadas with special reference to the genus *Magicicada*. J Comp Physiol 152: 197-204

II.2 Ecological Constraints for Sound Communication: From Grasshoppers to Elephants

Heiner Römer

Institut für Zoologie, Karl-Franzens-Universität, 8010 Graz, Austria

Abstract
Sensory ecology is mainly concerned with the mechanisms that enable an animal to produce or utilize signals within its specific environment, and how the information about identity or location of the sender is transmitted to the receiver(s). Biotic and abiotic factors constrain the transmission of an acoustic signal and the performance of receivers in detection and location. Since animals use acoustic signals in a wide range of carrier frequencies - from infrasound to high ultrasound - with corresponding wavelengths spanning a range from tens of meters to a few millimeters, the rate of attenuation and degradation of sound signals varies largely in different communication systems. As a result, the active range of a sound signal covers probably hundreds of kilometers for marine mammals, several kilometers for elephants, to less than 1 m in some grasshoppers. This chapter examines some mechanisms by which these different animals achieve a maximum range of communication. In particular, I will show how sound interference in stratified environments can dramatically influence communication distance, and how animals adapt their timing and location of the site of their signaling to these environmental conditions.

Key words Thermal gradients, sound transmission, active space

1 Introduction

This chapter focuses on one limited aspect of acoustic communication, by assuming that selection drives signals and signaling behavior towards greater range and conspicuousness. The assumption, however, that acoustic signals of all species should have design features to maximize the range of communication is not correct. Communication between mates or parents and their young often ocurrs over short ranges, and natural selection may favor a range of the signal over the minimal distance necessary. Similar arguments hold for acoustic signals involved in maintaining territories (review in Wiley and Richards 1982). Certainly, a conspicuous signal may be advantageous, for example in mate attraction, but it may also have costs. For example, predators and parasitoids are known to home in on acoustic signals. Parasitoid tachinid flies and passively listening bats use the acoustic signals to find their hosts (Cade 1975, 1981; Belwood and Morris 1987; Lakes-Harlan and Heller 1992,). We would thus expect a trade-off between the costs and benefits of signaling, and the degree of conspicuousness of a given signal would indicate the point of balance between the counteracting selective forces driving the evolution of the signal.

Leaving these costs aside and assuming that there are only benefits from a conspicuous advertisement call of a frog, bird or insect, how can the active space of the signal be increased (the space where the signal is above detection threshold of intended receivers)? This active space depends on: (1) the amplitude of the broadcast signal, (2) the properties of the transmission channel, (3) the amplitude and spectral features of ambient noise, and (4) the sensory performance of the receiver to detect signals and to discriminate between them. The main focus of this chapter is on the role of the transmission channel, and by comparing animals of very different sizes and in the different transmission media of air and water, I wish to demonstrate some of the general rules for sound transmission that apply to all of them.

2 Properties of the Transmission Channel

Natural habitats are not sound-proof rooms; they exhibit hard or soft boundaries, reflecting surfaces, scattering by objects and ambient noise which all affect the signal-to-noise ratio, temporal pattern, and frequency spectrum of the signal at the position of the receiver(s). There are several reviews dealing with physical and other aspects on sound propagation and acoustic communication outdoors, which are highly recommended for a deeper insight into the mechanisms relevant for sound propagation outdoors (Michelsen 1978; Wiley and Richards 1982; Michelsen and Larsen 1983; Spiesberger and Fristrup 1990; Embleton 1996; Römer 1998). In this chapter only some aspects will be briefly covered which are important for a signaler to increase the active space of a signal.

2.1 Excess Attenuation Through Scattering

If there is a selection pressure on a sender to be more conspicuous we would expect that either the signal or signaling behavior is adapted in a way which reduces any source for excess attenuation, in order to increase the signal-to-noise ratio at the site of the receiver. Atmospheric absorption, scattering by objects in the transmission channel, reflective layers such as the ground, and strata of vegetation can result in the attenuation of the signal in excess to that predicted from the spherical spread of sound alone, given by the 6 dB decrease for each doubling of distance. Absorption is not a major source of attenuation for most animals, but for frequencies as high as 100 kHz, attenuation due to absorption may amount to 10 dBm^{-1} (Griffin 1971). Such high frequencies are used by bats for echolocation, and high absorption coefficients have been shown to influence the evolution of echolocation signals of insectivorous bats which hunt in open space. Here bats detect their prey over much greater distances than those searching for prey close to, or within, vegetation, but in order to avoid strong absorption over these distances, their echolocation signals are much lower in carrier frequency (Neuweiler 1989). This can be interpreted as an adaptation to the greatly reduced range of high-frequency signals caused by absorption in open space.

Whereas the advantage of sound signals over visual signals is their usefulness at night and in habitats with complex vegetation, the latter nevertheless has attenuating effects on acoustic signals due to scattering. For insects or animals using high frequencies which call within, or close to, meadows, bushland, or forests, this can become an important source of excess attenuation, because such vegetation includes surfaces the size of which is in the same order of magnitude as the wavelength of sound. There is a strong correlation between sound frequency and the degree of excess attenuation due to scattering. Leaves in the order of 5 cm scatter sound with a wavelength of 33 cm (1 kHz) much less than sound of a wavelength of 3.3 cm (10 kHz).

The habitat therefore acts as a low-pass filter (Keuper and Kühne 1983). The technique with a "biological microphone" has been used to study such frequency-dependent sound propagation in insect habitats. It enables long-term recordings of afferent nervous activity from identified auditory interneurons of insects, and the portable device allows the study of the perception and coding of sound signals in any terrestrial habitat (Römer and Lewald 1992). Within a bushy thicket, pure tone pulses with a carrier frequency of 5 kHz were attenuated by 22 dB over a distance of 10 m, in contrast to 52 dB at ultrasound frequencies of 40 kHz. Similarly, as a result of frequency-dependent scattering, coniferous forests exhibit less excess attenuation than deciduous forests, these effects are more pronounced at higher frequencies (Morton 1975; Marten and Marler 1977).

Obvious solutions to counteract excess attenuation due to vegetation are (1) a reduction of the carrier frequency of the signal (lower frequencies and thus larger wavelengths are scattered less) or (2) by signaling from a position above the vegetation in open space, thus avoiding scattering of the signal. Both solutions are by no means easy options for most animals and are constrained by other factors. The carrier frequency of acoustic signals is in most cases negatively correlated with body size, which renders it difficult for small animals, in particular insects, to call at frequencies best adapted for transmission through vegetation. Calling on top of the vegetation would significantly reduce excess attenuation (Rheinlaender und Römer 1986; Römer und Lewald 1992), but such preferred broadcasting positions make a diurnal signaler in turn more conspicuous for visually hunting predators.

This tradeoff between conspicuousness of the acoustic signal and visual conspicuousness of the signaler is probably reflected in the calling behavior of the bog katydid *Metrioptera sphagnorum*. This small insect produces a continuous call at high sonic and ultrasonic frequencies of 17 and 34 kHz (Morris 1970, Fig. 1A), which are strongly attenuated when males call close to the ground within dense vegetation. Males prefer to sing during late afternoon on small spruce trees scattered around in the habitat, but their preferred height is not the top of these trees, but a site on average about 30-50 cm below (Fig. 1B). This reduces their visual conspicuousness (at least to the searching entomologist) dramatically, but has the effect that there is only little excess attenuation of the amplitude of their call.

Male bushcricket *T. viridissima*, however, establish positions which are less than ideal for maximal sound transmission. Males could have increased the detection range of their signal almost threefold by singing from even higher positions than those observed in the field (Arak and Eiriksson 1992). This probably reflects the common tradeoff between the costs of singing at higher, exposed elevations and the increased detection range of the signal.

2.2 Frequency Filtering

Metrioptera sphagnorum also provides an example for the frequency filtering effect of the transmission channel. Since higher frequencies with smaller wavelengths suffer from multiple scattering more strongly than lower frequencies, the transmission channel acts essentially as a low-pass filter. In the katydid *M. sphagnorum*, the amplitude of the ultrasonic component close to the source is 5 dB higher compared to the "audio"-component at 17 kHz, but after transmission over 10 meters in the bog habitat (source and receiver 50 cm above ground) its amplitude is below the detection threshold of a receiver, as shown by the activity of the omega-neuron at this distance (Fig. 1C). By contrast, the 17 kHz component is still perceived, resulting in a patterned response of the neuron at large distances.

Since the neuron is equally sensitive to both components, it does respond even more strongly to the ultrasonic component close to the source, because of its higher amplitude (Römer and Morris, unpublished). Since the 17 kHz component is the only one available at larger distances, but has been shown not to be sufficient for female phonotaxis, the lower frequency has been suggested to function in male spacing (Morris et al. 1975).

2.3 Directionality

Apart from frequency-dependent attenuation, the other effect of dense vegetation on the transmission channel is on the directionality of the signal. In the most extreme situation, multiple reflections and scattering will turn the sound field around a listening receiver into a diffuse sound field where sound waves arrive with equal amplitude from all directions. Directional hearing under these conditions would be strongly impaired.

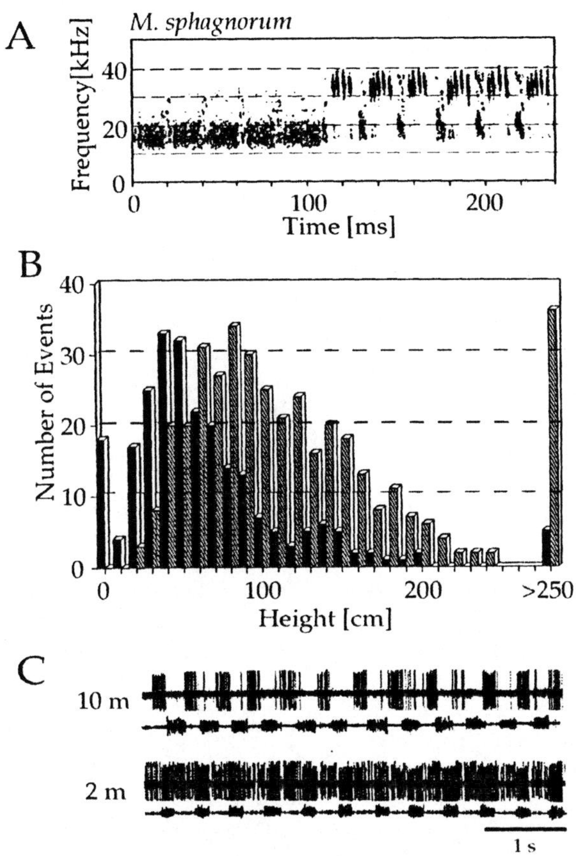

Fig. 1 A-C. A The calling song of the bog katydid *Metrioptera sphagnorum* is composed of an audio mode with frequencies around 17 kHz, and an ultrasonic mode at about 34 kHz, alternating continuously. **B** Males establish broadcast positions on small spruce trees within the bog, at an average height (*black bars*) about 50 cm below the maximum available height (top of the trees; *hatched bars*). **C** Discharge of the omega-neuron of a receiver at distances of 2 and 10 m, in response to a calling male. The calling song of a male was recorded with an ultrasonic detector (see *lower traces*), indicating the presence of the ultrasonic mode. The audio mode is always present in the pauses in between. Note that at 10 m the neuron only responds to the audio mode. (Römer and Morris, unpubl.)

A recent experimental demonstration for the grasshopper *Chorthippus biguttulus* was given by Gilbert and Elsner (2000) using the afferent activity of tympanal receptor fibres as a biological microphone in the field. They could show that in dense vegetation, but not on gravel or in sparse vegetation, binaural directional cues are strongly affected, a finding that is consistent with the observation that this grasshopper can be found either on gravel or sparse vegetation, but rarely in dense vegetation (for review see Michelsen 1998).

3 Shadow Zones and Enhanced Transmission Due to Environmental Gradients

Another means to increase the active space of a signal is to use spatial sound windows and to adapt to atmospheric conditions influencing long-distance animal vocalizations. Atmospheric conditions have rarely been considered here, particularly with respect to insects. In the following this singular aspect on sound communication outdoors will be considered. This will be achieved by comparing animals of very different sizes, from small insects to elephants and finback whales, communicating in different environments and microhabitats.

The wave-like nature of the propagation of sound results in a deflection when sound passes through environments in which the speed of sound varies. When sound crosses a boundary in air or water with different speed of propagation, part of the wave is reflected back into the first environment (like light being reflected from the surface of water). Vertical gradients of temperature (and of wind speed) develop during the day as a result of sun radiation. On a sunny day the air is warmer close to the ground (Fig. 2A), and this increases the sound speed as the ground is approached. Under these super-adiabatic conditions, in which temperature drops rapidly with height above ground, the gradient in the propagation speed of sound results in a bending of the sound rays toward the direction of lower speed, i.e., upwards (Fig. 2B).

This, in turn creates a shadow zone at some distance from the source, where no sound can penetrate directly. After sunset, the opposite occurs: the ground cools faster than the atmosphere; somewhere above the ground it is warmer compared to the surface, before at higher altitudes temperature decreases again (Fig. 2A). In such a temperature inversion the sound rays curve downwards during propagation into the layer of air with lower temperature. As a result, a sound channel can be created by repeated refraction towards the ground and reflection from the ground into the atmosphere. This channeling is even more effective if the ground is relatively smooth with little vegetation, so that sound is not scattered too much. Humans experience such a sound channel due to temperature inversion after sunset and at some distance from a highway, when we hear traffic noise some kilometers away. During the day this is not the case, because of the sound shadow zone. Are these conditions of any relevance for animal acoustic communication?

3.1 Elephants in the Savanna

Garstang et al. (1995) presented evidence that African elephants make use of atmospheric conditions conductive to low-frequency sound transmission.

Measurements of temperature at various height above the ground in a typical open African savanna show that strong temperature inversions form at the surface before sunset, and decay with sunrise (Fig. 2A). After sunset, the air is stratified near the ground and should lead to an enhancement of sound propagation for the infrasound signals between 15 and 50 Hz produced by elephants. Additional wind-speed measure-ments indicate that calm conditions do exist especially after sunset, with the consequence that such layers of air differing in propagation speed are not disturbed by topographic winds, because of turbulent mixing.

Such meteorological data has been used to calculate the effect of temperature gradients on the range of elephant vocalizations. The authors used idealized temperature profiles typical for midday superadiabatic conditions, late afternoon adiabatic lapse and evening inversions. They defined the elephants' calling range as the distance over which the call (117 dB SPL at a distance of 1 m) decreased by 67 dB to reach the hearing threshold (about 50 dB SPL) of receivers. This range changes from about 2 to 9 km between late afternoon and early evening, as a result of the sound channel created by a temperature inversion. Based on these calculations and temperature measurements over the African savanna for a period of 2 months, the same authors determined the calling area (i.e., the area over which the infrasound signal could possibly be detected by another elephant), which varies from less than 50 km^2 to values of 300 km^2. On any given day, the calling area undergoes an expansion and contraction, largest areas appear after sunset, or prevail at times of longer undisturbed weather conditions for most of the night (Fig. 2C).

The acoustic communication distances of up to 10 km estimated by Garstang et al. (1995) are larger than those of 2-4 km reported by Langbauer et al. (1991) using a playback technique, but the latter study was performed with elephants when they came to drink, without a control of the actual atmospheric conditions. Optimum conditions for maximizing calling area should also occur in other Savanna areas in Africa at dusk and at dawn, and if so, then these predictable conditions may have exerted a selective pressure on animals other than elephants to call at times when transmission is best. This is the case for lions (Schaller 1972) and coyotes (Laundré 1981, Larom et al. 1997).

3.2 Bladder Grasshoppers in Trees

Although the size of the animal in the following section contrasts greatly with that of an elephant, it makes use of environmental gradients for long-distance communication as well. Bladder grasshoppers (*Bullacris membracioides*) are endemic to southern Africa and the conspicuous nocturnal calls of males are audible to humans over several kilometers. The abdomen of the male is an air-filled space and it acts as a resonator set into vibration by rubbing a file on the hindleg against another row of teeth on the abdomen itself. This creates a very loud calling song with a carrier frequency of about 1.7 kHz and a SPL of 98 dB at a distance of 1 m.

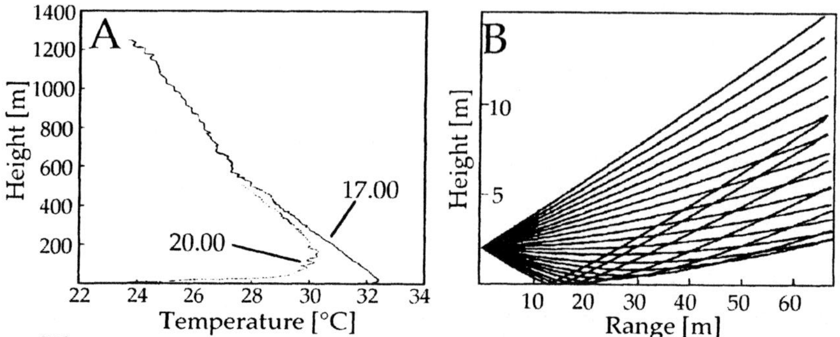

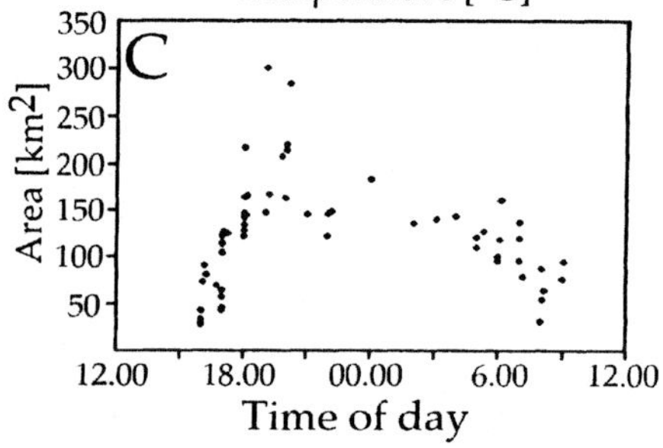

Fig. 2A-C. A Tethered balloon measurements of temperature profiles over the African savanna at Okaukuejo at 17.00 (*solid line*) and 20.00 (*dashed line*) local time. Note the temperature inversion after sunset (After Garstang et al. 1995). **B** The refraction of sound by daytime temperature gradients. Paths of rays emitted from a source at a height of 2 m, at intervals of 1° within 10° of horizontal, for a travel time of 200 ms. A receiver at the same (2m) and a distance of 70 m will be within the shadow zone and will not receive sound directly. **C** Area (in km^2) over which the elephant call could be detected by a conspecific, predicted from 65 tethered balloon measurements of the temperature profile over the African savanna, between 16 September and 4 October 1992 (After from Larom et al. 1997). Note the large variation in the active space of the signal within 24 h, with maxima occurring soon after sunset

Furthermore, the sensitivity of their hearing organ is very high and, combined with low ambient noise levels on the transmission channel, this sets the stage for an increased active range of the signal; but most spectacular is the use of atmospheric conditions with temperature gradients, favoring long-distance transmission, quite similar to the elephant example.

This was shown in field experiments performed in undisturbed habitat in KwaZulu-Natal, in South Africa. The area is characterized by steep-sided valleys, high humidity levels and vegetation comprising a mixed mosaic of bush and grassland. Thermal profiles reveal that superadiabatic conditions, in which temperature drops rapidly with height above the hot ground surface, prevail from mid-morning to mid-afternoon (van Staaden and Römer 1997). This gradient results in moderate turbulent mixing and the development of significant surface wind as the ground warms. By contrast, temperature increases with height from the ground at night. Strong temperature inversions form at the surface shortly after sunset and decay with sunrise, accompanied by calm wind conditions during early to late evening. The consequences of these contrasting conditions for sound transmission are summarized in Fig. 3.

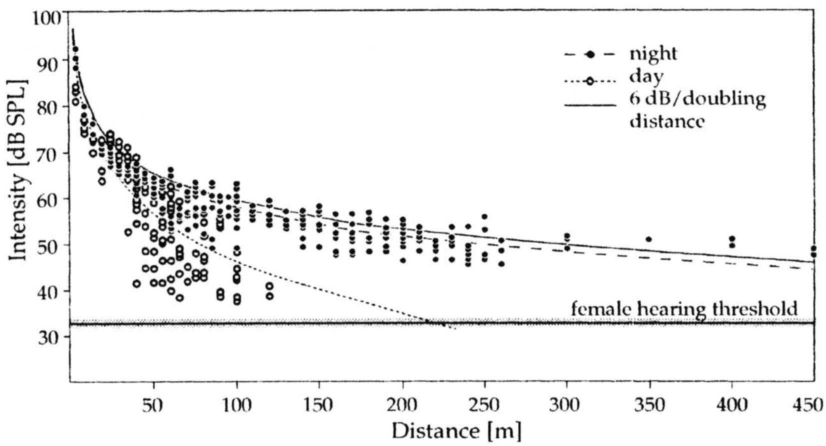

Fig. 3. Changes in sound pressure level of the male call of an African grasshopper, *Bullacris membracioides* over distance in open grassland indicate that night-time transmission (*solid circles*) is consistently close to ideal (6 dB/dd) and suffers little excess attenuation. During the day (*open circles*), however, measures are highly variable and decrease significantly at about 50 m, indicating a sound shadow zone as a result of temperature gradients of 4.5 °C/10 m above ground level. Extrapolating from the night trajectory and assuming a mean hearing threshold of a conspecific female of 32 dB SPL for the male call (*horizontal dashed line*), females can hear conspecific male calls at distances up to 1.4 to 1.9 km. (van Staaden and Römer 1997)

Playbacks of the male call during day and night (when *B. membracioides* actually communicates with sound) have been combined with measurements of the SPL at various distances. Daytime measurements showed substantial variability and a sudden drop in SPL of the male call starting at distances of about 40 m. This is consistent with the existence of a shadow zone. Almost all values fell below the theoretical value for geometric spreading beyond 50 m and below the hearing threshold of potential receivers beyond 100-150 m. Piercy et al. (1977) reported similar values of 20-30 dB excess attenuation as a result of the shadow zone, largely independent of sound frequency. By contrast, at night the male call suffers little or no excess attenuation. Under a wide variety of environmental conditions attenuation is about 6 dB/dd according to geometrical spreading of sound for distances up to 450 m. Extrapolating the male call transmission trajectory until the SPL of the transmitted call intersects the hearing threshold of a female gives a potential transmission distance to conspecific females ranging between 1400 and 1900 m.

The data and calculations in the elephant study predicted an enhancement of 12-15 dB for the infrasound frequencies, whereas the experimental data for night-time transmission of the grasshopper call show little evidence for a direct enhancement of the call, i.e., SPLs higher than predicted by geometrical spreading alone. However, the measured SPL at each distance is the additive effect of all enhancing and attenuating conditions affecting the sound wave between sender and receiver, and excess attenuation due to high grass vegetation on the transect most likely reduced the enhancing effect of temperature inversions. Moreover, compared with the infrasound frequencies of African elephants, enhancement effects should be reduced at the 1.7 kHz carrier frequency typical of the calls of this grasshopper.

Despite this difference, the design features of this communication system agree with all expectations of natural selection to maximize broadcast range while avoiding predation and intraspecific competition. For the female receiver the amplitude of the male signal is maximized relative to background noise by repetition of a temporally stereotyped signal, by an increased broadcast amplitude resulting from the abdominal resonator, by an increase of the transmitted signal amplitude using optimal weather conditions, and by (for an insect) extremely sensitive hearing.

3.3 Crickets on the Ground

If we go further down in the size of the animal, and close to the ground of terrestrial habitats, we find grasshoppers and crickets communicating by sound close to the ground, although physicists tell us that they should not do this. In fact, these insects face a complicated situation, where the sound waves they produce interact with the ground in an angle which is almost zero and sound waves travel almost parallel to the ground. The wave reflected from the ground will be out of phase with the direct wave at some distance, and when the two interact they cancel each other. Therefore sound propagation parallel to the ground is often referred to as the forbidden mode of propagation since it creates another type of sound shadow zone (for review see Embleton 1996).

Gryllus campestris is such a species living in and calling on the ground, and the broadcast range of a male signal would be limited to just 1 or 2 m, at least from theoretical calculations for ground waves. A field measurement confirmed attenuation in excess to geometric spreading (Popov et al. 1972), but the distance over which a female could probably hear the song was about 10-15 m and thus considerably larger than the forbidden mode of propagation would suggest. I suggest that microclimatic conditions very close to the ground could have an enhancing effect for sound propagation.

Layers of air in close contact with the ground are largely determined by sun radiation, with highest temperatures occurring where radiation is absorbed. In a meadow where the grass was about 30-40 cm high, we found a temperature profile as shown in Fig. 4 (Seither and Römer, unpubl.). Firstly, the data indicate an inverse relationship between temperature and humidity of the vegetation: in the early morning the meadow is wet with dew, and after sunrise temperature does not increase before most of the humidity has evaporated. Only then can temperature increase substantially. A very steep gradient of up to 10 °C within a height of only 30 cm developed, with the maximum occurring at the top of the vegetation. By comparison, the mean gradient in the free atmosphere in Middle Europe is 0.6 °C over 100 m (Geiger et al. 1995).

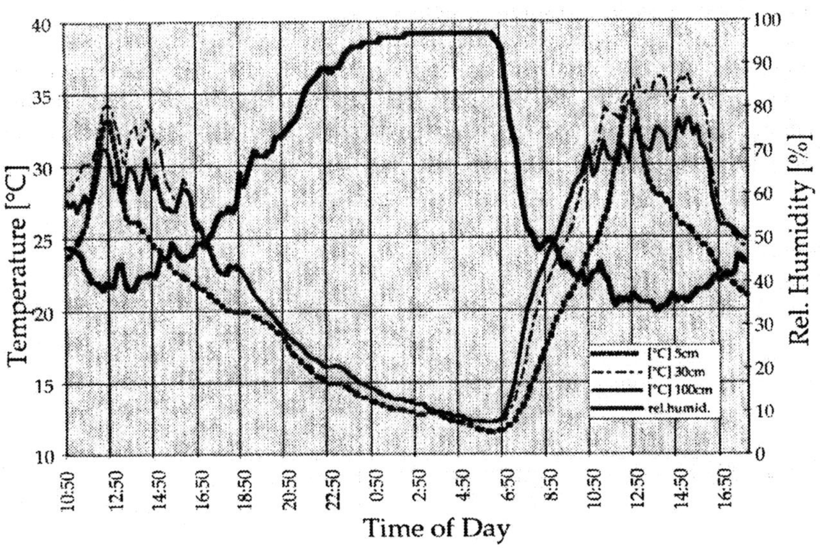

Fig. 4. Microclimatic conditions within and above a meadow on a summer day (12 and 13 August 1997), measured with three thermosensors and one sensor for humidity. Note that max. temperatures occur at the top of the grass, and that the gradient at about 16.00 h is 10 °C over only 30 cm. (Seither and Römer, unpubl.)

We therefore compared the transmission characteristics of such a meadow under a variety of meteorological conditions, for frequencies between 100 Hz and 15 kHz. Particular attention was given to the temperature gradient.

Figure 5 shows the result of a comparison between sound transmission on a sunny day with a steep temperature gradient, and transmission on a cloudy day with a weak gradient. The steep temperature gradient clearly favors the transmission of sound signals, at least in the frequency range of 300 to 6000 Hz. These frequencies are enhanced with increasing distance by as much as 17 dB at 32 m, compared to a situation with the low temperature gradient. The figure also shows the frequency dependency of this effect; there is no such influence at frequencies between 100 Hz and 300 Hz.

Although more experiments on sound transmission on or close to the ground are needed, the available findings tell us that a ground-dwelling cricket may still use this forbidden mode of propagation by exploiting a particular sound channel due to temperature gradients. The mechanism that brings about this gradient here is quite different from that underlying the two gradients reported from the African studies.

Whereas these result from night-time or evening inversions, the temperature gradient in a meadow results from the interaction of sun radiation with the top of grass vegetation during the day. I should emphasize, however, that for many other insects calling in such a meadow it does not matter whether they sing with or without a strong temperature gradient: many of the small grasshoppers in a summer meadow call so softly that a conspecific receiver will detect the signal at distances of 0.5 m to 1 m only (Lang, in press). Over such short distances and for the frequencies used by these grasshoppers, the temperature gradient has no enhancing effect at all.

D=h*Grad. strong* (f)-h*Grad. weak* (f)[dB]

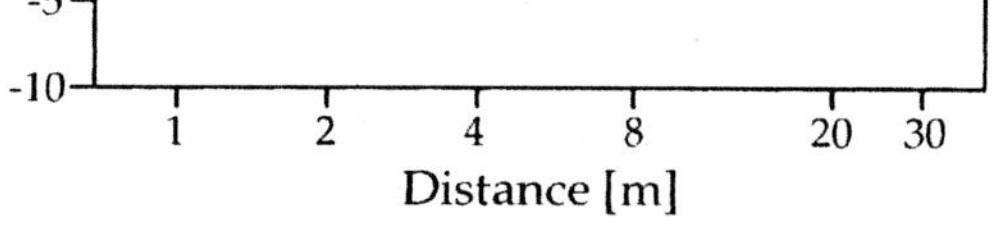

Fig. 5. Transmission characteristics of a meadow covered with grass about 30 cm high (cf. Fig. 4), for the frequency bands of 100-300 Hz and 300-6000 Hz, respectively. The height of both the speaker and microphone above ground was 2 cm. The *y-axis* gives the amount in decibels, by which the transmission with a strong temperature gradient was better (+), or worse (-) compared to a situation with a weak gradient. (Seither and Römer, unpubl.)

3.4 The SOFAR Channel in the Ocean

So far, the effect of gradients in the speed of airborne sound was compared at different scales, ranging from several kilometers to some centimeters. Considering the transmission of sound in water there is a real change in dimensions: we are no longer concerned with transmission over a few kilometers, but thousands of kilometers!

Sound travels about five times faster in water than in air, at a speed of about 1500 ms^{-1}. Propagation velocity is a function of the temperature, pressure and salinity of water. For example, the speed increases by 4.8 ms^{-1} per degree of temperature increase, whereas the effects of pressure and salinity are comparatively small. In aquatic environments horizontal stratifications of these parameters result in changes in sound velocity. This in turn redirects sound waves into the direction of lower velocity. A typical profile of sound velocity is shown in Fig. 6A.

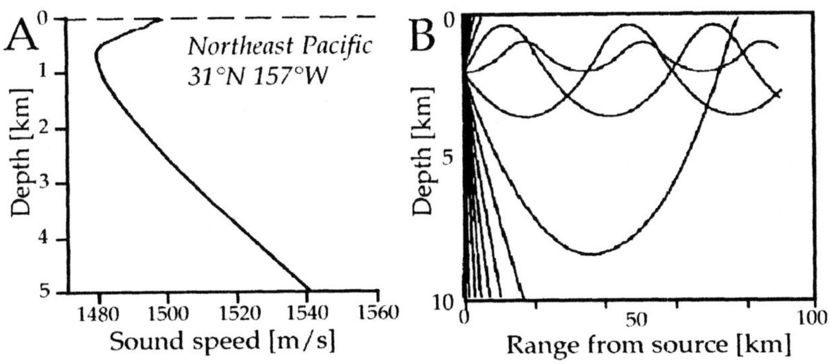

Fig. 6 A-B. **A** Profile of the sound propagation velocity of the ocean at a location in the northeast Pacific at 31°N 157°W (Spiesberger and Fristrup 1990). **B** The deep sound channel (SOFAR channel) in the ocean. Paths of rays separated by 10 ° intervals are shown for a travel time of the sound of 60 s after having left the source. The sound velocity profile was assumed to be similar to the one shown in A. Note that rays leaving the source over a range of at least 30° are trapped in the channel. (from Dusenbery 1992)

The minimum is at a depth of water of about 500 m, which is typical for the tropics and subtropics. Sound velocity increases at the surface because the water gets warmer, the increase towards the deep ocean is because pressure increases, whereas temperature stays almost constant at 4°C. The result of this minimum in sound velocity is that sound is trapped by repeated refraction into this layer, without suffering from absorption and scattering at the surface of the ocean, or on its ground (Fig. 6B). This deep channel for preferred sound propagation is called the SOFAR channel (Urick 1983). Animals within the SOFAR channel could, in theory, hear sound signals over tremendous distances, ranging up to thousands of kilometers.

Ironically, the best set of data result from the common human interest in global warming. Because the oceans play a crucial role in the temperature balance of the atmosphere, there is a need for measuring the average temperature and temperature changes over large ocean ranges. The idea was to use acoustic thermometry, with two simple considerations in mind: (1) the travel time of sound between 2 points in the ocean is a sensitive indicator of the ocean temperature in between, and (2) as mentioned above, the ocean is a good propagator of sound, and therefore the two points between which the sound travels can be thousands of kilometers apart. If we take 0.02 °C year^{-1} at the top of the ocean as a typical estimate of greenhouse warming, this decreases exponentially to 0.005 °C year^{-1} at a depth of 1 km, that is roughly at the depth of the SOFAR channel. If we take the latter change at the sound channel axis due to global warming effect, this would change the travel time for sound by 100 ms over a path of 10 000 km.

In 1991, the so-called Heard Island Feasability Test was carried out to determine the limits for such measurements (Munk and Forbes 1989; Munk et al. 1994). In this test, low-frequency acoustic signals were transmitted from a source near Heard Island in the southern Indian Ocean (Fig. 7). Fourteen receiver sites were installed in the Atlantic, Pacific, and Indian Oceans, with distances to the source of up to 18 000 km. The most impressive result was that a clear signal could be received at almost all receiver sites after transmission times of up to 3 h. The highest signal-to-noise ratios, sometimes exceeding 30 dB in a 1-Hz band, were obtained along a sound channel axis from the source ship to Ascension Island (A in Fig. 7) in the South Atlantic Ocean (Georges et al. 1994), which is a distance of more than 9 000 km.

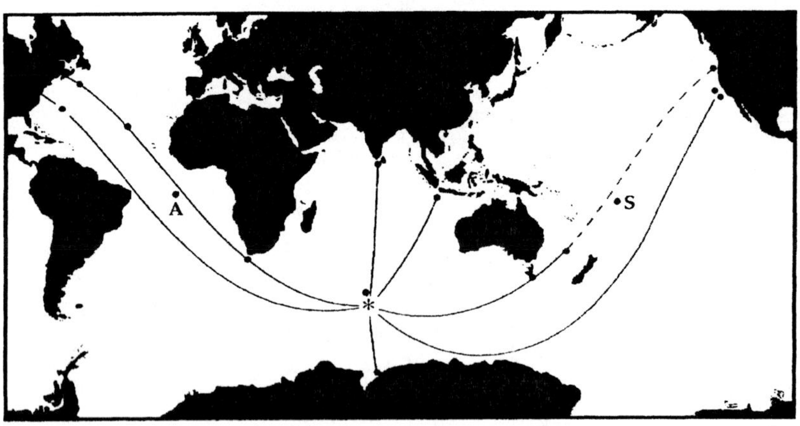

Fig. 7. The Heard Island Feasability Test. An underwater sound source at a depth of 300 m, positioned 50 km southeast of Heard Island (*asterisk*) broadcast low-frequency sound signals, which could be perceived at almost all receiver sites (*dots*) except for the one off Samoa (S). (After Munk et al. 1994)

The real-time spectral analysis of a recorded signal at this site reveals another interesting detail: during the test a signal was routinely received for at least 23 min after the direct signal had ended. This is due to scattering from the continental borders of South America, Africa, and Antarctica. If we imagine a finback whale signaling somewhere in the South Indian Ocean, a receiver along this sound channel will be able to first detect this signal after more than 90 min, and then continue to receive scattered signals from all these places for another 20 min: this is a scenario hard to imagine for us because of dimensions far beyond the signal ranges reported for terrestrial animals.

I do not argue that a whale in the ocean south of Australia communicates with another one in the Atlantic Ocean through the SOFAR channel. A realistic calculation for the detection range of a call of a finback whale is about 400 km (Spiesberger and Fristrup 1990). However, a consequence of the excellent sound propagation will be a considerable increase in noise levels, which in the frequency range from 1-100 Hz is dominated by ships and by the ocean's turbulence (Urick 1983). This can result in an increase in masking of the hearing thresholds of marine mammals and fish.

4 The "Noisy" Transmission Channel

The statement made at the beginning of this chapter that the sender - in order to increase the active space of its signal - should increase the amplitude of the signal is not absolutely precise. The sender should in fact increase the signal-to-noise ratio by whatever mechanism to enable easy signal detection by intended receivers. The problem with environmental noise, created by various sound sources such as wind, running water, or various individuals of different species, is its nonrandom nature and variation in spectral composition and temporal pattern.

The variation in noise is an indirect consequence of the temporal variation in sound transmission properties during day and night, as outlined above. This is because acoustically communicating animals adapt their broadcast times to these conditions. Diurnal animals tend to concentrate their singing activity either in the first hours of the day or in the evening. Well-known examples are the dawn choruses of birds and insects (Henwood and Fabrik 1979).

Nocturnal animals find more reliable conditions for sound transmission and often call throughout the night. However, the constraints for long-range sound transmission are similar for many species of mammals, amphibians, and insects, and the result is a complex sonic and ultrasonic background over which these animals must communicate. It is evident, then, that there is a tradeoff between the conditions optimal to broadcast one's signal and those for detecting and discriminating signals in the noise produced by the other species. Measures of the range of communication, or of sound discrimination abilities, are therefore of little relevance without considering the level of masking noise in the environment.

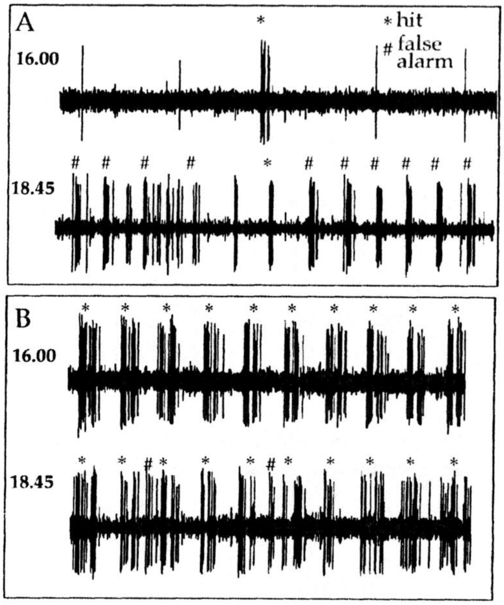

Fig. 8 A, B. **A** Signal detection in a spatially heterogenous habitat with ambient noise levels varying during the day. The preparation with recordings of extracellular activity of the omega-neuron of a katydid was placed within the rainforest and remained there for approximately 4.5 h. The activity was recorded at 16.00 h. under very low external noise conditions of 40 dB SPL and less (*upper line* in **A, B**) and at 18.45 h, when the noise level had reached more than 60 dB SPL (*lower line* in **A, B**). Under low-noise conditions, a short sound pulse of 20 ms duration presented only once every 5 s is still well perceived (many hits over the period of investigation, few false alarms), but after sunset, external noise result in a dramatic increase in the rate of false alarms. By contrast, a highly redundant signal (**B**) led to a high rate of hits under the low and high noise conditions, and less false alarms in high noise. (Lang and Römer unpubl.)

A sensory ecologist is used to describe the amplitude and frequency spectrum of the signal and then to compare it with the same properties of ambient noise, in order to evaluate the signal detection performance of a receiver. The actual receiver does it in a different way. A female listening to the call of a male does not know beforehand when a male is calling, her auditory system has to analyze continuously the afferent nervous activity within the auditory pathway and then somehow to decide, based on some discharge criterion, whether or not there had been a signal on the transmission channel.

Using the biological microphone with the omega-neuron activity mentioned before (Rheinlaender and Römer 1986), one can address this question directly in the noisy environment of the insect. Figure 8 shows representative periods of neural activity of the same preparation placed within the rainforest (Panama, Barro Colorado Island) at two times of day, at 16.00 and 18.30 h, and in response to a signal broadcast from a distance of 10 m. In the early afternoon, under ambient noise levels of 40 dB SPL and less, a burst of activity in this neuron was always associated with the broadcast signal (100% hits). It never happened that such a burst of activity occurred without a signal, i.e., a receiver using the burst criterion would never decide incorrectly on the presence of a signal (no false alarms). However, after another 2.5 h, shortly after sunset, the noise level had increased to more than 60 dB SPL. The same experiment now led to quite different results (Fig. 8A, lower line).

There is so much afferent activity due to noise from many other sources in the habitat that - based on the previous burst criterion - the receiver produced 1397 false alarms in a period of 5 min, indicating the presence of a signal where, in fact, there had been none. The signal detection performance deteriorates with increasing distance to the sender, then the rate of correct hits decreases, but at the same time the rate of false alarms stays high. In addition, at greater distances the number of cases where the neuron failed to generate a burst of action potentials in response to a signal increases (miss).

Results of such experiments also point to one solution to many communication systems to counteract noise, namely redundancy by signal repetition. When the signal is repeated at a high rate, both in low and high ambient noise levels the rate of hits is almost 100%, and there are only few false alarms even at high noise levels (Fig. 8B). Other solutions for avoiding mistakes in signal detection is the temporal segregation of calling activity of several signaling individuals, or the partitioning of the carrier frequencies of their calls, thus allowing for "private channels" of communication (Moore et al. 1989; Narins 1995; Riede 1995).

References

Arak A, Eiriksson T (1992) Choice of singing sites by male bushcrickets (*Tettigonia viridissima*) in relation to signal propagation. Behav Ecol Sociobiol 30:365-372

Belwood J, Morris GK (1987) Bat predation and its influence on calling behavior in neotropical katydids. Science 238:64-67

Cade WH (1975) Acoustically orienting parasitoids: fly phonotaxis to cricket song. Science 190:1312-1313

Cade WH (1981) Field cricket spacing, and the phonotaxis of crickets and parasitoid flies to clumped and isolated cricket songs. Z Tierpsychol 55:365-375

Dusenbery DB (1992) Sensory Ecology. WH Freeman and Company, New York

Embleton TFW (1996) Tutorial on sound propagation outdoors. J Acoust Soc Am 100: 31-48

Garstang M, Larom D, Raspet R, Lindeque M (1995) Atmospheric controls on elephant communication. J Exp Biol 198:939-951

Geiger R, Aron RH, Todhunter P (1995) The Climate Near the Ground. Textbook Meteorology. Vieweg, Braunschweig, Wiesbaden

Georges TM, Boden LR, Palmer DR (1994) Features of the Heard Island signals received at Ascension. JASA 96:2441-2447

Gilbert F, Elsner N (2000) Directional hearing of a grasshopper in the field. J Exp Biol 203:983-993

Griffin DR (1971) The importance of atmospheric attenuation for the echolocation of bats (Chiroptera). Anim Behav 19:55-61

Henwood K, Fabrik A (1979) A quantitative analysis of the dawn chorus: temporal selection for communicatory optimization. Am Nat 114:260-274

Keuper A, Kühne R (1983) The acoustic behavior of the bushcricket *Tettigonia cantans*. II. Transmission of airborne sound and vibration signals in the biotope. Behav Processes 8:125-145

Lakes-Harlan R, Heller KG (1992) Ultrasound sensitive ears in a parasitoid fly. Naturwissenschaften 79:224-226

Lang F (in press) Acoustic communication distances of a gomphocerine grasshopper. Bioacoustics

Langbauer WR, Payne KB, Charif RA, Rapaport L, Osborn F (1991) African elephants respond to distant playbacks of low-frequency conspecific calls. J Exp Biol 157:35-46

Larom D, Garstang M, Payne K, Raspet R and Lindeque M (1997) The influence of surface atmospheric conditions on the range and area reached by animal vocalizations. J Exp Biol 200: 421-431

Laundré JW (1981) Temporal variation in coyote vocalization. J Wildl Mgmt 45:767-769

Marten K, Marler P (1977) Sound transmission and its significance for animal vocalizations. I. Temperate habitats. Behav Ecol Sociobiol 2:271-290

Michelsen A (1978) Sound Reception in Different Environments In: Ali MA (ed) Sensory Ecology. New York, London, Plenum, pp 345-373

Michelsen A, Larsen ON (1983) Strategies for acoustic communication in complex environments. In: Huber F, Markl H (eds) Neuroethology and Behavioral Physiology. Berlin Heidelberg New York, Tokyo, Springer, pp 322-332

Michelsen A (1998) Biophysics of sound localization in insects. In: RR Hoy, AN Popper, RR Fay (eds) Comparative Hearing: Insects. New York, Berlin, Heidelberg, Springer, pp 18-62

Moore SW, Lewis EE, Narins PM and Lopez PT (1989) The call-timing algorithm of the white-lipped frog, *Leptodactylus albilabris*. J Comp Physiol A 164: 309-319

Morris GK (1970) Sound analysis of *Metrioptera sphagnorum* (Orthoptera: Tettigoniidae). Can Ent 102:363-368

Morris GK, Kerr, GE and Gwynne DT (1975) Calling song function in the bog katydid, *Metrioptera sphagnorum* (F.Walker) (Orthoptera, Tettigoniidae): Female phonotaxis to normal and altered song. Z Tierpsychol 37:502-514

Morton ES (1975) Ecological sources of selection on avian sounds. Am Nat 108:17-34

Munk W and Forbes AMG (1989) Global ocean warming: an acoustic measure. J Phys Oceanogr 19:1765-1778

Munk W, Spindel RC, Baggeroer A, Birdsall TG (1994) The Heard Island feasability test: JASA 96:2330-2342

Narins P (1995) Frog communication. Sci Am 273:62-67

Neuweiler G (1989) Foraging ecology and audition in echolocating bats. TEE 4: 160-166

Piercy JE, Embleton TFW, Sutherland LC (1977) Review of noise propagation in the atmosphere. J Acoust Soc Am 61:1402-1418

Popov AV, Shuvalov VF, Svetlogorskaya ID, Markovich AM (1972) Acoustic behavior and auditory system of insects. Rhein.-Westf. Akad Wiss 53:281-306

Rheinlaender J, Römer H (1986) Insect hearing in the field. I. The use of identified nerve cells as 'biological microphones'. J Comp Physiol A 158:647-651

Riede K (1995) Diversity of sound producing insects of a Bornean lowland rain forest. Proc Int Conf Trop Rainforest Res. Brunei, Daressalam

Römer H (1998) The sensory ecology of hearing in insects. In: RR Hoy, AN Popper, RR Fay (eds) Comparative Hearing: Insects. New York, Berlin, Heidelberg, Springer pp 63-96

Römer H, Lewald J (1992) High-frequency sound transmission in natural habitats: implications for the evolution of insect acoustic communication. Behav Ecol Sociobiol 29:437-444

Schaller GB (1972) The Serengeti Lion. Chicago, IL:University of Chicago Press. 480 pp

Spiesberger JL and Fristrup KM (1990) Passive localization of calling animals and sensing of their acoustic environment using acoustic tomography. Am Nat 135: 107-153

Staaden, van MJ, Römer H (1997) Sexual signalling in bladder grasshoppers: Tactical design for maximizing calling range. J Exp Biol 200: 2597-2608

Urick RJ (1983) Principles of Underwater Sound. New York, McGraw-Hill

Wiley RH, Richards DG (1982) Adaptations for acoustic communication in birds: Sound transmission and signal detection. In: Kroodsma DE, Miller EH, Quellet H (eds) Acoustic Communication in Birds. New York, Academic Press, pp 131-181

II.3 The Ecology of Avian Acoustical Signals

Hans Winkler

Konrad Lorenz-Institut für Vergleichende Verhaltensforschung der Österreichischen Akademie der Wissenschaften, Savoyenstraße 1A, 1160 Vienna, Austria

Abstract
The acoustical communication of birds is reviewed with an emphasis on ecological aspects. Sound propagation in the environment has specific consequences for signal transmission. The intensively studied ground effect and other influences offer good prediction for signal design in the frequency domain. The particular transmission properties observed in bird sounds are, in fact, at least as much governed by their temporal structure. Some principles of signal design are illustrated using woodpeckers as examples. They show how intimate and long-distance signals are structured, and how they relate to behavior and habitat use. The analysis of a set of species of antbirds of a rainforest community demonstrates that the temporal structure of the song is most important for explaining the transmission properties at singing post height. A review of the hypotheses proposed to explain the dawn chorus serves to point out the role of the receiver's demand for information played in the shaping of bird song. It is suggested that researchers should pay more attention to acoustical signals other than song, because these may offer better models for testing specific hypotheses about functional aspects. More and more evidence from different fields of research indicates that temporal characteristics of vocalizations need more attention. Finally, viewing acoustical communication more from the receiver's end may foster our understanding.

Key words Acoustic communication, bird song, habitat, sound propagation

1 Introduction

Birds rely extensively on sound as a means of communication. The diversity of sounds and vocal repertoires defies simple description. This chapter purports to review the ecological aspects of acoustical communication in birds. To that end I will concentrate on summarizing the current state of knowledge of basic aspects of sound transmission as far as they were not treated in the chapter by Römer, this Volume, ecological factors such as habitat and foraging, and sound reception. Some of the points made will be illustrated with examples from my own work. Recent reviews (e.g., Kroodsma and Miller 1996) show that our knowledge has progressed well in certain fields, but also that many gaps and open questions remain. Most of the examples given will deal with song and its analogues in non-songbirds. However, various types of calls and instrumental signals will be discussed as well. With respect to ecology I will focus on sound transmission in natural habitats. However, other aspects of ecology, such as predation, will be treated as well. Particular attention will be paid to the receivers' role.

2 Sound Propagation in Terrestrial Habitats

2.1 Attenuation

Sound attenuation by distance has several components, namely geometrical spreading, atmospheric absorption, effects of the ground, and miscellaneous environmental effects (noise, vegetation, see also Chapter II. 2). For a point source the spreading of sound is spherical with an attenuation equal to 6.02 dB. In certain habitats spreading of sound approaches two-dimensional cylindrical spreading with attenuation close to 3 dB per doubling of distance. Attenuation by air absorption is relevant for distances of more than several hundred meters only.

2.2 Attenuation over the Ground

Attenuation of sound by the ground has been extensively covered by theoretical and empirical studies (Piercy et al. 1977; Embleton et al. 1983; Attenborough 1985; Wempen 1991; Embleton 1996; Sutherland and Daigle 1997). The attenuation results from the interference between direct sound rays and those reflected from the ground. Ground attenuation strongly depends on the type of ground surface, the grazing angle, relative positions of sender and receiver, and the frequency of the sound. In addition to the plane wave reflection (Embleton 1996) there are the effects of the spherical wavefront. They differ depending on whether the signal is wide-band random noise or a pure tone (Chessell 1977; Sutherland and Daigle 1997). However, at medium distances (< 100 m) and moderate to large grazing angles ground attenuation is predominantly due to plane wave reflection (Sutherland and Daigle 1997). Increasingly harder ground shifts attenuation slightly towards higher frequencies, and turbulence mitigates the ground effect (Chessell 1977). If either the sender or receiver is not very close to the ground and if there are even moderate height differences, many transmission bands at lower frequencies and their multiples open up. Even small changes of position alter the base attenuation frequencies.

2.3 Refraction and Turbulence: Effects of Wind and Temperature

Vertical gradients of wind velocity and temperature are a main source for changes in sound levels. Only close to the ground can sound with reduced phase velocity and strong distance-dependent attenuation (creeping wave) penetrate the shadow region created by upward refraction. Downward refraction may cancel ground attenuation and excess attenuation may become zero (Embleton 1996, Sutherland and Daigle 1997).

Air turbulence, as it builds up over open areas during the day, scatters the sound energy. Turbulence fluctuations modulate mainly highly directed sound beams at medium distances and are insignificant for spherically expanding waves. Backscattering also attenuates sound negligibly (Sutherland and Daigle 1997). This contrasts with the deliberations of Wiley and Richards (1982), who considered possible severe attenuation of higher frequencies due to turbulence. Turbulence counteracts ground attenuation at high frequencies and long distances even at the moderate sound amplitude and phase fluctuations associated with temperature fluctuations of $\pm 1°C$ (Chessell 1977, Embleton 1996). It also offsets upward refraction by transferring sound energy into the shadow zone (Daigle et al. 1986) with scattering as the principal mechanism at frequencies above 500 Hz (McBride et al. 1992).

2.4 Reverberation and the Effects of Vegetation and Habitat

Signals can also be degraded in their temporal structure. Particularly pulsed signals and those with complex temporal structure are influenced by reverberations that cause trailing echoes (Richards and Wiley 1980; Brenowitz 1986). Reverberation depends chiefly on frequency and range. Richards and Wiley (1980) found that reverberations in forests are least in the 2-8 kHz range. Studies that relate to habitat influences mainly concentrate on gross habitat differences, such as those between open and closed (forest) habitats (Chappuis 1971; Hunter and Krebs 1979; Ryan and Brenowitz 1985). Generally, birds in open habitats produce higher-pitched signals. Particularly studies of neotropical passerines suggest that forest-dwelling species use lower frequencies at smaller bandwidths, and exhibit low repetition rates (Morton 1975; Richards and Wiley 1980). These features are largely shared by species of temperate American or subtropical Old World habitats (Wasserman 1979; Lemon et al. 1981; Wiley and Richards 1982; Sorjonen 1986; Wiley 1991; Smith and Yu 1992). Wiley (1991) in his comprehensive survey of eastern North American birds could demonstrate that habitat had much stronger effects on temporal features than on spectral ones. Suggestive evidence for habitat influences comes from variation within songs, within species and between closely related species (e.g., Nottebohm 1975b; Bowman 1979, Hunter and Krebs 1979; Handford 1981; Shy 1983, 1984;

Sorjonen 1983; Shy and Morton 1986), particularly in cases where at least some song characteristics match transmission properties of the habitats. Pertinent data also come from studies that compare the transmission of sounds between species-specific and atypical habitats (e.g., Shy and Morton 1986). In most cases, however, it remains unclear what exactly determines the acoustic climate of the habitat and the specific adaptive responses to it.

In closed habitats, tall grass, reed, foliage, branches and tree trunks may scatter sound. From empirical data and theory, the following picture of sound attenuation in forests close to the ground (1-2 m) emerges (Embleton 1963; Aylor 1971; Morton 1975; Linskens et al. 1976; Marten and Marler 1977; Marten et al. 1977; Wiley and Richards 1982; Price et al. 1988). Attenuation peaks at low frequencies due to ground effects, it is relatively low at frequencies from 0.5 to 2.5 kHz, and it steadily increases with frequency (between 3 and 6 dB per doubling of frequency) above approximately 1.5 kHz due to scattering. At these higher frequencies small additional peaks due to resonance, probably insignificant for avian communication in most cases (but see Heuwinkel 1982, 1990 for reed stands), may appear (Price et al. 1988). Scattering in dense growth, on the other hand disrupts ground interference (Price et al. 1988).

Wiley and Richards (1978) presented a model of sound propagation in forest canopies that considers mainly the effects of refraction and treats the canopy as a more or less even surface. The model certainly underestimates surface roughness, and wind induced turbulence generated by canopy and emergent trees (Fitzjarrald et al. 1990). The canopy also acts as a low-pass filter so that only large-sized eddies may penetrate into the forest interior (Gao and Li 1993). Waide and Narins (1988) suggest that a considerable proportion of sound from the canopy does not reach the ground.

Background noise in forests exhibits clear spectral patterns, diurnal variation and stratification. It peaks between 4 and 9 kHz and is about 10 dB higher than in grassland (Ryan and Brenowitz 1985). Noise of biological origin is substantial particularly in tropical forests. It concentrates in the canopy and peaks at dusk, and increases markedly above 4 kHz (Riede 1996). Noise generated by wind, on the other hand, declines with frequency and may constitute the main source for noise in temperate forests (Klump 1996). Waterfalls and torrents produce strong noise that decreases in dB level towards higher frequencies (Martens and Geduldig 1990; Riede 1996).

Open fields, such as grasslands exhibit strong ground attenuation effects and refraction caused by wind or temperature gradients exert powerful influences on sound propagation (Marten and Marler 1977; Römer, this Vol.). Noise levels seem generally to be lower than in forests, however. In the reed beds of the Old World, a habitat for many songbird species, attenuation is negatively correlated with height above ground, and is strong above 5 kHz. Amplitude modulation is present above 1 kHz. Resonance effects may contribute to low and even to negative attenuation below 4 kHz (Jilka and Leisler 1974; Heuwinkel 1982, 1990).

3 Communication

3.1 Birds as Receivers

Bird ears are most sensitive in the 1-5 kHz range. They can discriminate frequencies differing by only 1-2% (Konishi and Knudsen 1979; Dooling 1982; Saunders and Salvi 1993; Fay 1992, 1997; Langemann et al. 1998; Bradbury and Vehrencamp 1998). There are small but biologically significant species-specific differences with respect to frequencies of best hearing (Okanoya and Dooling 1988; Kreutzer et al. 1991) and frequency dependent critical noise masked auditory thresholds (Langemann et al. 1998). Birds have generally a narrower hearing range than mammals and are even less sensitive than humans at higher frequencies although the latter rank low among mammals in this respect. As in other vertebrates, the detectability of a signal increases with signal duration (Klump and Maier 1990; Fay 1992, 1997).

Birds discriminate among different signals not only on the basis of frequency-related features, such as position of center frequency, lowest and highest frequency present, signal bandwidth, and alternations in pitch between successive elements, but also on the basis of temporal patterns (Weary et al. 1986; Kreutzer et al. 1991; Loesche et al. 1992; Catchpole and Slater 1995). They especially discriminate between changes in signals of their own species (Dooling et al. 1992; Kreutzer et al. 1991; Loesche et al. 1992). Motivation affects discrimination as well. Johnsrude et al. (1994) showed, for example, that some signal aspects remain invariant across motivational contexts while others are more specific.

Birds recognize acoustical signals rather well even in the presence of high levels of noise (Schubert 1971; Brenowitz 1982; Klump 1996; Fay 1997, Langemann et al. 1998). Frequencies of best hearing in the presence of noise may be substantially different from those measured without noise, and may be better predictors of species-specific signal frequencies (Okanoya and Dooling 1988; Klump 1996; Langemann et al. 1998). Amplitude modulations reduce masking by noise (Klump 1996).

Birds can determine the direction of a sound (accuracies achieved by small birds in the range of 10-30°, Bradbury and Vehrencamp 1998; Nelson and Stoddard 1998) although their ears are very close together. The mechanism birds use as described by Lewis and Coles (1980), Coles et al. (1980), and Hill et al. (1980) is different from mammals and is based on interaural comparisons of differential sound pressures facilitated by a direct connection between the ears. Birds locate broad-band signals more accurately than pure tones, and frequency-modulated tones better than non-modulated ones (Brown 1982; Klump and Shalter 1984). The ability of the peripheral auditory system to process directional information also increases signal-to-noise ratios in perception (Klump 1996).

The ability of birds to determine not only the direction of a sound source, but also its distance has received great interest (Richards 1981b; Morton 1982; Morton et al. 1986, 1998; Naguib 1997a, b, 1998; Wiley and Godard 1996).

Morton (1986, 1998) proposed that birds have to learn these signals and familiarize themselves with the songs of territorial neighbors in order to be able to compare the incoming more or less degraded signal with some internal representation of the original one. Alternatively, they also may use general distance-dependent features for ranging and would not need to be familiar with the songs (Morton and Derrickson 1996; Wiley and Godard 1996; Naguib 1997a).

3.2 Signal Design and Broadcasting Tactics

The ultimate goal of research on the ecology of communication is to understand to what degree ecological factors contribute to the diversity of avian acoustical signals and what portion of the variance in their features they explain. Environmental influences on sound transmission, the requirements for signal reception and processing by the birds' sensory and cognitive systems, and the number of alternatives that must be coded by the signal set have to be respected if one wishes to come up with design rules for a set of signals (Morton 1980; Endler 1993; Bradbury and Vehrencamp 1998). Signal range and locatability are the ecologically most relevant aspects when discussing the specific design principles (Bradbury and Vehrencamp 1998).

The *range* of signals should be maximal for mate attraction, large to moderate for need signals, moderate for territorial defence, moderate to short for alarm calls, and short for courtship and other intimate signals (Bradbury and Vehrencamp 1998).

The range of territorial signals should increase with territory size and its "active space", that distance from the sender over which the signal can be detected by potential receiver's (Marten and Marler 1977; Brenowitz 1982), should at least include the territories of nearby neighbors (Bradbury and Vehrencamp 1998). If the sender does not care about the receiver's espionage or even has some interest to cooperate in that respect, it should mix degradable elements into its song (Sorjonen 1983), while in the former case a greater uniformity in song element features is predicted. To sum it up, it can be said that optimizing transmission distance is not necessarily equivalent to maximizing it (e.g., Lemon et al. 1981).

The following design principles can be deduced from present knowledge on sound transmission, if range maximising is assumed as in the acoustic adaptation hypothesis (Morton 1975; see also Römer, this Vol.): in terms of frequency, signals should be low pitched if the sender is very close to the ground. If signal-to-noise ratio is low, energy should be concentrated in a narrow frequency band and signal elements should become louder and longer (so that receivers can integrate over longer periods). Acoustic signals intended for close interactions only should be of low amplitude, unstructured, and involve frequencies that are quickly attenuated.

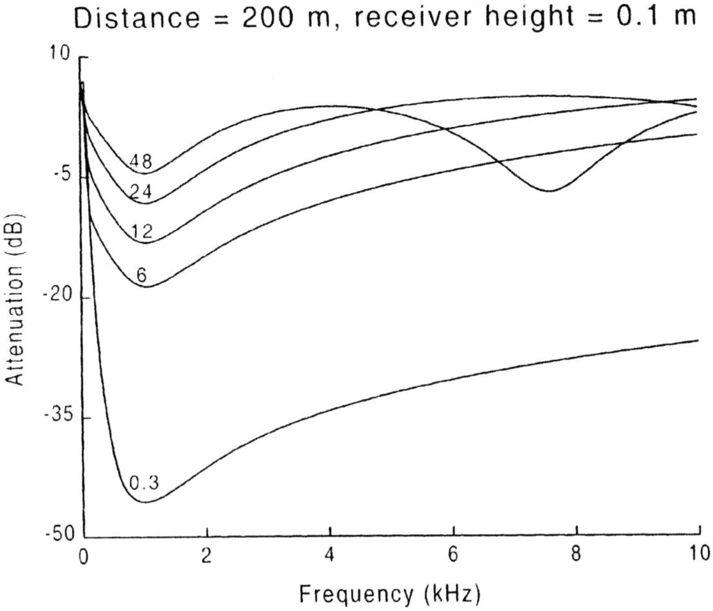

Fig. 1. Model calculations for frequency-dependent excess attenuation caused by ground effects. Distance between sender and receiver is 200 m, and receiver height is 0.1 m. *Different curves* apply to different sound source heights given in m (0.3, 6, 12, 24, 48). Based on the Delany-Bazley ground model with local reaction assumption and ground impedance 50 kPa (average forest floor) with a 0.1 m porous layer (see Sutherland and Daigle 1997). These particular calculations are representative for the habitat and behavior of thrushes, *Turdus*, that feed on the ground (receiver) and sing at various heights (6 – 48 m), or formicariid antbirds that receive and sing (0.3 m) close to the ground in a forest

Good *locatability* is required for mate attraction, long-range communication between social partners, and possibly for attracting members of mixed-species flocks and mobbing gangs. It should be low, however, for alarm calls that induce fleeing or hiding.

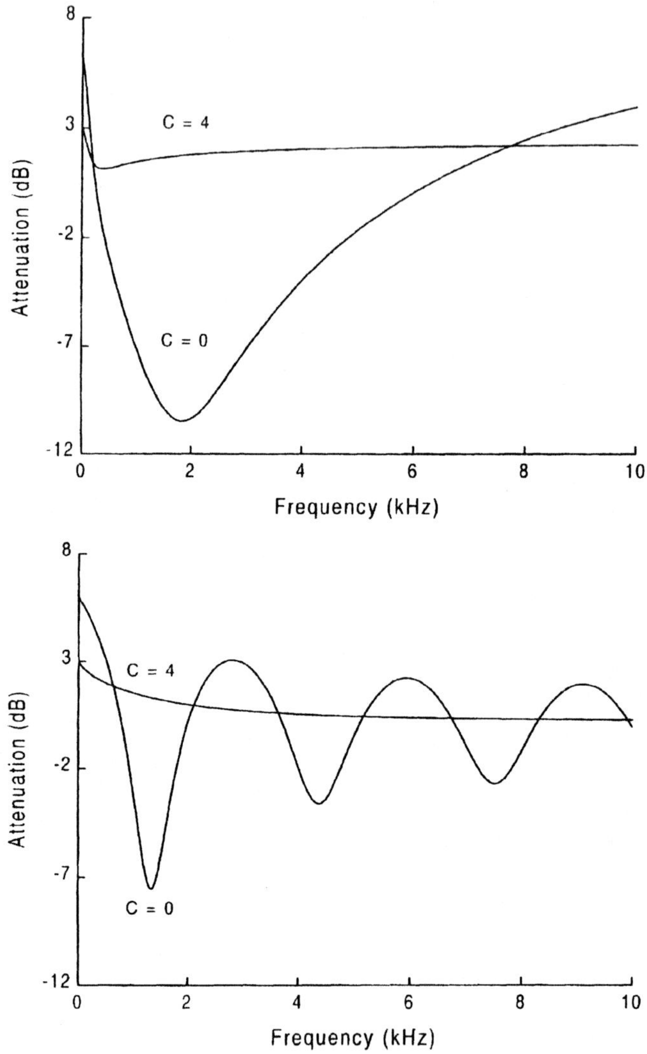

Fig. 2. Model calculations for frequency-dependent excess attenuation caused by ground effects. The turbulence constant C (Chessell 1977) differs for the two curves. Based on Delany-Bazley's ground model with local reaction assumption and a ground impedance of 200 kPa (grassland) with a 0.1 m porous layer (see Sutherland and Daigle 1997). These particular calculations are representative for skylarks, *Alauda,* that feed on the ground (receiver) and sing high above the ground in a song flight. **Top** Model for a bird singing at 12 m (about the minimum height observed, Glutz von Blotzheim and Bauer 1985), and **Bottom** for a bird singing at 120 m

Locatability is enhanced if signals are broad-band, frequency modulated, or abrupt in onsets and interruptions (Marler 1955, 1956; Brown 1982; Klump and Shalter 1984; Redondo and Arias de Reyna 1988). Conversely, pure tones (they also should be high-pitched because receivers may detect such signals more easily than the predator, Klump et al. 1986) with gradual onset camouflage the sender's position. Another aspect of locatability is the problem to direct the signal to a reduced set of receivers over long distance. Directionality is mainly a signaler's problem and cannot be discussed here for lack of room (see Witkin 1977; Larsen and Dabelsteen 1990 for references).

The receiver wants to maximize *signal detection* while reducing the likelihood of possibly costly false alarms (Richards 1981aM Wiley and Richards 1982). This depends very much on the uncertainty about the arrival of the signal, especially when it is sent rarely. Preambles that are followed by the signal proper at a predictable interval can solve this problem. Applied to bird songs, the introductory notes should be easily detected and consist of pure tones, lack species-specific characteristics, and they would be expected to precede signals rich in complex notes with high repetition rates that are very species-specific (Richards 1981a). Since motivation affects discrimination as well (Johnsrude et al. 1994), the preamble's response eliciting power should be invariant across motivational contexts.

The relative position of signaler and receiver with respect to each other and with respect to features in the environment that are significant for sound propagation has received some attention in the literature (Hunter 1980; Lemon et al. 1981; Wiley and Richards 1982; Wilczynski et al. 1989), but almost always treated from the signaler's perspective. Others (Dabelsteen et al. 1993; Mathevon et al. 1996), however, discuss the possibility that it might be more important for the signaler to optimize his receiving rather than his broadcasting range. It is most important to get high above the ground. Figures 1 and 2 present theoretical calculations based on the models and algorithms described in Chessell (1977), Price et al. (1988), Pirinchieva (1991), Wempen (1991), and Sutherland and Daigle (1997). Figure 1 might be relevant for antpittas (Sect. 3.2) that sing and receive on the ground and thrushes that forage on the ground (receivers) and sing in middle and upper tiers of the forest.

These calculations did not include the attenuation of higher frequencies through vegetation (Price et al. 1988). Moreover, with increasing signaler height the portion of the sound rays that travel through the understory is reduced. Figure 2 represents the situation for a skylark for which the receiver also is assumed to be at ground level. Here, I also included the effects of turbulence, following Chessell's (1977) approach. The most significant result of these model calculations is that singing high up during turbulent conditions, as would often occur in grasslands (but not in forests), may be advantageous.

The above design principles are subject to *constraints* by the other demands on the signals (Lambrechts 1996). The number of different classes of individuals (sex, age, kin, individual, species) which should be identified by the receiver, the modulation potential (within individual variation), emotional coding (Morton 1977, 1982), and signal honesty are paramount to social communication. Some further constraints derive from the morphology and physiology of the sender. The possibility of eavesdropping predators and their sensory performance certainly influences signal design as well (Klump and Shalter 1984; Klump et al. 1986). Although body size should theoretically constrain the ability to produce sound of low frequency, correlations between avian body size and pitch of the sounds produced are rather weak (Greenewalt 1968, Nottebohm 1975a).

Analyses related to this topic (Ryan and Brenowitz 1985; Lemon et al. 1981) did not, however, completely control for phylogenetic effects (Garland and Adolph 1994), which in the face of the observed low correlations might be rather relevant since body size in birds is strongly dependent on phylogeny (Owens et al. 1999).

4 Signal Design and Habitat Selection in the Acoustical Communication of Woodpeckers

The vocal repertoire of most species-rich group of woodpeckers, the genus *Picoides* (Winkler and Christie 1995), consists of more than ten categories (Winkler and Short 1978). Differences within repertoires are large among categories and rather consistent across species with species-specific differences mainly encoded in the two categories Call Note and Rattle Call (Winkler and Short 1978) that I will discuss further.

Call notes are the most frequent and general call. They are short (23-128 ms) in overall duration, with one exception (red-cockaded woodpecker, *P. borealis*, see below) simple chevron-like structures on the sonagram. Since sound energy is clearly concentrated in the raising and falling parts, effective signal duration is about 16 ms. Such a signal, transmitted from about 20 m height to a receiver at 10 m and at a distance of 100 m, is too short to be subject to attenuation through destructive interference. Its frequency is in the range of 1.7-3.6 kHz with an average of 2.7 ± 0.14 kHz (N = 19 species) and lies roughly in the frequency window of 1-2.5 kHz for sound transmission in forests (Morton 1986; Marten and Marler 1977; Marten et al. 1977; Price et al. 1988). These features indicate that this signal type is designed for good transmission.

The rattle is used over long distances and basically consists of strings of call notes in many species. In some species rattles consist of elements that are very different from ordinary call notes and they are typically uttered with leading call notes, thus representing composite signals (Fig. 3A) with an alerting and a message component. The only condition listed in Section 1.2.2 that is not met is the species-specific character of the leading call note (Winkler 1971, Winkler and Short 1978). Although pied woodpeckers cover a wide range of body sizes (13-160 g), there seems to be no relationship between body size and call pitch.

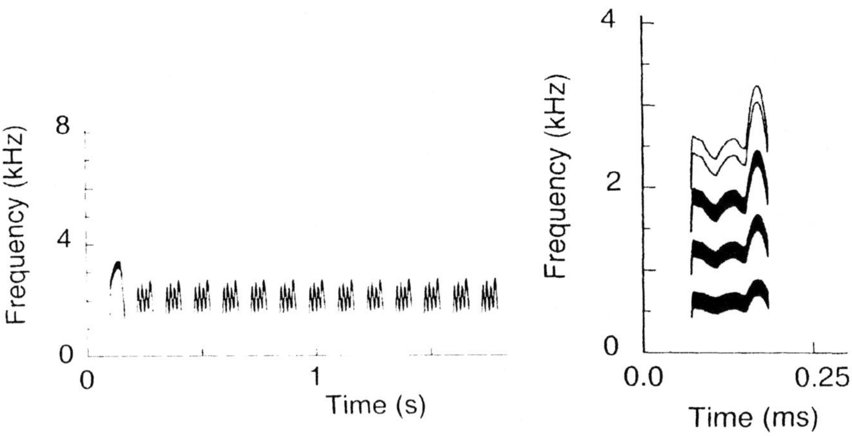

Fig. 3. Schematic representations of pied woodpecker signals. **Left**, Rattle call (long-distance communication), **Right**, Wad call (close distances)

In contrast, and as predicted, intimate signals are much less conspicuous. A soft call type (Wad call) that is used in the context of copulations and also during changeovers at the breeding hole contains frequencies between 0.65 and 1.85 kHz, and has a noisy, smacking quality (Fig. 3B).

The red-cockaded woodpecker is the only pied woodpecker that is a cooperative breeder and forages in groups (see Winkler and Christie 1995 for more information). The consequences for its acoustic communication are that those intimate calls that are used rarely by the other species are frequently uttered and are much more conspicuous (in the 1.5-4.3 kHz range), and highly variable. The call note duration (72-128 ms) is the longest in the genus and the only one that is regularly frequency modulated (that would allow for encoding of individuality, but this has never been tested).

Many models of the economics of signaling assume that there is at least some conflict between activities such as foraging and communication. Woodpeckers forage in certain strata and may also exhibit sexual dimorphism in microhabitat use (Wallace 1974; Peters and Grubb 1983; Matthysen et al. 1991; Winkler and Christie 1995). The long-range rattle calls are mainly broadcast from the canopy (Winkler et al. 1996).

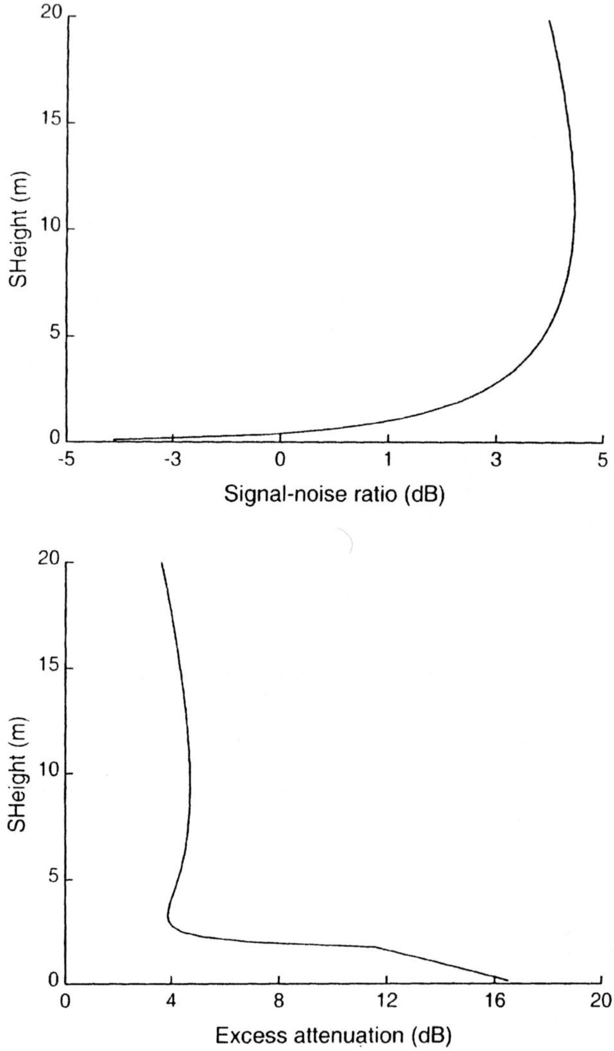

Fig. 4 A,B. Idealized representation of the vertical distribution of **A** top, signal-to-noise ratio and **B** bottom, excess attenuation in a tropical rain forest at a sender to receiver distance of 40 m. Data based on a regression of values measured in six species of antbirds (five Thamnophilidae and one Formicariidae)

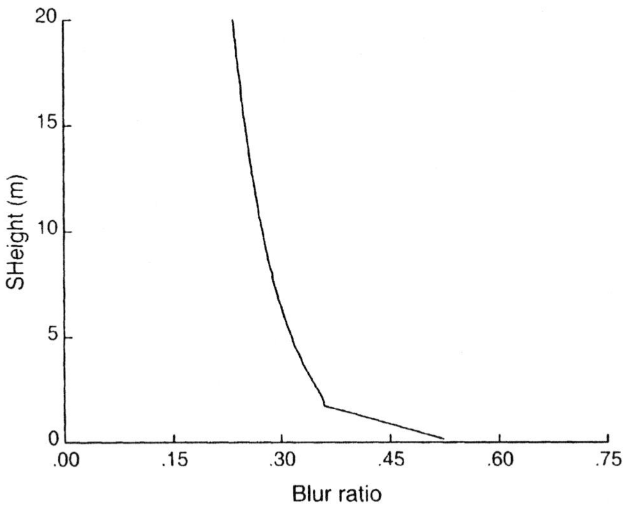

Fig. 4 C. Blur ratio in a tropical rain forest at a sender to receiver distance of 40 m. Data based on a regression of values measured in six species of antbirds (five Thamnophilidae and one Formicariidae)

In the Arizona woodpecker *Picoides stricklandi* 71% of these signals are sent at treetop height, while foraging birds are found with only 36% probability at this level (Winkler 1979). This significant difference hints at a potential conflict between signaling and foraging. It may contribute to the ecological sexual dimorphism in some species such as the downy woodpecker *P. pubescens*. Males of this species forage at higher levels (Peters and Grubb 1983; Matthysen et al. 1991), and rattles are sent mainly from tree tops (84% pers. obs.).

While there is no significant difference between calling and foraging habitat in males, the difference for females is significant. This vocalization is associated with dominance and males invest more into territorial behavior and signaling than females (Winkler and Short 1978), and frequent switching between signaling and foraging habitat may be costly.

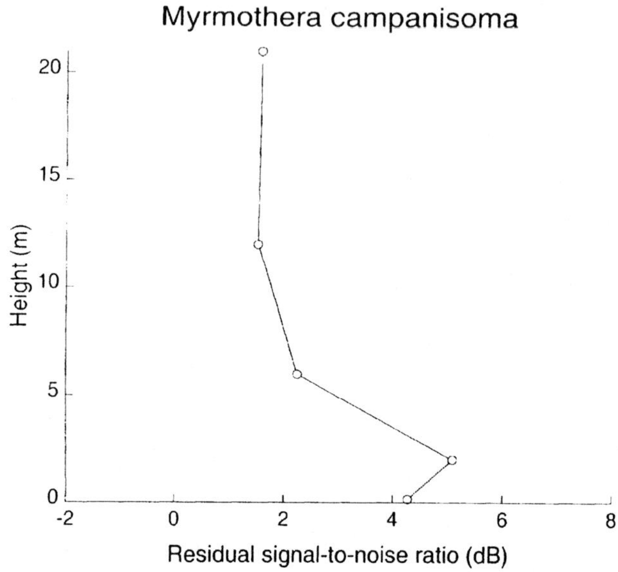

Fig. 5. Residual signal-to-noise ratio of thrush-like antpitta, *Myrmothera campanisoma*, songs transmitted at different heights in a rainforest. Shown are the residuals from the regression depicted in Fig. 4. This species sings near the ground

5 Propagation of Antbird Songs in a Rainforest

Antbirds in the wide sense are a neotropical group of suboscines that comprise the families Thamnophilidae and Formicariidae. They do not learn their song (Kroodsma and Konishi 1991) and in many species both sexes sing. We studied the propagation of antbird songs in a lowland rainforest in Southern Venezuela (Nemeth et al. 1997; Nemeth and Winkler in prep.). The study site represents an undisturbed, not very tall forest with moderate reverberation: echoes decay to -60 dB in about 0.9 s, about half the value of a full, medium sized concert hall, with a tendency to higher values near the ground, and with background noise lowest between 550 Hz and 3.5 kHz, indicating the gap between low-frequency wind generated and high-frequency biogenic noise. The noise is lowest early in the morning (less than 40 dB re 20μPa) and highest in the evening (about 55 dB re 20μPa), mainly due to insects (N Ellinger pers. comm.). We found a source of amplitude modulation at about 0.2 Hz that cannot be attributed to atmosphaeric turbulence, but we could not detect amplitude modulations in the frequency range described by Richards and Wiley (1980). We recorded the song of several species at close range and later transmitted them in the forest along a transect at various distances and heights with the speaker and microphone always at the same height. The data was analyzed for excess attenuation and signal-to-noise ratio. Furthermore, we computed the blur ratio, the ratio between the energy of the observed signal and the sum of the energy in the bird vocalization and the channel noise. These measures represent the changes in most of the information-carrying parameters in the sounds (Dabelsteen et al. 1993).

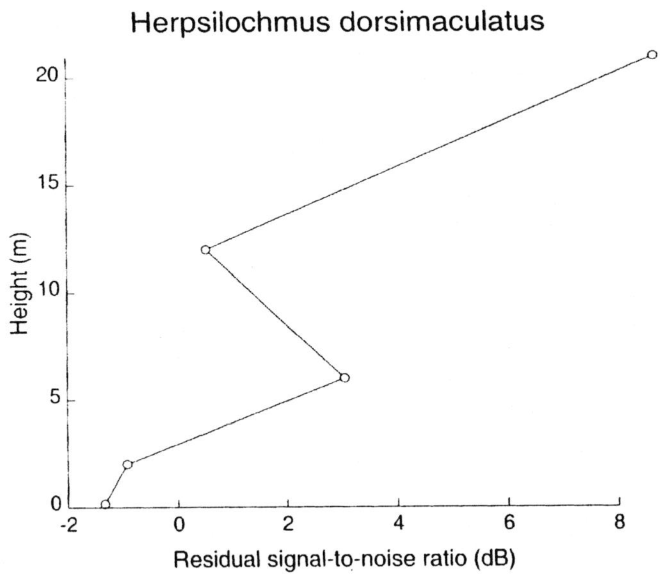

Fig. 6. Residual signal-to-noise ratio of spot-backed antwren, *Herpsilochmus dorsimaculatus*, songs transmitted at different heights in a rainforest. Shown are the residuals from the regression depicted in Fig. 4. This species sings in the forest canopy

Figure 4 shows the average fate of antbird signals expressed in the aforementioned parameters. Multiple regression analysis with height above the ground and distance as predictors was used to calculate the average relationships-to-blur ratio, signal-to-noise ratio and excess attenuation for all species together. Residual values with respect to these curves were taken as indication for adaptation to long-range transmission. Figures 5 and 6 show the respective values for the ground-dwelling antpitta *Myrmothera campanisoma* (Formicariidae) and the canopy-inhabiting spot-backed antwren *Herpsilochmus dorsimaculatus* (Thamnophilidae). The analysis suggests that signal properties are tuned to the conditions for sound transmission at the typical height of the singing post of these species. Previous studies measured habitat influences in the frequency domain only (Marten and Marler 1977).

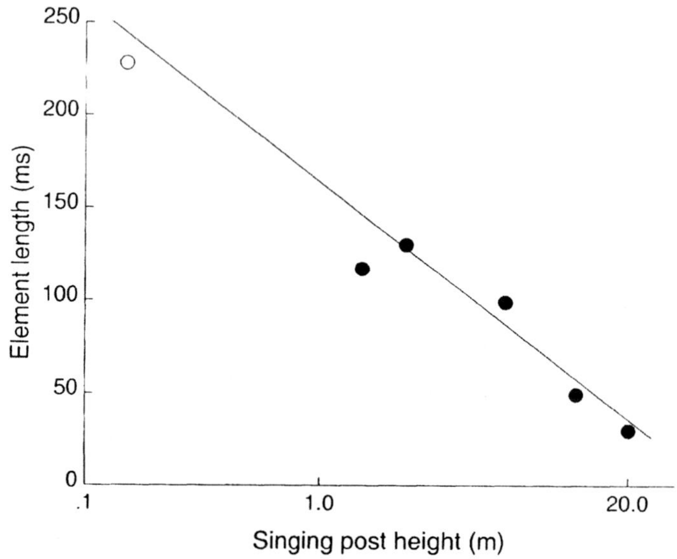

Fig. 7. Relationship between element duration and singing height of six species of antbirds. Regression on log singing post height (m) and element length (ms)

The results presented here stress the role of acoustical time patterns. In fact, there is a strong negative correlation between element length, and the logarithm of singing post height (Fig. 7). With respect to frequency, the antpitta's song is remarkable because it sings in the frequency band (approximately 1 kHz) that is attenuated most according to theory (see Fig. 1).

However, loudness and concentration of sound energy in a very narrow frequency band ensure good transmission over larger distances (see Sect. 2.3) and its length ensures detection by temporal summation (Klump and Maier 1990; Sect. 2.1). Completely unrelated species that use this singing habitat in Asia produce similar songs (Smith and Yu 1992).

These data lend support to the acoustic adaptation hypothesis that states that long-distance signals are structured for maximal transmission (Morton 1975) which had not been supported by other studies (e.g. Fotheringham et al. 1997). They also suggest that they may be more easily detected if they are tuned to the receiver's mechanisms of perception.

6 Dawn Chorus: Ecology, Sex, or Updating?

Diel patterns in acoustic signaling are universal among animals in general, and in birds in particular. Many species of birds have a main peak of song at dawn and a minor one at dusk. Many explanations have been proposed to explain this pattern. In their review, Staicer et al. (1996) list 12 hypotheses and group them into those that rest on physiology or the internal state of the singer, those that emphasize environmental and other ecological constraints, and those that invoke social necessities. Since song may have several functions, these hypotheses are not mutually exclusive.

It has been known for some time that song influences reproductive behavior and hormonal levels of females (Wingfield and Farner 1993; Cheng et al. 1998), and may even control gene expression in certain parts of the brain (Ball and Gentner 1998; Jarvis et al. 1998). Singing may also stimulate the signaler's own hormones. The self-stimulating hypothesis posits that singing changes the hormone levels of the singer, and that this is most effective at dawn. The daily cycle hypothesis proposes a circadian fluctuation of hormone levels with the dawn chorus as being a nonfunctional consequence (Staicer et al. 1996). Thorough testing of these possibilities must wait until methods exist which allow one to monitor and manipulate hormones in the course of a day while the birds are singing.

At low light levels neither day- nor night-active hunters may find ideal conditions. Hence, dawn and dusk would be ideal periods for singing. The predation hypothesis implies that males could sing from more exposed locations at dawn and dusk than at other times (Staicer et al. 1996). Neither observational nor experimental evidence supporting the predation hypothesis seems to exist. Foraging and singing may interleave, as in vireos (*Vireo*, Vireonidae), may constrain each other to some extent, or represent activities clearly separated with respect to microhabitat (e.g. Hunter 1980) and timing. Experiments on captive birds show that the birds are able to forage only at luminances that are reached after the time of the dawn chorus (Kacelnik 1979). Besides this, the hypothesis of singing at periods when foraging is inefficient has received little support. A more complex hypothesis derived from dynamic programming models (Hutchinson et al. 1993, Houston and McNamara 1999) shares with the former the assumption that foraging and singing are mutually exclusive activities, and the complete neglect of factors related to the receiver. These models predict that, if there is uncertainty about daily energy requirements, time for singing is optimally allocated at dawn and dusk, Hutchinson et al. (1993) assume that females deem continuous songs most attractive and that singing is costly (mainly in terms of lost opportunity for other activities). Although there is little empirical support for this hypothesis, it is attractive because it also explains the dusk chorus. Like the previous hypothesis it suggests that the less singing is separated from foraging, the less it should be restricted to dawn and dusk. Environments differ in the way in which atmospheric conditions change in the course of the day. For instance, in the rainforest studied by us, temperature gradients and turbulence build up only slowly, yet, songs of most birds are clearly concentrated or confined to very brief periods at dawn and dusk. In other words: no ecological causes can be invoked as explanation.

Signals should be honest and uncheatable, particularly if the interest of sender and receiver are different. As stated in the much-discussed handicap principle, signals are reliable indicators of the signaler's quality provided they are costly (Zahavi 1977, Grafen 1990). A signal that indicates only the presence of the signaler can be cost-free and reliable at the same time when there is no conflict of interests (Maynard Smith 1994). As an explanation for the dawn chorus, the handicap principle would require that singing is costly, and that these costs are particularly high at dawn. It would predict not only that there is a strong correlation between the quality of the singer and dawn song output, but also that the variance among singers is highest at this time of day (Staicer et al. 1996). Note that ecological hypotheses assumed that singing at dawn is more efficient and presumably less costly. High predation risk in the morning is incompatible with the predation hypothesis, but would represent just the type of handicap honest signaling and sexual selection models would require.

Mate-guarding is one important means by which males can prevent their mates from engaging in extrapair copulations (Birkhead and Møller 1992). Singing in the dawn may ward off potential intruders and also may also persuade the female not to leave for other males. It has been believed for some time that extrapair copulations are most successful at dawn because a fertility window would be open at that time. This early fertility window seems, however, not to be a general phenomenon (Sheldon and Burke 1994). Singing and accompanying the female, the most effective way of guarding her, may not be compatible activities. This, and the fact that in most species males do not sing in the fertile period of their females do not render paternity assurance as a plausible cause for the dawn chorus (Staicer et al. 1996; Gil et al. 1999).

The territory defence hypothesis rests on the assumption that territorial intrusion peaks at dawn (Staicer et al. 1996). It also implies that listeners use the information about presence and status of the singer immediately. Thus, it would predict that the timing of singing matches the timing of territory take-overs, and suggests corresponding playback experiments (Staicer et al. 1996).

Staicer et al. (1996) propose a new hypothesis that stresses the role of singing for adjusting social relationships among territorial neighbors. Among other things, they suggest that males who assert their status at dawn benefit from that during social interactions during the day. The main prediction of the social dynamics hypothesis is that variation in dawn singing is correlated with the instability of social relationships. The higher the instability, the longer or intense the dawn chorus would be (Staicer et al. 1996). Factors that contribute to this instability would include overnight mortality, frequent extrapair copulations, and dynamic territorial boundaries. This hypothesis is attractive because it combines ecologically relevant factors (mortality) with aspects of communication; but, it is the latter aspect that remains somewhat elusive, for it is rather unspecific about how song contributes to the stability of the social relationships. As in the territory defence hypothesis, effects of singing are immediate. So, for example, males should sing when fertile females ready to engage in an extrapair copulation are within earshot (Staicer et al. 1996). In this view information translates into female decisions without much delay. Hence, the timing and intensity of signaling must match those of the actions of the receiver. I suggest that the main issue in social dynamics is the singer's uncertainty about the status of the other members of the social network. Singing then is as much information-gathering through provoking the signals of others, as it constitutes valuable information for the receiver. Since uncertainty is maximal after the night, this information is most valuable in the morning, and it should influence all later decisions.

7 Concluding Remarks

Bird song is arguably the most intensively studied aspect of avian biology. Nevertheless, bird calls and the songs of nonoscines remain little understood. Bird songs are complex, carry many messages, have several functions, and are subject to many different influences, none the least of which is the fact that they are learned (Catchpole and Slater 1995). The acoustic communication of other birds, such as woodpeckers and antbirds, may represent much better models for studying general principles of communication. Since the auditory system codes natural sounds at higher information rates and efficiencies than white noise (Rieke et al. 1995) and transmission properties may also be sensitive to complex signal features, especially in temporal dimensions, more experiments that involve naturalistic sounds are needed.

We need measurements of real sender-receiver interactions under biologically relevant conditions. They should cover more signal types and more species or clades. The acoustical properties of many natural habitats are still unknown. Clearly, more realistic models and empirical measurements of sound propagation in forest canopies are urgently needed, particularly because much, if not most, of avian acoustic signals in forests originate in this layer.

Much could certainly be gained if the requirements for information-gathering were more respected. Most signalers also need to receive information and their broadcasting may have the main purpose of eliciting information from others. With the exception of song learning studies, practically all laboratory and field experiments deal with immediate responses of birds to acoustic signals. To gain more insights, we have to extend this approach and compare informed versus uninformed receivers, not only at the time they become informed but also after they have been informed. Very much along the same line, I would also argue for a stronger emphasis on studying the need for information by the receiver, how that influences the design and performance of its senses, the signaling behavior of conspecifics and other organisms, and its own information-gathering efforts through signaling. After all, information is the currency of communication.

Acknowledgments. I thank F.G. Barth for his invitation to write this chapter. E. Nemeth provided the data for the analyses on antbirds. For this, his productive discussion, and great enthusiasm during our field work, I am most grateful. The work at the Surumoni crane site was supported by the Fonds zur Förderung der Wissenschaftlichen Forschung, project P11563. J. Szarzynski provided valuable results and comments related to his extensive micrometeorological studies at the Surumoni study site. Special thanks go to T. Dabelsteen and his coworkers, who introduced us into their elaborate techniques of sound analysis. W. Deutsch helped in many ways in procuring the technical literature. Work on woodpeckers had been supported by the F.M. Chapman Fund, AMNH, New York.

References

Attenborough K (1985) Acoustical impedance models for outdoor ground surfaces. J Sound Vibr 99:521-544

Aylor D (1971) Noise reduction by vegetation and ground. J Acoust Soc Am 51:197-205

Ball GF, Gentner TQ (1998) They're playing our song: gene expression and birdsong perception. Neuron 21:271-274

Birkhead T, Møller AP (1992) Sperm Competition in Birds. Academic Press, London, 282 p

Bowman RI (1979) Adaptive morphology of song dialects. J Ornithol 120:353-380

Bradbury JW, Vehrencamp SL (1998) Principles of Animal Communication. Sinauer, Sunderland MS

Brenowitz EA (1982) The active space of red-winged blackbird song. J Comp Physiol A 147:511-522

Brenowitz EA (1986) Environmental influences on acoustic and electric animal communication. Brain Behav Evol 28:32-42

Brown CH (1982) Ventriloquial and locatable vocalizations in birds. Z Tierpsychol 59:338-350

Catchpole CK, Slater PJB (1995) Bird Song: Biological Themes and Variations. Cambridge University Press, Cambridge, 248 p

Chappuis C (1971) Un exemple de l'influence du milieu sur les émissions vocales des oiseaux: L'évolution des chants en forêt équatoriale. Terre et Vie 2:183-202

Cheng MF, Peng JP, Johnson P (1998) Hypothalamic neurons preferentially respond to female nest coo stimulation: demonstration of direct acoustic stimulation of luteinizing hormone release. J Neurosci 18:5477-5489

Chessell CI (1977) Propagation of noise along a finite impedance boundary. J Acoust Soc Am 62:825-834

Coles RB, Lewis DB, Hill KG, Hutchings ME, Gower DM (1980) Directional hearing in the Japanese quail (*Coturnix coturnix japonica*). II. Cochlear physiology. J Exp Biol 86:153-170

Dabelsteen T, Larsen ON, Pedersen SB (1993) Habitat-induced degradationof sound signals: Quantifying the effects of communication sounds and bird location on blur ratio, excess attenuation, and signal-to-noise ratio in blackbird song. J Acoust Soc Am 93:2206-2220

Daigle GA, Embleton TFW, Piercy JE (1986) Propagation of sound in the presence of gradients and turbulence near the ground. J Acoust Soc Am 79:613-627

Dooling R (1982) Auditory perception in birds. In: Kroodsma DE, Miller EH (eds) Acoustic Communication in Birds, Vol 1. Academic Press, New York, pp 95-130

Dooling RJ, Brown SD, Klump GM, Okanoya K (1992) Auditory perception of conspecific and heterospecific vocalizations in birds: evidence for special processes. J Comp Psychol 106:20-28

Embleton TFW (1963) Sound propagation in homogeneous deciduous and evergreen woods. J Acoust Soc Am 35:1119-1125

Embleton TFW (1996) Tutorial on sound propagation outdoors. J Acoust Soc Am 100:31-48

Embleton TFW, Piercy JE, Daigle GA (1983) Effective flow resistivity of ground surfaces determined by acoustical measurements. J Acoust Soc Am 74:1239-1244

Endler JA (1993) Some general comments on the evolution and design of animal communication systems. Phil Trans R Soc Lond B 340:215-225

Fay RR (1992) Structure and function in sound discrimination among vertebrates. In: Webster DB, Fay RR, Popper AN (eds) The Evolutionary Biology of Hearing. Springer-Verlag, New York, pp 229-263

Fay RR (1997) Hearing and sound perception among vertebrate animals. In: Crocker MJ (ed) Encyclopedia of Acoustics, Vol 4. Wiley, New York, pp 1789-1797

Fitzjarrald DR, Moore KE, Cabral O, Scolar J, Manzi AO, de Abreu Sá LD (1990) Day-time turbulent exchange between the Amazon forest and the atmosphere. J Geophys Res 95:16825-16838

Fotheringham FR, Martin PR, Ratcliffe L (1997) Song transmission and auditory perception of distance in wood warblers (Parulinae). Anim Behav 53:1271-1285

Gao W, Li, BL (1993) Wavelet analysis of coherent structures at the atmosphere-forest interface. J Appl Meteorol 32:1717-1725

Garland T Jr, Adolph SC (1994) Why not do two-species comparative studies?: limitations on inferring adaptation. Physiol Zool 67:797-828

Gil D, Graves JA, Slater PJB (1999) Seasonal patterns of singing in the willow warbler: evidence against the fertility announcement hypothesis. Anim Behav 58:995-1000

Glutz von Blotzheim UN, Bauer KM (1985) Handbuch der Vögel Mitteleuropas, Vol. 10/I. Aula-Verlag, Wiesbaden, 507p

Grafen A (1990) Biological signals as handicaps. J Theor Biol 144:517-546

Greenewalt CH (1968) Bird Song: Acoustics and Physiology. Washington DC, Smithsonian Press, 194 p

Handford P (1981) Vegetational correlates of variation in the song of *Zonotrichia capensis*. Behav Ecol Sociobiol 8:203-206

Heuwinkel H (1982) Schalldruckpegel und Frequenzspektren der Gesänge von *Acrocephalus arundinaceus*, *A. scirpaceus*, *A. schoenobaenus* und *A. palustris* und ihre Beziehungen zur Biotopakustik. Ökol Vögel (Ecol Birds) 4:85-174

Heuwinkel H (1990) The effect of vegetation on the transmission of songs of selected European Passeriformes. Acta Biol Benrodis 2:133-150

Hill KG, Lewis DB, Hutchings ME, Coles RB (1980) Directional hearing in the Japanese quail (*Coturnix coturnix japonica*). I. Acoustic properties of the auditory system. J Exp Biol 86:153-170

Houston A, McNamara J (1999) Models of Adaptive Behaviour. Cambridge University Press, Cambridge, 378 p

Hunter ML (1980) Microhabitat selection for singing and other behaviour in great tits, *Parus major*: some visual and acoustical considerations. Anim Behav 28:468-475

Hunter ML, Krebs JR (1979) Geographical variation in the song of the great tit (*Parus major*) in relation to ecological factors. J Anim Ecol 48:759-785

Hutchinson JMC, McNamara JM, Cuthill IC (1993) Song, sexual selection, starvation and strategic handicaps. Anim Behav 45:1153-1177

Jarvis ED, Scharff C, Grossman MR, Ramos JA, Nottebohm F (1998) For whom the birds sing: context-dependent gene expression. Neuron 21:775-788

Jilka A, Leisler B (1974) Die Einpassung dreier Rohrsängerarten (*Acrocephalus schoenobaenus*, *A. scirpaceus*, *A. arundinaceus*) in ihre Lebensräume in Bezug auf das Frequenzspektrum ihrer Reviergesänge. J. Ornithol. 115:192-212

Johnsrude IS, Weary DM, Ratcliffe LM, Weisman RG (1994) Effect of motivational context on conspecific song discrimination by brown-headed cowbirds (*Molothrus ater*). J Comp Psychol 108:172-178

Kacelnik A (1979) The foraging efficiency of great tits (*Parus major* L.) in relation to light intensity. Anim Behav 27:237-241

Klump GM (1996) Bird communication in the noisy world. In: Kroodsma DE, Miller EH (eds) Ecology and Evolution of Acoustic Communication in Birds. Cornell Univ. Press, Ithaca, NY, pp 321-338

Klump GM, Maier EH (1990) Temporal summation in the European starling (*Sturnus vulgaris*). J Comp Psychol 104:94-100

Klump GM, Shalter MD (1984) Acoustic behaviour of birds and mammals in the predator context. Z Tierpsychol 66:189-226

Klump GM, Kretzschmar E, Curio E (1986) The hearing of an avian predator and its avian prey. Behav Ecol Sociobiol 18:317-323

Konishi M, Knudsen EI (1979) The oilbird: hearing and echolocation. Science 204:425-427

Kreutzer, ML, Dooling RJ, Brown SD, Okanoya K (1991) A comparison of song syllable perception by five species of birds. Int J Comp Psychol 4:141-155

Kroodsma DE, Konishi M (1991) A suboscine bird (Eastern Phoebe, *Sayornis phoebe*) develops normal song without auditory feedback. Anim Behav 42:477-487

Kroodsma DE, Miller EH (eds 1996) Ecology and Evolution of Acoustic Communication in Birds. Cornell Univ. Press, Ithaca, NY, 587 p

Lambrechts MM (1996) Organization of birdsong and constraints on performance. In: Kroodsma DE, Miller EH (eds) Ecology and Evolution of Acoustic Communication in Birds. Cornell Univ. Press, Ithaca, NY, pp 305-320

Langemann U, Gauger B, Klump GM (1998) Auditory sensitivity in the great tit: perception of signals in the presence and absence of noise. Anim Behav 56:763-769

Larsen ON, Dabelsteen T (1990) Directionality of blackbird vocalization. Implications for vocal communication and its further study. Ornis Scand 21:37-45

Lemon RE, Struger J, Lechowicz MJ, Norman RF (1981) Song features and singing heights of American warblers: Maximization or optimization of distance? J Acoust Soc Am 69:1169-1176

Lewis B, Coles R (1980) Sound localization in birds. Trends Neurosci 3:102-105

Linskens HF, Martens MJM, Hendriksen HJGM, Roestenberg-Sinnige AM, Brouwers WAJM, van der Staak ALHC, Strik-Jansen AMJ (1976) The acoustic climate of plant communities. Oecologia 23:165-177

Loesche P, Beecher MD, Stoddard PK (1992) Perception of cliff swallow calls by birds (*Hirundo pyrrhonota* and *Sturnus vulgaris*) and humans (*Homo sapiens*). J Comp Psychol 106:239-247

Marler P (1955) Characteristics of some animal calls. Nature 176:6-8

Marler P (1956) Über die Eigenschaften einiger tierlicher Rufe. J Ornithol 97:220-227

Marten K, Marler P (1977) Sound transmission and its significance for animal vocalization I, Temperate habitats. Behav Ecol Sociobiol 2:271-290

Marten K, Quine D, Marler P (1977) Sound transmission and its significance for animal vocalization II. Tropical forest habitats. Behav Ecol Sociobiol 2:291-302

Martens J, Geduldig G (1990) Acoustic adaptations of birds living close to Himalayan torrents. Proc. Int. 100. DO-G Meeting:123-131

Mathevon N, Aubin T, Dabelsteen T (1996) Song degradation during propagation: importance of song post for the wren *Troglodytes troglodytes*. Ethology 102:397-412

Matthysen E, Grubb TC, Cimprich D (1991) Social control of sex-specific foraging behaviour in Downy Woodpeckers, *Picoides pubescens*. Anim Behav 42:515-517

Maynard Smith J (1994) Must reliable signals always be costly? Anim Behav 47:1115-1120

McBride WE, Bass HE, Raspet R, Gilbert KE (1992) Scattering of sound by atmospheric turbulence: Predictions in a refractive shadow zone. J Acoust Soc Am 91:1336-1340

Morton ES (1975) Ecological sources of selection on avian sounds. Amer Nat 109:17-34

Morton ES (1977) On the occurrence and significance of motivation-structural rules in some bird and mammal sounds. Amer Nat 111:855-869

Morton ES (1980) The ecological background for the evolution of vocal sounds used at close range. Acta XVII Congr Int Ornithol: 737-741

Morton ES (1982) Grading, discreteness, redundancy and motivation-structural rules. . In: Kroodsma DE, Miller EH (eds) Acoustic Communication in Birds, Vol 1. Academic Press, New York, pp 183-212

Morton ES (1986) Prediction from the ranging hypothesis for the evolution of long distance signals in birds. Behaviour 99:65-86

Morton ES (1998) Degradation and signal ranging in birds: memory matters. Behav Ecol Sociobiol 42:135-137

Morton ES, Derrickson KC (1996) Song ranging by the dusky antbird, *Cercomacra tyrannina*: ranging without song learning. Behav Ecol Sociobiol 39:195-201

Morton ES, Gish SL, van der Voort M (1986) On the learning of degraded and undegraded songs in the Carolina Wren. Anim Behav 34:815-820

Naguib M (1997a) Ranging of songs with the song type on use of different cues in Carolina wrens: effects of familiarity. Behav Ecol Sociobiol 40:385-393

Naguib M (1997b) Use of song amplitude for ranging in Carolina Wrens, *Thryothorus ludovicianus*. Ethology 103:723-731

Naguib M (1998) Perception of degradation in acoustic signals and its implications for ranging. Behav Ecol Sociobiol 42:139-142

Nelson BS, Stoddard PK (1998) Accuracy of auditory distance and azimuth perception by a passerine bird in natural habitat. Anim Behav 56:467-477

Nemeth E, Winkler H, Dabelsteen T (1997) Adaptations in bird songs in a neotropical rainforest. Advances in Ethology 32, 119

Nottebohm F (1975a) Vocal behavior in birds. In: Farner DS, King JR, Parkes KC (eds) Avian Biology, Vol 5, New York, Academic Press, pp 287-332

Nottebohm F (1975b) Continental patterns of song variability in *Zonotrichia capensis*: some possible ecological correlates. Amer Nat 109:605-624

Okanoya K, Dooling RJ (1988) Hearing in the swamp sparrow, *Melospiza georgiana*, and the song sparrow, *Melospiza melodia*. Anim Behav 36:726-732

Owens IPF, Benett PM, Harvey PH (1999) Species-richness among birds: body size, life history, sexual selection or ecology? Proc R Soc Lond B 266:933-939

Peters WD, Grubb TC (1983) An experimental analysis of sex-specific foraging in the Downy Woodpecker, *Picoides pubescens*. Ecology 64:1437-1443

Piercy JE, Embleton TFW, Sutherland LC (1977) Review of noise propagation in the atmosphere. J Acoust Soc Am 61:1403-1418

Pirinchieva RK (1991) Model study of sound propagation over ground of finite impedance. J Acoust Soc Am 90:2678-2682
Price MA, Attenborough K, Heap NW (1988) Sound attenuation through trees: Measurements and models. J Acoust Soc Am 84:1836-1844
Redondo T, Arias de Reyna L (1988) Locatability of begging calls in nestling atricial birds. Anim Behav 36:653-661
Richards DG (1981a) Alerting and message components in songs of rufous-sided towhees. Behaviour 76:223-249
Richards DG (1981b) Estimation of distance of singing conspecifics by the Carolina wren. Auk 98:127-133
Richards DG, Wiley RH (1980) Reverberations and amplitude fluctuations in the propagation of sound in a forest. Implications for animal communication. Am Nat 108:17-34
Riede K (1996) Diversity of sound-producing insects in a Bornean lowland rain forest. In: Edwards DS et al. (eds), Tropical Rainforest Research - Current Issues. Kluwer Academic Publisher pp. 77-84
Ryan MJ, Brenowitz EA (1985) The role of body size, phylogeny, and ambient noise in the evolution of bird song. Am Nat 126:87-100
Rieke F, Bodnar DA, Bialek W (1995) Naturalistic stimuli increase the rate and efficiency of information transmission by primary auditory afferents. Proc R Soc Lond B 262:259-265
Saunders SS, Salvi RJ (1993) Psychoacoustics of normal adult chickens: thresholds and temporal integration. J Acoust Soc Am 94:83-90
Schubert M (1971) Untersuchungen über die reaktionsauslösenden Signalstrukturen des Fitis-Gesanges, *Phylloscopus t. trochilus* (L.), und das Verhalten gegenüber arteigenen Rufen. Behaviour 38:250-288
Sheldon BC, Burke T (1994) Copulation behaviour and paternity in the chaffinch. Behav Ecol Sociobiol 34:149-156
Shy E (1983) The relation of geographic variation in song to habitat characteristics and body size in North American tanagers (Thraupinae: *Piranga*). Behav Ecol Sociobiol 12:71-76
Shy E (1984) The structure of song and its geographical variation in the scarlet tanager (*Piranga olivacea*). Am Midl Nat 112:119-130
Shy E, Morton ES (1986). Adaptation of amplitude structure of songs to propagation in field habitat in song sparrows. Ethology 72:177-184
Smith JI, Yu H-T (1992) The association between vocal characteristics and habitat type in Taiwanese passerines. Zool Sci 9:659-664
Sorjonen J (1983) Transmission of the two most characteristic phrases of the song of the thrush nightingale *Luscinia luscinia* in different environmental conditions. Ornis Scandinavica 14:278-288
Sorjonen J (1986) Factors affecting the structure of song and the singing behaviour of some northern European passerine birds. Behaviour 98:286-304
Staicer CA, Spector DA, Horn AG (1996) The dawn chorus and other diel patterns in acoustic signaling In: Kroodsma DE, Miller EH (eds) Ecology and Evolution

of Acoustic Communication in Birds. Cornell Univ Press, Ithaca, NewYork, pp 426-453

Sutherland LC, Daigle GA (1997) Atmospheric sound propagation. In: Crocker MJ (ed) Encyclopedia of Acoustics. John Wiley & Sons, NewYork, pp 341-365

Waide RB, Narins PM (1988) Tropical forest bird counts and the effect of sound attenuation. Auk 105:296-302

Wallace RA (1974) Ecological and social implications of sexual dimorphism in five melanerpine woodpeckers. Condor 76:238-248

Wasserman FE (1979) The relationship between habitat and song in the white-throated sparrow. Condor 81:424-426

Weary DM, Lemon RE, Date EM (1986) Acoustic features used in song discrimination by the veery. Ethology 72:199-213

Wempen J (1991) Schallausbreitung über Erdboden. Bibliotheks- und Informationssystem der Universität Oldenburg, Oldenburg, 102 p

Wilczynski W, Ryan MJ, Brenowitz EA (1989) The display of the blue-black grassquit: the acoustic advantage of getting high. Ethology 80:218-222

Wiley RH (1991) Association of song properties with habitats for territorial oscine birds of eastern North America. Am Nat 138:973-993

Wiley RH, Godard R (1996) Ranging of conspecific songs by Kentucky warblers and its implications for interactions of territorial males. Behaviour 133:81-102

Wiley RH, Richards DG (1978) Physical constraints on acoustic communication in the atmosphere: Implications for the evolution of animal vocalizations. Behav Ecol Sociobiol 3:69-94

Wiley RH, Richards DG (1981) Adaptations for acoustic communication in birds: sound transmission and signal detection. In: Kroodsma DE, Miller EH (eds) Acoustic Communication in Birds, Vol 1. Academic Press, New York, pp 131-181

Wingfield JC, Farner DS (1993) Endocrinology of reproduction in wild species. In: Farner DS, King JR, Parkes KC (eds) Avian Biology, Vol 9. Academic Press, London, pp 163-327

Winkler H (1971) Die artliche Isolation des Blutspechts *Picoides (Dendrocopos) syriacus*. Egretta 14:1-20

Winkler H (1979) Foraging ecology of Strickland's Woodpecker in Arizona. Wilson Bull 91:244-254

Winkler H, Christie DA (1995) Woodpeckers. A Guide to the Woodpeckers, Piculets and Wrynecks of the World. Pica Press, Sussex, 406 p

Winkler H, Newton AV, Newton SF (1996) On the ecology and behaviour of the Arabian woodpecker *Picoides dorae*. Zool Middle East 12:33-45

Winkler H, Short LL (1978) A comparative analysis of acoustical signals in pied woodpeckers (Aves, Picoides). Bull Am Mus Nat Hist 160:1-109

Witkin SR (1977) The importance of directional sound radiation in avian vocalization. Condor 79:490-493

Zahavai A (1977) The cost of honesty (further remarks on the handicap principle). J Theor Biol 67:603-605

III.1 The Motion-Sensing Hairs of Arthropods: Using Physics to Understand Sensory Ecology and Adaptive Evolution

J. A. C. Humphrey[1], F. G. Barth[2] and K. Voss[3]

[1]Department of Mechanical and Aerospace Engineering, University of Virginia, Charlottesville, Virginia 22904, USA
[2]Biocenter, Zoology Institute, Vienna University, Althanstr. 14, 1090 Vienna, Austria
[3]Department of Mathematics, College of Arts and Sciences, Bucknell University, Lewisburg, Pennsylvania 17837, USA

Abstract.
In an effort to understand the "design" principles of arthropod filiform hairs, it is desirable to explain the effects that ecology and, therefore, natural selective pressures may have had on the adaptive evolution of these medium motion sensors in terrestrial and aquatic habitats, respectively. This is a complex interdisciplinary problem involving various facets of biology, mechanics, and mathematics. Using a simplified, physically approximate theoretical analysis, the present work places on a more general foundation the understanding of the observed variations of the maximum angular deflection and maximum angular velocity, and their respective resonance frequencies, of filiform hairs as a function of the physical parameters that affect these four quantities. Calculated results obtained using the approximate analysis are compared with corresponding results from a more exact physical analysis and with previous measurements and calculations to show that all qualitative aspects of hair behavior are correctly captured by the simplified theory. The theory is then used to explain the dependence of hair motion on the physical parameters that affect it, and to explore the sensitivity of hair motion to small changes in these parameters.

Key words Trichobothria, viscosity, inertia, sensory ecology, modeling

1 Introduction

1.1 Purpose and Background of This Study

Among the many types of mechanoreceptive cuticular hairs of arthropods, the filiform hairs of insects and the analogous trichobothria of the arachnids are the most spectacular ones regarding the sensitivity of their response to the slightest movement of air (Shimozawa and Kanou 1984; Barth et al. 1993). In a series of earlier investigations, several authors have derived and applied physical models to describe the behavior of filiform cuticular hairs responding to oscillatory movements of the surrounding fluid medium in both terrestrial and aquatic arthropods; see Fletcher (1978), Tautz (1979), Shimozawa and Kanou (1984), Humphrey et al. (1993, 1997), Barth et al. (1993, 1995), Devarakonda et al. (1996), Shimozawa, et al. (1998), and Barth (2000). The theoretical component of that work is numerical in nature. It requires the judicious calculation and plotting of numerous individual cases in order to elucidate the effects on hair maximum angular deflection and velocity, and their corresponding resonance frequencies, resulting from changes in hair diameter and length, the mechanical spring and damping constants of the hair, and the medium density, viscosity, and velocity.

In spite of its accuracy, the numerical solution approach is not entirely satisfactory. This is because it does not provide easily usable and interpretable analytical formulas from which to derive generally applicable conclusions concerning the expected performance or evolutionary adaptation of filiform hairs as the result of small changes in hair properties or the environment. The purpose of this study is to derive such formulas subject to simplifications that render the mathematics tractable but which, nevertheless, retain the essential physics of the problem and from which correct qualitative results can be obtained.

The combination of increased understanding and predictive capability allowed by the present highly integrative analytical approach squarely addresses a major challenge in sensory ecology explained elsewhere in this book by David Dusenbery. In the case of arthropod filiform motion-sensing hairs it can be stated as follows. Biomechanical and physiological studies clearly show that these hairs are tuned to preferred frequencies (or frequency ranges) and require minimal deflection, velocity, and acceleration thresholds to detect fluid medium-borne deflection stimuli originating from other animals (mates, prey, and predators) and the environment. It seems reasonable to assume that over evolutionary time periods the hairs have adapted to selective pressures in response to changes in the frequencies and/or deflection velocities and accelerations detected. As Dusenbery explains, two types of constraints influence this evolution: the history of previous evolution through genetic encoding, and the physical-chemical laws that all systems, animate or inanimate, must obey. We note that the first constraint evolves with time while the second has been fixed since the beginning of time. It would advance understanding to determine how the fixed physical-chemical constraints influence and guide evolutionary adaptation. We explore this problem here for the case of arthropod filiform hairs.

1.2 Review of Earlier Work

This section summarizes some of the main experimental and numerical findings of the earlier studies referred to above with emphasis on the trichobothria of spiders. In the case of *Cupiennius salei*, the trichobothria appear in clusters of 2-30 hairs ranging in length from 100 to 1400 μm and in base diameter from 5 to 15 μm. Aerodynamic drag is favored by the feathery structure of the hairs, which are tuned to best frequency ranges between 40 Hz and 600 Hz. The variation of hair length in a cluster fractionates both the intensity and frequency range of a stimulus (Barth et al. 1993) and, as a result, trichobothria can function as displacement, velocity, and acceleration sensors over different frequency ranges. Primarily because of the effects of dynamic viscosity on their effective inertia, the resonance frequencies of filiform hairs in water tend to be much lower than for hairs in air (for hairs of similar geometrical and mechanical characteristics). Hairs in water appear to work as low-frequency (1-150 Hz) sensors and, whereas in some receptors physiological threshold values follow water velocity, they follow acceleration in others (Bleckmann 1994). Also, the most typical motion-sensing hairs in water tend to be very short, correlating with the smaller values of the boundary layer thickness in water compared to air at the same frequency (Devarakonda et al. 1996).

Building mainly on the work of Fletcher (1978) and Shimozawa and Kanou (1984), Humphrey et al. (1993, 1997) derived a physical model that describes the motion of a hair protruding from a solid substrate where it is driven by the oscillating movement of a viscous fluid (air or water). Using Stokes' (1851) theory, their model accounts for the important effect of the "added" or "virtual" mass force acting on the hair, and the dependence of this force and that of the viscous drag on frequency. In their study, straight hairs are modeled as constant diameter cylinders of length L and effective diameter d. Subsequently, Shimozawa et al. (1998) considered the variation of hair diameter with length for calculating the moment of inertia and the added mass and drag forces. The bulk of these effects can be captured using a length-averaged effective diameter for the hair.

Humphrey et al. (1993, 1997) show that for a fluid oscillating at f (Hz) parallel to the longitudinal axis of a cylindrical substrate of diameter D (m), Stokes' (1851) velocity distribution for a fluid oscillating parallel to a flat surface is accurate provided $fD^2/\nu > 20/\pi$, where ν (m^2s^{-1}) is the fluid kinematic viscosity. In such cases, the thickness of the viscous-dominated boundary layer near the substrate is given by $\delta = 4.5\ (\nu/\pi f)^{1/2}$ and it is possible to obtain a closed form analytical solution for hair motion (Humphrey et al. 1993). In agreement with experiments (Barth et al. 1993), this solution shows that hairs oscillate at the frequency of the oscillating fluid medium but possess resonance frequencies, $\omega_{res(\theta)}$, that decrease with increasing hair length. The solution also shows that the hair maximum deflection angle increases linearly with increasing far field fluid velocity amplitude, U_o, and approximately linearly with increasing hair length; although the lengths of the trichobothria of the spider *Cupiennius salei* infrequently exceed the substrate boundary layer thickness (Barth et al. 1993). While a corresponding analytical solution is not obvious for the case of a fluid oscillating perpendicular to the cylindrical substrate, results based on rigorous numerical calculation (Humphrey et al. 1993) reveal significantly higher values of hair deflection angle, velocity, and acceleration compared to a parallel flow for the same values of U_o and f. This is because in the perpendicular orientation substrate surface curvature has the effect of accelerating the flow near the substrate. In contrast, maximum deflection angle and velocity resonance frequencies do not depend on flow magnitude or direction.

Humphrey et al. (1993, 1997) and Barth et al. (1993) have demonstrated two very different ways to derive values for the hair torsional restoring constant, S, and damping constant, R, from related experimental data. In one method the phase difference between hair and air motion must be known as a function of the hair (or flow) oscillation frequency. The other is based on knowing the variation of hair maximum deflection angle with hair (or flow) oscillation frequency. Both methods yield reasonable values of S and R and a variation of the second has been applied by Shimozawa et al. (1998) to derive values of S and R as a function of hair length for crickets. All the results support a dependence of S and R on hair length of the form $y = a\, L^b + c$, where a, b, and c are experimentally determined positive constants that differ for S and R, respectively. Both analytical and numerical results show that increasing S or decreasing R works to increase hair resonance frequency.

2 Physical and Analytical Considerations

2.1 The Equation of Angular Momentum for a Filiform Hair

The starting point for the present analysis is Eq. (13) in the study by Humphrey et al. (1993, 1997) for the conservation of angular momentum of a hair attached to a substrate and immersed in a fluid oscillating at frequency ω $(=2\pi f)$ radians per second with far field velocity amplitude U_o. It is written here for a straight, cylindrically shaped hair of length L, effective diameter d and density ρ_h immersed in a fluid medium of density ρ, dynamic viscosity μ, and velocity V_F at location y along its length. For these conditions, the equation is:

$$(I + I_\rho + I_\mu)\ddot{\theta} + (R + R_\mu)\dot{\theta} + S\theta = 4\pi\mu G \int_0^L V_F y\,dy + \left(\frac{\pi\rho d^2}{4} - \frac{\pi^2 \mu G}{g\omega}\right)\int_0^L \dot{V}_F y\,dy, \quad (1)$$

where θ denotes the angular deflection of the hair and the dotted superscripts denote differentiation with respect to time. The quantities I, R, and S are, respectively, the moment of inertia, the damping constant, and the torsional restoring constant of the hair. These are mechanical properties inherent to the hair; the first can be determined from the hair shaft geometry but the latter two must be determined experimentally, and all three are constant for a given hair. The quantities I_ρ, I_μ, and R_μ denote additional contributions to the moment of inertia and the damping constant of the hair associated with the fluid medium density and viscosity, respectively. The following definitions apply (Humphrey et al. 1993):

$$I = \frac{\pi\rho_h d^2}{48}\left(4L^3 + \frac{3}{4}d^2 L\right) \quad (2) \qquad\qquad I_\rho = \frac{\pi\rho d^2 L^3}{12} \quad (3)$$

$$I_\mu = -\frac{\pi^2 \mu G L^3}{3 g \omega} \qquad (4) \qquad\qquad R_\mu = \frac{4}{3} \pi \mu G L^3. \qquad (5)$$

In Eqs. (4) and (5), the quantities G and g are given by

$$G = -\frac{g}{(g^2 + \pi^2/16)} \qquad (6) \qquad\qquad g = 0.577 + \ln\left(\frac{d^2 \omega \rho}{16\mu}\right)^{1/2}. \qquad (7)$$

In the present study, the velocity of the fluid medium driving hair motion is taken from Stokes (1851) and assumed to be

$$V_F = U_o(-\cos(\omega t) + \cos(\omega t - \beta y)e^{-\beta y}). \qquad (8)$$

This equation corresponds to the velocity at time t and position y of a fluid oscillating at frequency ω and far field amplitude U_o above a fixed flat substrate supporting the hair. The quantity $\beta = (\omega\rho/2\mu)^{1/2}$ is related to the oscillating flow boundary layer thickness, δ, defined above, according to

$$\delta = \frac{4.5}{\beta} = 6.4\left(\frac{\mu}{\rho\omega}\right)^{1/2}. \qquad (9)$$

With V_F given by Eq. (8), the terms on the right hand side of Eq. (1), representing the flow-induced torques that drive hair motion, are completely defined.

For small deflection angles ($\theta < 10°$), Eq. (1) admits an exact analytical steady-periodic solution. This is given in Appendix 1 of Humphrey et al. (1993) and the result is

$$\theta = C_1 \cos(\omega t) + C_2 \sin(\omega t), \qquad (10)$$

where

$$C_1 = \frac{P(S - I_t\omega^2) - Q\omega R_t}{(S - I_t\omega^2)^2 + (\omega R_t)^2} \qquad (11) \quad \text{and} \quad C_2 = \frac{P\omega R_t + Q(S - I_t\omega^2)}{(S - I_t\omega^2)^2 + (\omega R_t)^2} \qquad (12)$$

with the following quantities as defined:

$$I_t = I + I_p + I_\mu \qquad (13) \qquad\qquad R_t = R + R_\mu \qquad (14)$$

$$P = A2\pi\mu G U_o L^2 + B\left(\frac{\pi\rho d^2 \omega U_o L^2}{8} - \frac{\pi^2 \mu G U_o L^2}{2g}\right) \qquad (15)$$

$$Q = B2\pi\mu GU_oL^2 + A\left(\frac{\pi^2\mu GU_oL^2}{2g} - \frac{\pi\rho d^2\omega U_oL^2}{8}\right) \tag{16}$$

$$A = e^{-\beta L}[-(1/\beta L)\cos(-\beta L) - (1/\beta L + 1/(\beta L)^2)\sin(-\beta L)] - 1 \tag{17}$$

$$B = e^{-\beta L}[(1/\beta L)\sin(-\beta L) - (1/\beta L + 1/(\beta L)^2)\cos(-\beta L)] + 1(\beta L)^2. \tag{18}$$

The details concerning the derivation of Eq. (1) and its exact analytical solution, Eq. (10), and a discussion of the conditions for which this solution applies, are provided in Humphrey et al. (1993). Suffice it to say here that Eq. (10) is based on Stokes' (1851) analysis through the assumptions embedded in Eqs. (1) and (8), and that it provides an excellent theoretical basis for describing all major known aspects of filiform hair motion in air or water, especially for hairs on cylindrically shaped substrates where the flow field oscillates parallel to the longitudinal axis of the cylinder (Humphrey et al. 1993; Barth et al. 1993); see these references for examples of measured and calculated maximum deflection angles versus frequency.

In this study the result given by Eq. (10) is referred to as the physically exact analytical solution. This solution is very powerful because it explicitly states the quantitative dependence of the hair deflection angle, θ, and, by differentiation, its first- and second-order time derivatives (hair angular velocity and acceleration) on all the physical parameters that affect these three quantities (d, L, R, S, ρ, μ, ρ_h, U_o, ω, t). In principle, it should be possible to obtain from Eq. (10) analytical expressions for the hair maximum deflection angle, θ_{res}, and maximum velocity, $V_{res} = \partial\theta/\partial t|_{res}$, at their corresponding resonance frequencies, $\omega_{res(\theta)}$ and $\omega_{res(V)}$, and for the resonance frequencies themselves. In practice, however, the nonlinear frequency dependencies embedded in the P, Q, I_t, and R_t terms that contribute to C_1 and C_2 in Eq. (10) preclude obtaining closed form analytical solutions for these quantities. The problem can be bypassed by solving Eq. (10) numerically for a variety of physical situations in order to determine from the results obtained the quantitative dependencies of $\omega_{res(\theta)}$, θ_{res}, $\omega_{res(V)}$, and V_{res}, on the physical parameters that affect them. While this is certainly an accurate way to proceed, such a numerical approach is unsatisfying because of its lack of universality; meaning that it does not allow the formulation of generally applicable analytical conclusions leading to the kind of broad understanding sought here.

2.2 The Physically Approximate Analytical Solution

Inspection of the terms that compose P, Q, I_t, and R_t in Eq. (10) shows that they all depend on ω: the first three *explicitly* [see Eqs. (4), (15), and (16)] and all four *implicitly*, through the inclusion of ω in the logarithmic term in Eq. (7) and through the dimensionless quantity βL in Eqs. (17) and (18). However, calculations of these terms for typical conditions corresponding to hairs in air and water reveal that they are essentially constant for frequencies exceeding 50-75 Hz. In order to proceed, we assume the terms are constant for the purpose (only) of deriving *physically approximate but analytically exact solutions* for the resonance frequencies, $\omega_{res(\theta)}$ and $\omega_{res(V)}$, and their corresponding maximum angular deflection, θ_{res}, and maximum angular velocity, V_{res}, respectively. The validity of the approach is determined via *a posteriori* verification.

The derivation of the desired expressions is provided in the Appendix and the results are:

$$\omega_{res(\theta)} = \left[\frac{S}{I_t} - \frac{R_t^2}{2I_t^2}\right]^{1/2} \quad (19a) \qquad \theta_{res} = \frac{2I_t}{R_t}\left[\frac{P^2+Q^2}{4I_tS - R_t^2}\right]^{1/2} \quad (19b)$$

$$\omega_{res(V)} = \left[\frac{S}{I_t}\right]^{1/2} \quad (20a) \qquad V_{res} = \frac{1}{R_t}\left[P^2+Q^2\right]^{1/2}, \quad (20b)$$

where all quantities are in SI units, $\omega_{res(\theta)}$ and $\omega_{res(V)}$ are in rad s^{-1}, θ_{res} is in rad and V_{res} is in rad s^{-1}. These are familiar solutions describing the behavior of a forced, damped, harmonic rod-like oscillator; except that here the oscillator consists of a hair and the fluid medium (air or water) immediately around the hair that moves with it. The latter is the so-called added or virtual mass effect, originating in Stokes' (1851) analysis (Humphrey et al. 1993). From Eq. (13) it is clear that the added or virtual mass contribution to the total moment of inertia, I_t, arises through separate contributions associated with the density and viscosity of the fluid, respectively. In contrast, from Eq. (14) we see that the corresponding contribution to the total damping constant, R_t, is due only to the fluid viscosity.

The only acceptable physical solutions for the set of Eqs. (19a, 19b) and (20a, 20b) correspond to conditions for which the quantities under the ½ exponents are positive. The most stringent condition is imposed by Eq. (19a) and requires $2I_t S - R_t^2 > 0$. This constraint is the outcome of the constant-terms approximation made in the analysis.

It is reasonable to suppose that there will be cases involving combinations of physical conditions for which the constraint is not observed and to which, therefore, the physically approximate analytical results do not apply. For these cases it is preferable to determine $\omega_{res(\theta)}$, θ_{res}, $\omega_{res(V)}$, and V_{res} directly from Eq. (10) and its time derivative or, for more general velocity distributions than those given by Eq. (8) on which Eq. (10) is based, from numerical solutions of the original differential equation, Eq. (1).

2.3 Evaluation of Physically Approximate Analytical Solution

2.3.1 Filiform Hairs in Air

Prior to using the physically-approximate analytical results given by Eqs. (19a, 19b) and (20a, 20b) to discuss and interpret the dependencies of $\omega_{res(\theta)}$, θ_{res}, $\omega_{res(V)}$, and V_{res} on the physical parameters that affect them, we first establish their physical accuracy. For this, we perform calculations corresponding to the experimental conditions of Barth et al. (1993). These authors measured maximum deflection angles and resonance frequencies for a number of individual hairs of four selected groups of trichobothria on the spider *Cupiennius salei* under controlled oscillating flow conditions. In particular, for the MeD1 group of hairs reported they found the following fits for maximum deflection angle and resonance frequency: $\theta_{res} = 3.3 + 7\ 10^{3}\ L$ and $f_{res(\theta)} = 4.25\ 10^{-3}\ L^{-1.316}$, where θ_{res} is in degrees, $f_{res(\theta)}$ in Hz and L in m. They also determined values for the damping constant R and the torsional restoring constant S for two of these metatarsal hairs. For a short hair they obtained: $L = 250$ μm, $d = 5$ μm, $R = 0.27\ 10^{-15}$ N m s rad^{-1}, and $S = 0.62\ 10^{-12}$ N m rad^{-1}; and, for a long hair: $L = 750$ μm, $d = 7$ μm, $R = 2.20\ 10^{-15}$ N m s rad^{-1}, and $S = 5.77\ 10^{-12}$ N m rad^{-1}. (Although the long hair had an additional curved portion 250 μm long, it was not considered in the calculations of their Fig. 19 with which we compare here.) Curve fits of these data to the form $y = a\ L^{b}$ yield: $d = 6.343\ 10^{-5}\ L^{0.306}$, $R = 2.031\ 10^{-9}\ L^{1.909}$, and $S = 1.272\ 10^{-5}\ L^{2.03}$, where L is in m. Because these fits are based on only two points each, they are subject to (unknown) uncertainties. Similar fits for R and S have been obtained from larger data sets by Shimozawa et al. (1998) for the filiform hairs of crickets. In spite of the care taken by these authors to obtain accurate measurements, the scatter in the data ranges from ± 20 to ± 80% for S and ± 7 to ± 38% for R, approximately, over the range of hair length explored.

Except for where otherwise noted, present calculations for hairs in air based on the physically approximate solution assume $d = 7$ μm, $R = 1\ 10^{-15}$ N m s rad^{-1}, and $S = 4\ 10^{-12}$ N m rad^{-1} regardless of hair length. For hairs in water the same value of d is used, but with different values for R and S, as discussed in Section 2.3.2. The calculations for hairs in air based on the physically exact solution given by Eq. (10) use the above fits for d, R, and S. All calculations assume fluid medium physical properties at 27 °C: air ($\rho = 1.177$ kg m^{-3}, $\mu = 1.846\ 10^{-5}$ kg m^{-1} s^{-1}), water ($\rho = 995.8$ kg m^{-3}, $\mu = 8.6\ 10^{-4}$ kg m^{-1} s^{-1}). The hair density used is $\rho_h = 1100$ kg m^{-3} and the far field velocity amplitude is set to $U_o = 50$ mm s^{-1}. All calculations are performed in SI units. To solve Eq. (10) directly for $\omega_{res(\theta)}$ we proceed as follows. A value of ω is fixed in the equation and t is varied until the maximum deflection angle, θ_{max}, is found. Sequentially varying ω yields values of θ_{max} from which the value of $\omega_{res(\theta)}$, corresponding to the largest value of θ_{max} ($\equiv \theta_{res}$), can be obtained. This approach retains all the ω dependencies embedded in the P, Q, I_t and R_t terms which the physically approximate analytical solution neglects. When applied to the time derivative of Eq. (10), the same approach yields $\omega_{res(V)}$ and V_{res}. To solve Eq. (19a) for $\omega_{res(\theta)}$ and Eq. (20a) for $\omega_{res(V)}$, we

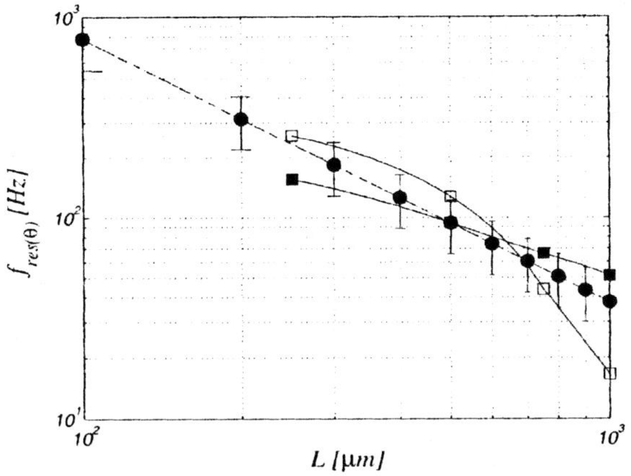

Fig. 1. Maximum displacement angle resonance frequency versus hair length. Physically approximate solution, Eq. (19a) *empty squares* $d = 7$ μm, $R = 1\ 10^{-15}$ N m s rad^{-1}, $S = 4\ 10^{-12}$ N m rad^{-1}. Exact physical solution, Eq. (10) *filled squares d, R,* and *S* prescribed from fits to experimental data; see text. Best fit to experimental data of Barth et al. (1993): *filled circles connected by dashed line* (*error bars* reflect scatter in the original measurements); see text. $U_o = 50$ mm s^{-1} in all cases

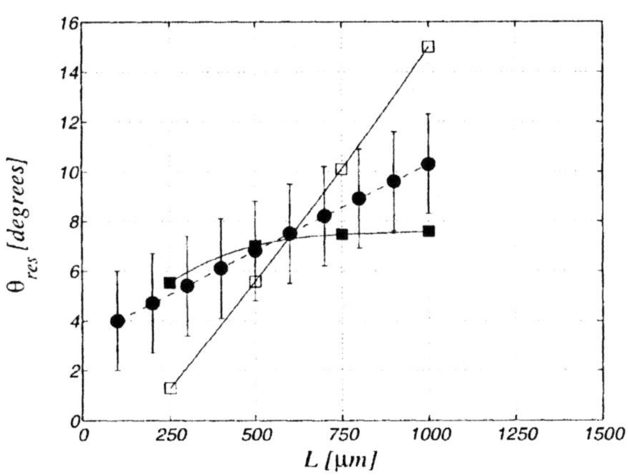

Fig. 2. Maximum displacement angle at resonance frequency versus hair length. Physically approximate solution, Eq. (19b) *empty squares* $d = 7$ μm, $R = 1\ 10^{-15}$ N m s rad^{-1}, $S = 4\ 10^{-12}$ N m rad^{-1}. Exact physical solution, Eq. (10) *filled squares d, R,* and *S* prescribed from fits to experimental data; see text. Best fit to experimental data of Barth et al. (1993): *filled circles connected by dashed line* (*error bars* reflect scatter in the original measurements); see text. $U_o = 50$ mm s^{-1} in all cases

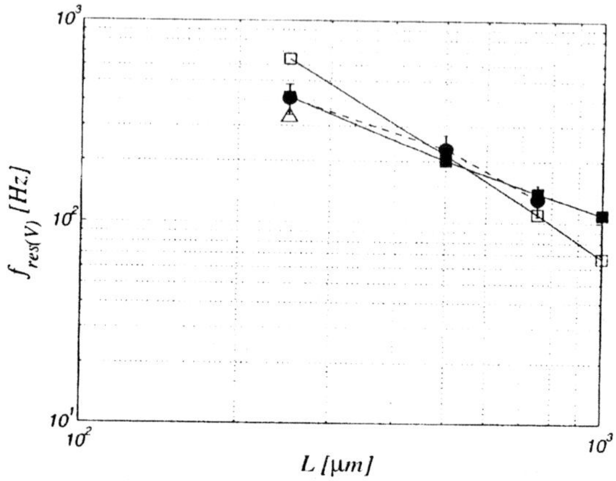

Fig. 3. Maximum angular velocity resonance frequency versus hair length. Physically approximate solution, Eq. (20a) *empty squares* d = 7 μm, R = 1 10^{-15} N m s rad^{-1}, S = 4 10^{-12} N m rad^{-1}. Exact physical solution, Eq. (10) *filled squares d, R,* and *S* prescribed from fits to experimental data; see text. Best fit to calculations of Barth et al. (1993): *filled circles connected by dashed line* (*error bars* reflect plot reading error); see text. U_o = 50 mm s^{-1} in all cases. Physically approximate solution for L = 250 μm obtained using experimental values of *d, R,* and *S*: *open triangle*

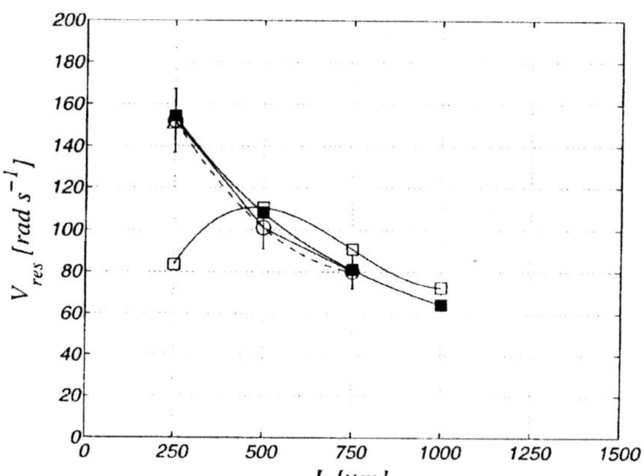

Fig. 4. Maximum angular velocity at resonance frequency versus hair length. Physically approximate solution, Eq. (20b) *empty squares* d = 7 μm, R = 1 10^{-15} N m s rad^{-1}, S = 4 10^{-12} N m rad^{-1}. Exact physical solution, Eq. (10) *filled squares d, R,* and *S* prescribed from fits to experimental data; see text. Best fit to calculations of Barth et al. (1993): *filled circles connected by dashed line* (*error bars* reflect plot reading error); see text. U_o = 50 mm s^{-1} in all cases. Physically approximate solution for L = 250 μm obtained using experimental values of *d, R,* and *S*: *open triangle*

substitute best-guess values for ω into the right-hand side of the equations and compare with the values calculated from the equations. The results obtained when the substituted and calculated values agree are $\omega_{res(\theta)}$ and $\omega_{res(V)}$, respectively. When resubstituted into their corresponding Eqs. (19b) and (20b), these quantities yield θ_{res} and V_{res}.

The results for hair maximum deflection resonance frequency and maximum deflection angle at resonance frequency are given in Figs. 1 and 2, respectively. (Note that in the figures we plot $f_{res(\theta)} = \omega_{res(\theta)}/2\pi$.) The approximate solution yields results in good qualitative agreement with the experimental fits. Because the exact solution accounts for the experimental variations of d, R and S with L, it shows better overall agreement for both $f_{res(\theta)}$ and θ_{res}. Results for hair velocity resonance frequency and the maximum velocity at that frequency are plotted in Figs. 3 and 4 where comparisons are made with values determined numerically by Barth et al. (1993). Again, both the exact and the approximate analytical solutions yield results in good agreement with the earlier calculations of $f_{res(V)} = \omega_{res(V)}/2\pi$ by Barth et al. (1993). Although the physically approximate result for V_{res} corresponding to $L = 250$ µm deviates markedly from the present exact solution and the earlier numerical calculation, Fig. 4 shows that the discrepancy is removed if the physically approximate calculation is repeated using the experimental values for d, R, and S.

2.3.2 Filiform Hairs in Water: Determination of *R* and *S*

The evaluation performed in Section 2.3.1 is for hairs in air, for which values of the damping constant R and the torsional restoring constant S are known. Unfortunately, corresponding values of R and S are not known for motion-sensing hairs in water. On this topic the literature is very sparse and most measurements do not provide the accuracy or precision needed. In many or even most cases, the hair is directly coupled to a vibrating stylus as opposed to being driven by fluid motion, and in many other cases the control of the water movement used as a stimulus is inadequate. The best we can say at this point is that, for comparable mechanical sensitivities, hairs that are sensitive to hydrodynamic stimuli are probably much shorter, in general, than their aerodynamic counterparts (Barth et al. 1993; Bleckmann 1994; Devarakonda et al. 1996), with resonance frequencies lying below 150 Hz (as determined electrophysiologically through cell response). We do not know of any investigation in which the mechanical response of a hydrodynamic receptor hair has been studied adequately enough to draw any conclusion regarding its values of R and S.

Notwithstanding, it is possible to obtain estimates of R and S for hairs in air *and* in water as follows. Equation (19a) can be rearranged so that it is explicit in S. The result is

$$S = I_t \omega^2_{res(\theta)} + \frac{R_t^2}{2I_t}. \tag{19a'}$$

We can now fix all known physical properties, as well as the length, diameter and resonance frequency of a particular hair, and use Eq. (19a′) to calculate S as a function of R. The result is a series of (R, S) pairs all of which satisfy the values fixed in the equation. Figure 5 shows the result of doing this for the trichobothria of Section 2.3.1. Each curve connecting a series of filled symbols corresponds to a particular hair in air. A striking feature of these curves is that they overlap in the neighborhood of $R = 1\ 10^{-15}$ N m s rad^{-1} and $S = 4\ 10^{-12}$ N m rad^{-1} (the values used for the physically approximate calculations shown in Figs. 1-4). This is not a coincidence. While R and S are expected to vary with hair length, they are mechanical properties of the hair-supporting apparatus which consists of cuticular material (chitin fibers embedded in a protein matrix). Thus, we expect R and S to vary, but within limited ranges (of 1 - 2 orders of magnitude) with limits fixed by the molecular structure of the cuticular material.

Support for this hypothesis is provided by the data for R and S in Barth et al. (1993). Figure 5 shows three points (filled diamonds) connected by a dotted curve. The two end points correspond to the 250- and 750-μm hairs investigated by these authors and the one in the center is the result of an interpolation for a 500-μm hair. The correspondence is good between the theoretically predicted overlap region and the experimental results for R and S.

We apply the above principle to hairs in water. From the limited information available we expect to find shorter hairs responding to lower frequency stimuli and, for illustration purposes, we assume two hairs of the following characteristics: $d = 5$ μm, $L = 250$ μm and $f_{res(\theta)} = 50$ Hz; and $d = 6$ μm, $L = 500$ μm, and $f_{res(\theta)} = 10$ Hz. Using these values to calculate S as a function of R yields the two curves connecting the unfilled symbols in Fig. 5. As for the hairs in air, these two curves also cross, very near to the point given by $R = 1\ 10^{-14}$ N m s rad^{-1} and $S = 2\ 10^{-11}$ N m rad^{-1}. Even if approximate, these results are sufficiently different from those for hairs in air to conclude that for hairs to perform as motion sensors in water, with sensitivities comparable to those found in air, they must possess larger values of R and S than required in air.

3 Discussion

3.1 Parameter Dependence of Hair Motion

It is possible to solve Eq. (1) numerically for very general conditions and analytically for the special case of the velocity profile given by Eq. (8), the result being Eq. (10) (Humphrey et al. 1993). However, even this analytical solution requires numerical evaluation to determine hair deflection angles, velocities, and frequencies. Hence the motivation for the physically approximate solution derived here, given by Eqs. (19a, 19b and 20a, 20b), which explains several of the observed dependencies of hair behavior on the physical parameters that affect that behavior and allows us to uncover new ones. For example, inspection of Eqs. (19a, 19b) shows that, in principle, the damping constant, R, and the torsional restoring constant, S, both affect the values of the hair maximum deflection resonance frequency, $\omega_{res(\theta)}$, and the maximum hair deflection at resonance frequency, θ_{res}. However, to the approximations made in the analysis, Eqs. (20a, 20b) show that the effects of R and S are completely decoupled for $\omega_{res(V)}$ and V_{res}! It is also clear that both $\omega_{res(\theta)}$ and $\omega_{res(V)}$ depend on S to the 1/2 power and on L to the -3/2 power, and that both θ_{res} and V_{res} depend linearly on U_o.

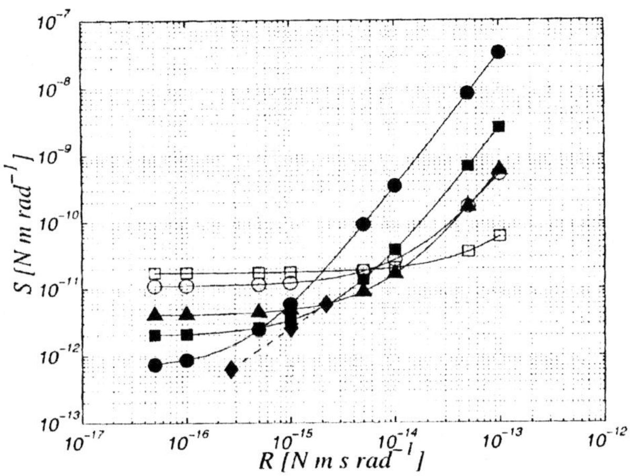

Fig. 5. Physically approximate calculations of (R, S) pairs in air (*filled symbols*) and water (*open symbols*) that yield a specific maximum displacement angle resonance frequency for a hair of fixed diameter and length. In air: d = 5 µm, L = 250 µm, $f_{res(\theta)}$ = 234 Hz (*circles*); d = 6 µm, L = 500 µm, $f_{res(\theta)}$ = 94 Hz (*squares*); d = 7 µm, L = 750 µm, $f_{res(\theta)}$ = 55 Hz (*triangles*); d and $f_{res(\theta)}$ determined from fits to the experimental data in Barth et al. (1993), see text. In water: d = 5 µm, L = 250 µm, $f_{res(\theta)}$ = 50 Hz (*circles*); d = 6 µm, L = 500 µm, $f_{res(\theta)}$ = 10 Hz (*squares*); values proposed for illustration purposes, see text. The *three filled diamonds* correspond to (R, S) pairs fitted to the data in Barth et al. (1993) for L = 250, 500, and 750 µm, respectively; each of these three points corresponds to a particular d, L, and $f_{res(\theta)}$; see text

It is possible to perform an order of magnitude analysis of the terms in Eqs. (19a, 19b) and (20a, 20b) consistent with the approximations made to obtain these equations. For this we note that evaluations of the various terms contributing to the total moment of inertia, I_t, and the total damping constant, R_t, typically yield $I \cong I_\mu >> I_\rho$ and $R_\mu > R$ in air, and $I_\mu >> I \cong I_\rho$ and $R_\mu > R$ in water. Using these findings and noting that the frequencies of stimuli in air tend to be larger than the corresponding values in water, it is possible to derive the approximate physical parameter dependencies presented in Table 1.

The tabulated results for air show the expected explicit power dependencies of $\omega_{res(\theta)}$ and $\omega_{res(V)}$ on hair density, diameter, length, and torsional restoring constant, respectively; compare, for example, the predicted $L^{-3/2}$ dependence of $\omega_{res(\theta)}$ in air to fits to the data in Barth et al. (1993) ($\omega_{res(\theta)} \sim L^{-1.316}$) and Kumagai et al. (1998) ($\omega_{res(\theta)} \sim L^{-0.92}$), respectively. The lack of dependence of $\omega_{res(\theta)}$ and $\omega_{res(V)}$ on air density or viscosity is notable. In water, both resonance frequencies vary with L^{-3} and S but do not depend on hair density or diameter. Instead, they depend on water viscosity according to μ^{-1}.

Two relations are provided for hair maximum deflection angle and velocity, respectively, depending on the value of the damping constant R relative to the viscous damping μL^3. In air, for example, if $R >> \mu L^3$ we see that $\theta_{res} \sim L^{13/4} S^{-3/4}$

Table 1. Approximate functional dependencies (given as products of the relevant physical parameters raised to their respective powers) of hair maximum deflection angle, θ_{res}, and maximum velocity, V_{res}, and of their associated resonance frequencies $\omega_{res(\theta)}$ and $\omega_{res(V)}$. The forms of the dependencies are for the cases where $R \cong \mu L^3$ and $R << \mu L^3$, respectively. Numerical coefficients affecting the tabulated relations (ranging in value between 0.1 and 10, approximately) are omitted for clarity

	Air	Water
$\theta_{res} \sim$	$\rho_h^{3/4} d^{3/2} L^{13/4} S^{-3/4} \mu^{3/2} \rho^{-1/2} U_o/[R+\mu L^3]$ $\rho_h^{3/4} d^{3/2} L^{1/4} S^{-3/4} \mu^{1/2} \rho^{-1/2} U_o$; $(R << \mu L^3)$	$L^{11/2} S^{-3/2} \mu^3 \rho^{-1/2} U_o/[R+\mu L^3]$ $L^{5/2} S^{-3/2} \mu^2 \rho^{-1/2} U_o$; $(R << \mu L^3)$
$V_{res} \sim$	$\rho_h^{1/4} d^{1/2} L^{7/4} S^{-1/4} \mu^{3/2} \rho^{-1/2} U_o/[R+\mu L^3]$ $\rho_h^{1/4} d^{1/2} L^{-5/4} S^{-1/4} \mu^{1/2} \rho^{-1/2} U_o$; $(R << \mu L^3)$	$L^{5/2} S^{-1/2} \mu^2 \rho^{-1/2} U_o/[R+\mu L^3]$ $L^{-1/2} S^{-1/2} \mu \rho^{-1/2} U_o$; $(R << \mu L^3)$
$\omega_{res(\theta)} \sim$ and $\omega_{res(V)} \sim$	$\rho_h^{-1/2} d^{-1} L^{-3/2} S^{1/2}$; $(R << \mu L^3$ for $\omega_{res(\theta)})$	$L^{-3} S \mu^{-1}$; $(R << \mu L^3$ for $\omega_{res(\theta)})$

R^{-1} and $V_{res} \sim L^{7/4} S^{-1/4} R^{-1}$, while for $R << \mu L^3$ we have $\theta_{res} \sim L^{1/4} S^{-3/4}$ and $V_{res} \sim L^{-5/4} S^{-1/4}$. Similar results are obtained for θ_{res} and V_{res} in water and, in all cases, the physically approximate and exact calculations in Figs. 1-4 yield power dependencies on L falling within these theoretical ranges. In this regard, recall that calculations based on the physically approximate solution assume constant values of R and S while those based on the exact solution allow for the variation of R and S with L. Thus, in both air and water, the final dependencies of θ_{res} and V_{res} (and of $\omega_{res(\theta)}$ and $\omega_{res(V)}$) on L are functions of the *implicit* dependencies of R and S on L.

The expressions in the table allow the evaluation of ratios. For hairs of the same d and L in different oscillating media with the same value of U_o, substitution of typical values yields $[(\theta_{res})_{wat}]/[(\theta_{res})_{air}] > 1$ and $[(\omega_{res(\theta)})_{wat}]/[(\omega_{res(\theta)})_{air}] < 1$, in agreement with the numerical calculations of Devarakonda et al. (1996). It can also be shown that $[(V_{res})_{wat} / (V_{res})_{air}] \cong 1$, suggesting that hairs in air and water have comparable maximum angular velocities at resonance frequency. Although the values of the numerical coefficients embedded in the relations in Table 1 have been omitted for clarity, it can further be shown that $\omega_{res(V)} > \omega_{res(\theta)}$ in both media, as expected from inspection of Eqs. (19a) and (20a).

A measure of hair sensitivity to changes in fluid medium velocity is given by LdV_{res}/dU_o. This dimensionless quantity characterizes the rate of change of hair tip velocity with respect to a change in the amplitude of the far field fluid velocity and is easily approximated using the entries in Table 1. It is of interest to plot this measure of stimulus sensitivity against hair length normalized by the boundary layer thickness, $\delta_{res(V)} \sim (\mu/\rho\omega_{res(V)})^{1/2}$; that is, with respect to $L/\delta_{res(V)}$. Again, this dimensionless quantity is also easily obtained from Table 1. Results for hairs in air and in water are shown in Fig. 6 for three cases corresponding to $R << \mu L^3$, $R >> \mu L^3$, and $R \cong \mu L^3$, respectively. For the calculations in air, values of d, R, and S are specified from the fits to the data of Barth et al. (1993) provided in Section 2.3.1. For water, R and S are fixed to the pair of values obtained in Section 2.3.2. A general observation is that all three cases for hairs in water collapse onto a single line with slope −1, approximately, for small values of $L/\delta_{res(V)}$, and that all three cases for hairs in air converge towards a single line (not shown) also with slope -1, approximately, for large values of $L/\delta_{res(V)}$.

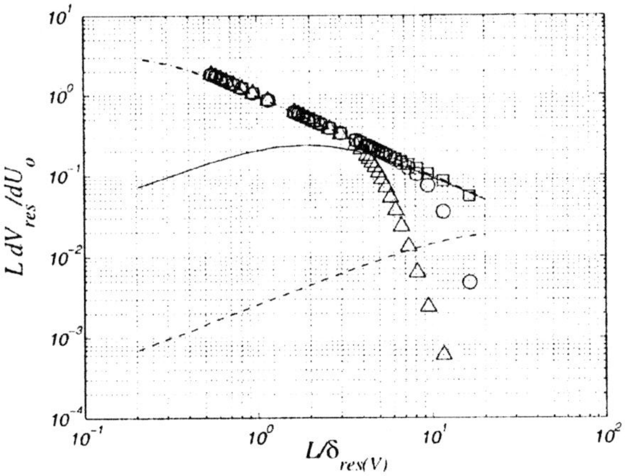

Fig. 6. Sensitivity of maximum hair velocity to changes in fluid medium far field velocity amplitude plotted as a function of hair length, both in dimensionless units. $R << \mu L^3$ (air *dot-dashed line*; water *squares*); $R \cong \mu L^3$ (air *continuous line*; water *circles*); $R >> \mu L^3$ (air *dashed line*; water *triangles*). See text for calculation details

The plots for hairs in air show clearly that when $R \cong \mu L^3$ hair sensitivity maximizes at a value $L/\delta_{res(V)} \sim O(1)$ (meaning of order 1 since the approximations made do not allow a more accurate claim than this). This finding agrees nicely with the experiments of Barth et al. (1993), who find that $L/\delta_{res(\theta)} \sim O(1)$, where $\delta_{res(\theta)} \sim (\mu/\rho\omega_{res(\theta)})^{1/2}$, for the trichobothria of *Cupiennius salei*. The results for $R << \mu L^3$ and $R >> \mu L^3$ in air show that increases in the damping factor significantly decrease hair sensitivity, but that this effect is much more pronounced for $L/\delta_{res(\theta)} < O(1)$ (hairs embedded in the boundary layer) than for $L/\delta_{res(\theta)} > O(1)$ (hairs poking through the boundary layer). Bearing in mind that we have used constant values for R and S for hairs in water, the sensitivity results for $L/\delta_{res(V)} \sim O(10)$ show the same dependence on the damping constant as for hairs in air. However, in water the dependence vanishes as $L/\delta_{res(V)}$ decreases. Over the entire parameter range investigated in water, sensitivity always increases with decreasing $L/\delta_{res(\theta)}$.

3.2 What Can Physics Say About Adaptive Evolution?

It is of special interest to explore the sensitivity of hair maximum deflection angle resonance frequency, $\omega_{res(\theta)}$, with respect to the physical parameters that affect it. (The same can be done for $\omega_{res(V)}$, θ_{res}, and V_{res}.) Mathematically speaking, we seek the values of the partial derivatives in the expression

$$d\omega_{res(\theta)} = \sum_i (\partial\omega_{res(\theta)} / \partial x_i) dx_i, \tag{21}$$

where the x_i are $x_1 = d$, $x_2 = L$, $x_3 = R$, $x_4 = S$, $x_5 = \mu$, and $x_6 = \rho$. Call $S_i = \partial\omega_{res(\theta)}/\partial x_i$ the absolute sensitivity of $\omega_{res(\theta)}$ with respect to the parameter x_i. Values of the different S_i can be determined numerically via finite difference approximation ($S_i \cong \Delta\omega_{res(\theta)}/\Delta x_i$) using the physically approximate solution. More relevant, however, is to compare relative sensitivities. For this, we rewrite Eq. (21) to read

$$d\omega_{res(\theta)} = \sum_i x_i (\partial\omega_{res(\theta)} / \partial x_i)(dx_i / x_i), \tag{22}$$

where the $R_i = x_i\ (\partial\omega_{res(\theta)}/\partial x_i) \cong x_i\ (\Delta\omega_{res(\theta)}/\Delta x_i)$ are the relative sensitivities. Values for the R_i are plotted in Fig. 7 for two hairs of equal dimensions ($d = 7$ μm, $L = 500$ μm) using $R = 1\ 10^{-15}$ N m s rad^{-1} and $S = 4\ 10^{-12}$ N m rad^{-1} for the hair in air and $R = 1\ 10^{-14}$ N m s rad^{-1} and $S = 2\ 10^{-11}$ N m rad^{-1} for the hair in water. (Note that in the plot the sign of the relative sensitivity associated with the increase of a particular physical parameter is indicated in parentheses after that parameter.) The plot shows for the conditions calculated that the relative sensitivities in air are larger by an order of magnitude or more than the corresponding values in water, and that they are largest for the parameters L, R, S, and μ.

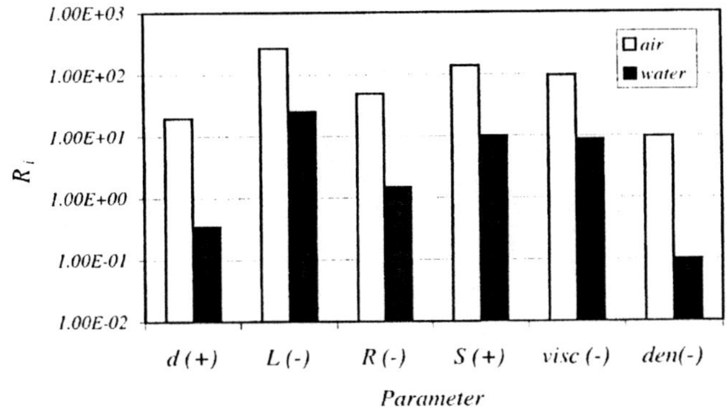

Fig. 7. Relative sensitivities, R_i, of a hair in air or in water for each of the parameters shown. The sign for R_i associated with an increase in a parameter is contained in the parentheses following that parameter. Calculations are based on the physically approximate solution and are centered about $d = 7$ μm, $L = 500$ μm, $R = 1\ 10^{-15}$ N m s rad^{-1}, $S = 4\ 10^{-12}$ N m rad^{-1} (air), and $d = 7$ μm, $L = 500$ μm, $R = 1\ 10^{-14}$ N m s rad^{-1}, $S = 2\ 10^{-11}$ N m rad^{-1} (water)

It is clear that two parameters (x_i, x_j) have the same relative sensitivity, meaning that they induce the same change in $\omega_{res(\theta)}$, if $R_i\ (dx_i\ /x_i) = R_j\ (dx_j\ /x_j)$. Again, finite differences yield $(\Delta x_i /x_i) = (R_j /R_i)\ (\Delta x_j /x_j)$ and it is possible to show that in air $\Delta L/L \cong - 0.50\ \Delta S/S$, $\Delta L/L \cong 0.20\ \Delta R/R$, and $\Delta L/L \cong - 0.10\ \Delta d/d$. Thus, for example, for a hair 500 μm long in air, the fractional increase in hair length required to induce a decrease $\Delta\omega_{res(\theta)}$ in hair resonance frequency is 0.50 the fractional decrease required of S, or 0.20 the fractional increase required of R. The third result, connecting ΔL with Δd, can be further manipulated to show that the *mass* increase required to raise the resonance frequency of this hair by increasing its diameter is about 20 times larger than the corresponding mass decrease required to achieve the same result by decreasing its length. Clearly, relative to changes in hair diameter, changes in hair length produce variations in resonance frequency much more economically in terms of the required mass changes!

Even more interesting is the possibility that changes in hair resonance frequency induced by fractional changes in the torsional restoring constant, S, or the damping constant, R, could, in principle, be due to evolutionary alterations of the molecular structure of the cuticular material composing the membrane suspending the filiform hair apparatus (chitin fibers embedded in a protein matrix). This is because long-term evolutionary changes in S and/or R resulting from genetically guided rearrangements of the cuticular material molecular structure are expected to involve very small changes in mass and, therefore, might be less costly relative to the mass changes required to alter hair length or diameter.

A model describing hair motion can be coupled to an optimization procedure, such as a genetic algorithm, to calculate possible evolution scenarios resulting in optimally performing hairs: meaning optimal values of d, L, R, and S for a particular stimulus frequency or sensitivity in a fluid medium. However, what is "optimal" has to be decided (and may be a moving target) and an associated "cost function" must be defined. In addition, to obtain realistic results, biological, physical-chemical, mechanical, and ecological constraints must be imposed on such an optimization problem. This is the subject of continuing research by the authors.

Acknowledgments. The authors acknowledge helpful discussions with Aaron Spak and thank him for his assistance in preparing the figures. FGB was supported by a grant from the Austrian Science Foundation (FWF P 12 192 BIO) and JACH by a grant from the National Science Foundation (9907203).

References

Barth FG (2000) How to catch the wind: spider hairs specialized for sensing the movement of air. Narturwissenschaften 87:51-58

Barth FG, Humphrey JAC, Wastl U, Halbritter J, Brittinger W (1995) Dynamics of arthropod filiform hairs. III. Flow patterns related to air movement detection in a spider *(Cupiennius salei* Keys.). Phil Trans R Soc London, B 347:397-412

Barth FG, Wastl U, Humphrey JAC, Devarakonda R (1993) Dynamics of arthropod filiform hairs. II. Mechanical properties of spider trichobothria *(Cupiennius salei* Keys.). Phil Trans R Soc London, B 340:445-461

Bleckmann H (1994) Reception of Hydrodynamic Stimuli in Aquatic and Semiaquatic Animals. Stuttgart, Jena, New York: G. Fischer

Devarakonda R, Barth FG, Humphrey JAC (1996) Dynamics of arthropod filiform hairs. IV. Hair motion in air and water. Phil Trans R Soc London, B 351:933-946

Fletcher NH (1978) Acoustical response of hair receptors in insects. J Comp Physiol 127:185-189

Humphrey JAC, Devarakonda R, Iglesias I, Barth FG (1993) Dynamics of arthropod filiform hairs. I. Mathematical modelling of the hair and air motions. Phil Trans R Soc London, B 340:423-444

Humphrey JAC, Devarakonda R, Iglesias I, Barth FG (1997) Errata for Dynamics of arthropod filiform hairs. I. Mathematical modelling of the hair and air motions. Phil Trans R Soc London, B 352:1995

Kumagai T, Shimozawa T, Baba Y (1998) Mobilities of the cercal wind-receptor hairs of the cricket, *Gryllus bimaculatus*. J Comp Physiol 183:7-21

Shimozawa T, Kanou M (1984) The aerodynamics and sensory physiology of range fractionation in the cercal filiform sensilla of the cricket *Gryllus bimaculatus*. J Comp Physiol A 155:495-505

Shimozawa T, Kumagai T, Baba Y (1998) Structural scaling and functional design of the cercal wind-receptor hairs of a cricket. J Comp Physiol A 183:171-186

Stokes GG (1851) On the effect of the internal friction of fluids on the motion of pendulums. Trans Camb Phil Soc 9:8ff. (Reprinted in *Mathematical and physical papers*, vol. III, 1-141. Cambridge University Press, 1901)

Tautz J (1979) Reception of particle oscillation in a medium – an unorthodox sensory capacity. Narturwissenschaften 66:452-461

Appendix

Beginning with Eqs. (10-12) in the text, we seek the maximum angular deflection θ_{res} and its associated resonance frequency $\omega_{res(\theta)}$. By Eq. (10), θ is a function of t and ω, and at its maximum value satisfies $\partial\theta/\partial t = 0$ and $\partial\theta/\partial\omega = 0$. The calculation of $\partial\theta/\partial t$ yields

$$\frac{\partial\theta}{\partial t} = -C_1\omega\sin(\omega t) + C_2\omega\cos(\omega t)\,. \tag{A-1}$$

Setting this to zero and solving for t results in

$$t = t(\omega) = \frac{1}{\omega}\tan^{-1}\left(\frac{C_2}{C_1}\right). \tag{A-2}$$

Resubstituting this result into Eq. (10) for θ leads to

$$\theta[\omega, t(\omega)] = (C_1^2 + C_2^2)^{1/2} \tag{A-3}$$

and using Eqs.(11) and (12) for $C_1=C_1(\omega)$ and $C_2=C_2(\omega)$ produces

$$\theta[\omega, t(\omega)] = \left[\frac{P^2+Q^2}{(S-I_t\omega^2)^2+R_t^2\omega^2}\right]^{1/2}. \tag{A-4}$$

For the moment assume that P, Q, I_t, and R_t are all independent of ω. Since the value of ω that maximizes θ also maximizes θ^2, we can differentiate θ^2 with respect to ω to obtain

$$\frac{d\theta^2}{d\omega} = -\frac{(P^2+Q^2)2\omega(2I_t^2\omega^2+R_t^2-2SI_t)}{((S-I_t\omega^2)^2+R_t^2\omega^2)^2}. \tag{A-5}$$

This expression is zero for the trivial nonphysical case when $\omega = 0$ and when

$$\omega = \omega_{res(\theta)} = \left[\frac{S}{I_t} - \frac{R_t^2}{2I_t^2} \right]^{1/2}, \tag{A-6}$$

which is the resonance oscillation frequency corresponding to the maximum angular deflection, θ_{res}. The value of θ_{res} is found by substituting Eq. (A-6) into Eq. (A-4) to obtain

$$\theta_{res} = \frac{2I_t}{R_t} \left[\frac{P^2 + Q^2}{4I_t S - R_t^2} \right]^{1/2}. \tag{A-7}$$

Equation (A-1) gives the angular velocity of a hair, $V=\partial\theta/\partial t$, at any time t and frequency ω. It is possible to proceed the same way as above to obtain an expression for the resonance frequency, $\omega_{res(V)}$, at which the velocity maximizes, and for the value of the velocity itself, V_{res}, at this resonance frequency. Setting the time derivative of Eq. (A-1) equal to zero and solving for t yields

$$t = t(\omega) = \frac{1}{\omega} \tan^{-1}\left(-\frac{C_1}{C_2}\right). \tag{A-8}$$

Resubstituting this result into Eq. (A-1) for V leads to

$$V[\omega, t(\omega)] = \frac{\partial\theta[\omega, t(\omega)]}{\partial t} = \omega (C_1^2 + C_2^2)^{1/2} \tag{A-9}$$

and using Eqs.(11) and (12) for $C_1 = C_1(\omega)$ and $C_2 = C_2(\omega)$ produces

$$V[\omega, t(\omega)] = \omega \left[\frac{P^2 + Q^2}{(S - I_t\omega^2)^2 + R_t^2\omega^2} \right]^{1/2}. \tag{A-10}$$

We again assume that P, Q, I_t, and R_t are all independent of ω and, since the value of ω that maximizes V also maximizes V^2, we can differentiate V^2 with respect to ω to obtain

$$\frac{dV^2}{d\omega} = \frac{(P^2 + Q^2) 2\omega (2I_t^2\omega^2 + R_t^2 - 2SI_t)}{[(S - I_t\omega^2)^2 + R_t^2\omega^2]^2}. \tag{A-11}$$

This expression is zero for the trivial nonphysical case when $\omega = 0$ and when

$$\omega = \omega_{res(V)} = \left[\frac{S}{I_t}\right]^{1/2}, \tag{A-12}$$

which is the resonance oscillation frequency corresponding to the maximum angular velocity, V_{res}. The value of V_{res} can be found by substituting Eq. (A-12) into Eq. (A-10) to obtain

$$V_{res} = \left.\frac{\partial\theta}{\partial t}\right|_{res} = \frac{1}{R_t}\left[P^2 + Q^2\right]^{1/2}. \tag{A-13}$$

III.2 Vibration Communication in Vertebrates

Peter M. Narins

Department of Physiological Science, UCLA, PO Box 160606, Los Angeles, California 90095-1606, USA

Abstract
In this chapter I shall attempt to explore the ways in which several vertebrate groups utilize vibrational or seismic signals. Specifically, I hope to encourage the study of animals other than those used in the standard mammalian preparations by emphasizing that an animal's natural acoustic behavior often leads to insights concerning the underlying physiological mechanisms. Applied across a wide range of taxa, this neuroethological approach has yielded a great deal of information about vertebrate behaviors including bat echolocation, sound localization by owls, electrolocation and object avoidance in electric fishes, song learning in passerine birds, and frog acoustic and vibrational communication. Among the vertebrates, strong evidence for seismic signal detection and/or communication has been documented for only a handful of cases. These animals appear to rely on surface (Rayleigh) waves to transmit vibrational signals. In *Leptodactylus* and perhaps in *Crotalus*, the apparatus for detecting these signals resides both in the sensory epithelium of the inner ear and in its associated sensory structures and appears to take advantage of the multimodal response properties of the sensory hair cells. Methods used to explore these responses and new behavioral evidence for seismic cue detection in several animal groups are discussed.

Key words Amphibian, mammal, seismic, hearing, inner ear

1 Introduction

Some animals possess sensory capacities so fundamentally different from our own that it is especially intriguing to imagine how they must perceive the world. An electric fish in a muddy river navigates, avoids obstacles, and recognizes kin by small perturbations in the local electric field surrounding its body; a horseshoe bat determines the character of its potential prey by analyzing the ultrasonic echolocation pulses reflected by the flapping wings of a passing insect; a pigeon in eastern United States, in theory, can hear the infrasound produced by wind streaming over the Rocky Mountains in the west, and presumably use this cue for orientation on its homeward flight; the sunfish can detect the plane of polarized light reflected from otherwise camouflaged prey. These and many other cases teach us that the essence of what is human is in a real sense defined by contrast to what is not human.

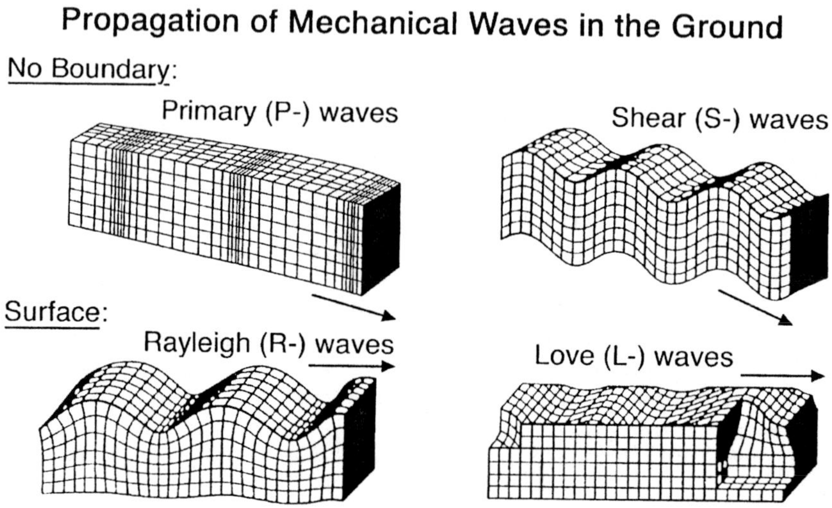

Fig. 1. Propagation of mechanical waves in elastic solids. *Primary* and *Shear* waves are found within a three-dimensional medium, whereas *Rayleigh* and *Love* waves propagate only at the interface between two media with different impedances (e.g., air and ground). Males of the white-lipped frog (*Leptodactylus albilabris*) in Puerto Rico generate Rayleigh waves during calling that are used for communication between conspecific males. In this form of surface wave, substrate particles trace out a retrograde, elliptical orbit. The *arrows* indicate the direction of wave propagation. (After Aicher and Tautz 1990)

What is the difference between sound and vibration? We think of sound as a sensation we hear with our ears and vibration as a sensation we feel with our bodies. Sound is usually associated with a long-distance pressure wave propagated through an elastic medium, whereas vibrations in solid objects are of two basic types: body waves and surface waves (Narins 1990). Body waves travel through the earth and are either of the high-velocity compressional (P) type or the low-velocity shear (S) type. Velocities through granite, for example, of 5.5 kms^{-1} and 3.0 kms^{-1} are typical for P-waves and S-waves, respectively. Moreover, since these waves are being dissipated within the earth, the major source of amplitude attenuation is via spherical spreading, just as for sound waves in free space. In other words, the amplitude of a compressional or shear wave falls off as 1/r, where r is the distance from the source.

The second basic class of elastic waves consists of those propagated disturbances restricted in their motion to the boundary layer at the earth's surface.

Two types of well-studied surface waves are Love waves and Rayleigh waves, which differ in both their plane of polarization and their velocity. The plane of polarization of a seismic wave is that plane containing the particle motion, and the direction (or axis) of propagation is the direction of energy flow for the wave. Displacements of the substrate particles in Love waves are perpendicular to, and in Rayleigh waves parallel to, the vertical plane containing the axis of wave propagation. Love waves cause particle displacement to occur in a direction transverse to the axis of energy flow. In contrast, Rayleigh waves result in no transverse vibration, but impart a strong vertical component to individual substrate particles. Both categories of propagated mechanical waves are illustrated in Fig. 1. Primary and Shear waves are high-velocity (body) waves that travel through the earth but are not known to be important for animal signaling. Only surface (R- and L-) waves have been shown to be used in animal communication signals- L-waves for desert-dwelling scorpion signaling and R-waves for signaling not only by the white-lipped frog, but for a variety of other terrestrial vertebrates as well.

Some variation in these ideal polarization patterns occurs due to both local inhomogeneities in the substrate and proximity to the source. Surface waves dissipate their energy in an ever expanding circular pattern concentric with the source; as a result, the amplitude of a Rayleigh or Love wave decreases as $1/r^{1/2}$, where r is the distance from the source. Thus, far from the source, the major remaining component of the seismic disturbance is the surface wave. Rayleigh waves propagate significantly slower than body waves; typical Rayleigh wave velocities in moist soil are 100 ms^{-1} (Lewis and Narins 1985) and in loose sand, 40-50 ms^{-1} (Brownell 1977).

Recent studies in our laboratory and in others have revealed that, at least in one group of animals (frogs), a population of inner ear sensors that respond to low-frequency sound are also exceptionally sensitive to whole-body vibrations. Upon reflection, this is not surprising, given the striking similarity of the anatomy and physiology of the sensory cells of the auditory and vestibular systems (Lewis and Narins 1999). Thus, for some of the inner ear sensors in the frog, sound and vibration appear to converge. Clearly, in this case, we might adopt the view that auditory and vibratory signals differ principally in the pathway traveled to the inner ear sensors. Thus, the tympanic auditory pathway describes vibratory stimuli reaching the inner ear receptors via the tympanic membrane [and thence the middle ear ossicle(s)]. This pathway is often used for the detection of airborne or water-borne vibrations, then referred to as sound. The extra-tympanic auditory pathway conducts signals to the inner ear through any route in which energy does not directly excite the tympanic membrane (e.g., most modes of bone conduction). This pathway is often used for the detection of vibrations traveling in a solid substrate, for example, soil (= seismic channel).

A totally independent somatosensory pathway, mediated by specialized cutaneous mechanorereceptors, may also be used to detect seismic communication signals (Nevo et al. 1991). It follows, therefore, that the few vertebrate organisms known or presumed to exploit vibratory communication signals all inhabit media that are capable of both providing an alternate pathway and supporting seismic waves, such as underground tunnels, muddy soil, a dense mat of reeds, etc. It is the relationship between habitat, signal structure and communication behavior in seismically active animals that I wish to explore and understand.

2 Definition of Seismic Signaling

Terrestrial seismic communication is the exchange of information between individuals using self-generated vibrational signals transmitted via a solid substrate such as soil, a plant stem, or a blade of grass. Communication using substrate-borne vibrations demands three physical prerequisites: (1) A sender must have the structural apparatus to generate a seismic signal and couple it efficiently to the substrate. This could take the form of an impulsive thump, delivered, for example, to the ground using a limb, a vocal sac, or a specialized accessory structure. (2) The existence of a channel for conducting the signal. Once produced, the seismic disturbance travels outward from the source perhaps as a surface or bending wave. Its propagation characteristics depend intimately on the physical properties of the channel, for example, the substrate density, water content of the soil, obstacles encountered, etc. (3) The seismic receiver must have the structural and neural systems to detect the vibrational signal, extract it from the ambient noise, and decode it for subsequent interpretation. In addition, the signal must be capable of producing a reliable and quantifiable change in the recipient's behavior for communication to have occurred.

This modality affords many advantages for terrestrial animals; seismic communication signals propagate without light during day or night, are most effective at close range, and have only short persistence. Thus, they are well suited for information exchange between neighboring individuals - there is little danger of their detection by distant predators (Narins 1990).

3 The Seismically Sensitive Vertebrates

Although anecdotes abound, evidence for the use of seismic signals for intraspecific communication of biologically significant information has been gathered in only three species of vertebrates to date: the white-lipped frog (*Leptodactylus*) of Puerto Rico, the blind mole-rat (*Spalax*) of Israel, and the Cape mole-rat (*Georychus*) of South Africa. The western rattlesnake (*Crotalus*) of the United States, the sandsfish lizard (*Scincus*) of the Sahara Desert, and the Namib Desert golden mole (*Eremitalpa*) of Namibia are thought to use prey-generated seismic cues for passive location, whereas spadefoot toads (*Scaphiopus*) in the western United States respond to environmentally induced substrate vibrations. Recently, it has been shown that the veiled chameleon (*Chamaeleo*) from Yemen produces vibratory signals in response to both visual and tactile stimuli, whereas the bannertail kangaroo rat (*Dipodomys*) of the southwestern United States uses a vibrational signal as an intermediate component of its communication system. The list of potential vertebrate seismic communicators includes Bushveld rain frogs (*Breviceps*) from the Republic of South Africa and the common Malaysian treefrog (*Polypedates*) among others.

3.1 A Test Case - *Leptodactylus albilabris* - the White-Lipped Frog

Males of most species of frogs and toads produce airborne species-specific advertisement calls to signal their reproductive readiness and locations to potential mates. In addition, several species of frogs are known to produce substrate-borne vibrational signals which may also communicate biologically significant information.

The white-lipped frog of Puerto Rico, *Leptodactylus albilabris*, is the first vertebrate for which anatomical, neurophysiological, and behavioral evidence has been brought to bear implicating the use of seismic signals in intraspecific communication (Narins and Lewis 1984; Lewis and Narins 1985).

This animal has the physical means for generating, the sensory apparatus for detecting, and access to a channel for transmitting vibratory signals between conspecifics. Males of this species vocalize while partially buried in the muddy soil. During the call, the vocal sac expands explosively, striking the substrate as it inflates.

The resulting thump generates a vertically polarized Rayleigh wave (Fig. 1) which propagates in the muddy soil outward in all directions from the source at a velocity of roughly 100 ms^{-1}, or about 1/3 the velocity of sound in air at sea level.

3.1.1 How Does a Frog Detect Extremely Small Motions of the Substrate?

It has been known for many years that the inner ear of anuran amphibians (frogs and toads) is highly sensitive to substrate-borne vibrations (Ashcroft and Hallpike 1934), but only more recently has this sensitivity been quantified (Koyama et al. 1982; Narins and Lewis 1984; Lewis and Narins 1985; Christensen-Dalsgaard and Jørgensen 1988; Jørgensen and Christensen-Dalsgaard 1991).

Anatomical and physiological studies have shown that the vibration sensitivity of terrestrial frogs resides principally in the sacculus of their inner ear (Koyama et al. 1982; Lewis et al. 1982; Christensen-Dalsgaard and Jørgensen 1988; Christensen-Dalsgaard and Narins 1993). The sacculus is an otolithic organ containing a membranous sac filled with a slurry of dense calcium carbonate crystals.

The 3000 sensory hair cells in the sacculus of the North American bullfrog (*Rana catesbeiana*), are concentrated in a macula with their stereocilia in contact with the overlying otoconial mass which can weigh 20 mg in adult frogs (Lewis and Lombard 1988). When the animal undergoes an acceleration, the vibrations reach the saccular macula, where they result in a shearing of the stereocilia of the sensory hair cells, due to the differential motion between the sensory epithelium and the otoconial mass. This leads directly to a modulation of the resting discharge rate in the afferent fibers innervating the saccular hair cells.

To investigate the neural basis of the remarkable behavior of *L. albilabris*, the responses of single fibers in the eighth (acoustico-vestibular) cranial nerve to whole-body vibrations were studied (Narins and Lewis 1984). A male frog was placed on a platform which was sinusoidally vibrated over a wide range of frequencies and amplitudes. To properly isolate the animal from ambient microseismic vibrations, it was necessary to enclose the entire recording apparatus and the animal in a specially designed vibration-damped room, in which the vibration noise floor was reduced to at least an order of magnitude lower than any stimulus being applied. In the case of the white-lipped frog, a vibration noise floor of 0.0001 cms^{-2} was obtained, enabling accurate measurements of accelerations as low as 0.001 cms^{-2}.

Single saccular fibers in the eighth cranial nerve were impaled with KCl-filled glass micropipettes and the responses to whole-body vibrations were recorded. Vibration-sensitive fibers show a characteristic tuned response; that is, each fiber responds most vigorously (highest discharge rate) to a best vibratory frequency (BVF), analogous to the best excitatory frequency of auditory fibers. Two populations of fibers were found that responded selectively to whole-body vibrations: a highly sensitive group with BVFs below 160 Hz, and a second, less-sensitive group with BVFs between 200-300 Hz.

The most sensitive fibers in this species respond to peak whole-body accelerations on the order of about 0.001 cm/s^{-2} (Narins and Lewis 1984). This represents seismic sensitivity of about an order of magnitude greater than saccular fibers in the North American bullfrog, *Rana catesbeiana* (Koyama et al. 1982), and two orders of magnitude more sensitive than has been reported for mammalian inner ear organs.

3.1.2 Bimodal Fiber Response Properties

In fact, most low-frequency axons in the frog inner ear are acutely sensitive to both substrate-borne vibrational (seismic) and sound stimuli (Yu et al. 1991; Christensen-Dalsgaard and Narins 1993; Christensen-Dalsgaard and Jørgensen 1996a); these are called bimodal fibers. Lewis et al. (1982a) dye-filled fibers from various regions of the eighth nerve in the bullfrog, and found that a fiber's distal origin could often be predicted based on its location within the nerve.

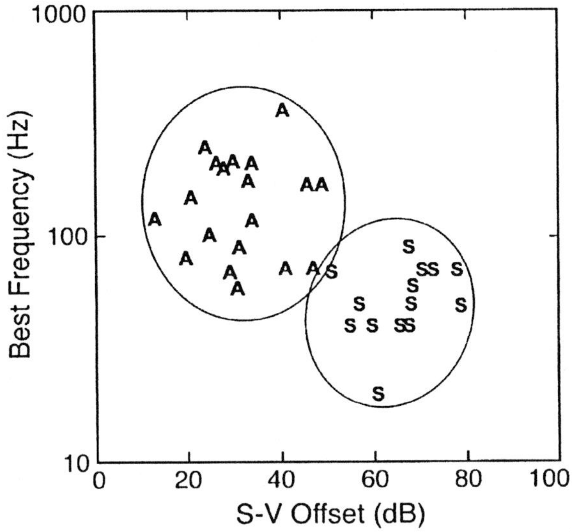

Fig. 2. Scatterplot of S-V offsets and BFs in 38 eighth nerve fibers in *R. p. pipiens* and *L. albilabris*. Elements in the *A* and *S* cluster were partitioned according to their S-V offset using a K-means clustering algorithm (Systat, v. 5.1). (After Christensen-Dalsgaard and Narins 1993)

Christensen-Dalsgaard and Narins (1993) demonstrated that the bimodal low-frequency (CFs<0.5 kHz; CF or characteristic frequency of an eighth nerve fiber is defined as that frequency for which a threshold response may be eleicited with the lowest intensity stimulus) eighth nerve fibers from the northern leopard frog, *Rana p. pipiens*, and the white-lipped frog, *Leptodactylus albilabris*, could be said to originate in either the sacculus or amphibian papilla (AP), based on their position within the eighth nerve, their spontaneous discharge rate and their relative sensitivity to sound and vibration. Moreover, the thresholds for whole-body vibrational stimuli in both populations are lower than those for airborne sound, when directly compared using particle velocity as a common metric.

Bimodal fibers had identical CFs for both sound and vibration. For each bimodal fiber encountered, a rate-level function (RLF- a plot of the spike rate vs. stimulus level) for each modality was determined, and the horizontal displacement between the two RLFs was measured in dB. This horizontal displacement (sound-vibration, or S-V offset) for the putative saccular fibers was 50-80 dB, whereas S-V offsets for the putative AP fibers ranged from 20-50 dB (Fig. 2). The conspicuous S-V offset differ ences between the response modalities promises to be a reliable, on-line physiological indicator of a bimodal fiber's anatomical origin that could help further characterize the fiber without the need for histological tracing methods; however, for this promise to be realized, putative origins based on S-V offset must first be confirmed for bimodal fibers by dye-filling methods.

3.1.3 How do we Characterize Bimodal Fibers?

Most studies of tuning (Frishkopf et al. 1968; Feng et al. 1975; Capranica 1976;Capranica and Moffat 1980), adaptation (Megela and Capranica 1981; Megela 1984), two-tone suppression (Liff and Goldstein 1970; Benedix et al. 1994), one-tone suppression (Christensen-Dalsgaard and Jørgensen 1996) and dynamic range (Feng 1980, 1982; Narins 1987; Dunia and Narins 1989) for amphibian low-frequency auditory fibers have been accomplished using traditional methods of analysis. These include frequency-threshold curves (FTCs) or tuning curves, PST, ISI, and period histograms in response to pure tones, clicks, and natural sounds, and rate-level functions. Nevertheless, this plethora of experimental techniques has not yielded a model of auditory unit function capable of accurately predicting the temporal response of these units to novel stimuli. Recently, there has been rekindled interest in characterizing the filtering properties of acoustic units in terms of Wiener kernel analysis (Wiener 1958). This analysis is applicable to the investigation of nonlinear systems with unknown characteristics, and is inherently a stochastic one- often relying on the use of Gaussian white noise as the test input ensemble. The experimental possibilities of this analysis were much advanced by Lee and Schetzen's discovery that the Wiener kernels could be derived by cross-correlation of the system's output with its Gaussian white noise input (Lee and Schetzen 1965; Marmarelis and Marmarelis 1978).

The Wiener series is capable of predicting the response of spontaneously active auditory units to novel and arbitrarily complex sounds (first-order series: de Boer and de Jongh 1978; Wolodkin et al. 1997; second-order series: van Dijk et al. 1994, 1997a,b; Yamada and Lewis 1999). Thus, the Wiener kernels embody properties of the auditory filter such as suppression, adaptation, separable AC and DC responses, and spectro-temporal filtering (Yamada 1997; Yamada and Lewis 2000). The kernels can be analyzed to describe these properties. For example, the amplitude of the discrete Fourier transform (DFT) of the first-order Wiener kernel (K_1) of an auditory unit with a low CF will approximate that unit's pure-tone-derived tuning curve or FTC (Evans 1977), while the DFT of the significant singular vectors of the second-order Wiener kernel (K_2) of a fiber with a high CF will approximate its FTC (Yamada 1997). K_2s have been gaining favor as the preferred method for characterization of auditory system non-linearities; as such we are attempting to apply them to the understanding of cross-modality interaction in the eighth nerve.

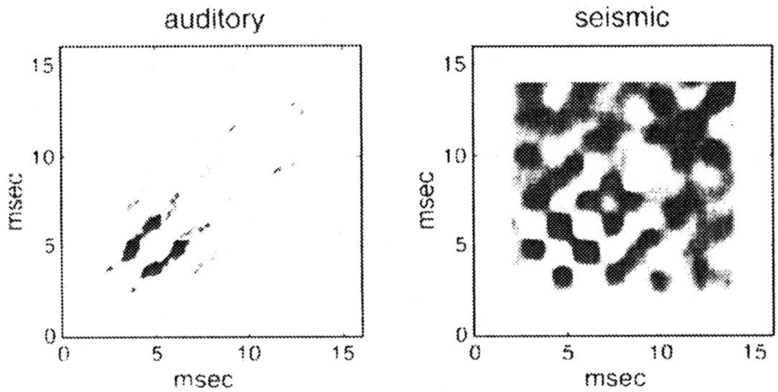

Fig. 3. Auditory (*left*) and seismic (*right*) components of the second-order Wiener kernel for bimodal unit no. 4-1499-1 from the leopard frog eighth nerve. In these plots, the origin represents the time of spike occurrence in response to broadband noise (left) or broadband, whole-body vibration (*right*). Both the *abscissa* and the *ordinate* indicate time prior to the spike, and the distance from the origin to the diagonal pattern is equivalent to the latency from the end of the stimulus to the resulting spike. These plots are an indication of the temporal relationships in the broadband stimuli that, on average, evoke a spike

In preliminary experiments we have examined a fiber's response to bimodal (auditory-plus-seismic) stimuli using second-order Wiener kernels. This is a powerful new technique that holds great promise for auditory scientists in that it has been successful in incorporating an accurate description of the temporal response of a fiber to an arbitrary auditory stimulus (Yamada 1997; Yamada and Lewis 1999). We are presently using this method for studying cross-modality suppression. To accomplish this, both stimuli (e.g., broadband airborne sound and broadband substrate vibration) are applied simultaneously to obtain the cross-modality response properties of the fiber. The seismic and auditory components of the second-order Wiener kernel for a single bimodal eighth nerve fiber in the leopard frog are shown in Fig. 3.

Our long-term goal is to use these kernels to predict the temporal response (sensu Yamada and Lewis 1999) to a mixed, auditory-plus-seismic stimulus (e.g., the species-specific call of the white-lipped frog, *L. albilabris*, which contains both auditory and seismic components: Lewis and Narins 1985; Narins et al. 1985). This is accomplished by digitizing both components of the advertisement call, convolving the digitized waveform with the K_1 and K_2 of the fiber (calculating the second-order Wiener series for this waveform), and comparing the resulting output of that simulation with the PSTH of the unit's response to the stimulus. Predicting the temporal response of a frog eighth nerve fiber in this way has been reported only once to our knowledge (Yamada and Lewis 1999) strictly in the auditory domain. Our current work seeks to extend these exciting results to bimodal units.

3.2 *Spalax ehrenbergi* - the Blind Mole-Rat

Blind mole rats (*Spalax ehrenbergi*) are highly solitary and extremely aggressive and live in individual tunnel systems where they rarely encounter each other except during the mating season (Nevo 1961). In a landmark study, Rado et al. (1987) demonstrated both with captive and wild blind mole rats that head-drumming responses could be elicited by finger tapping and scratching by the experimenter on the substrate. In a subsequent study, it was suggested that (1) the unique morphology of the middle ear of *Spalax* and of the articulation between the lower jaw and the skull, and (2) a peculiar jaw listening behavior enable substrate-borne vibrations to be transmitted to the inner ear mainly by bone conduction (Rado et al. 1989). This might be seen as compensating for the restricted range of airborne signals to which this animal is sensitive (Bruns et al. 1988; Heffner and Heffner 1992).

More recently, Rado et al. (1998) showed that the seismic signals that mole rats use for long-distance communication are processed primarily by the auditory rather than the somatosensory system. Clearly, these fossorial mammals have used the properties of seismic signals to their advantage for intraspecific communication. As we shall see, *Spalax* is not alone among the mammals in its exploitation of substrate-borne signals.

3.3 *Georychus capensis* - the Cape Mole-Rat

African mole-rats in the family Bathyergidae comprise five genera, living in a wide range of habitats. Three of the genera are solitary for some part of their lives, the most extreme case being the Cape mole-rat, *Georychus capensis*, which spends the vast majority of its life under Africa, alone in its burrow, save for a short period during each breeding season. This monotypic genus appears to be most closely related to two other solitary mole-rat genera, *Heliophobius* and *Bathyergus*, and less closely related to the social *Cryptomys* and *Heterocephalus* within the Bathyergidae (Honeycutt et al. 1991).

Males and females each construct complex burrow systems of approximately 130 m length and mean diameter 10 cm in sandy clay soils (Du Toit et al. 1985; Narins et al. 1992). Burrow systems of *Georychus* approach within a few meters of each other and are rather evenly spaced (Davis and Jarvis 1986; Bennett and Jarvis 1988a). Individuals of *Georychus* produce sexually dimorphic patterns of foot-druming in the laboratory and in the field, and it is possible that these signals are used for maintenance of spacing and communication between individuals. It is also likely that mate attraction is entirely mediated by seismic signaling between individuals in burrows.

Foot-drumming signals contain both auditory and seismic components. Measurements within, and at various distances from a burrow revealed that only the seismic component of the foot-drumming signal is detectable in the substrate at distances corresponding to natural interburrow distances (3-4 m); the amplitude of the acoustic component attenuates into the noise level within 1 m of the source (Narins et al. 1992).

Moreover, using an artificial thumper (constructed from the solenoid of an electric typewriter), it has been established that the horizontal and vertical components of the surface (Rayleigh) wave produced by foot-drumming attenuate differentially as they propagate through the sandy soil inhabited by *Georychus* (Fig. 4). By comparing amplitudes of the two surface wave components, a Cape mole-rat could, in principle, determine the distance to the signaling individual, as has been suggested to occur in the scorpion *Paruroctonus* (Brownell and Farley 1979a,b; Brownell 1987).

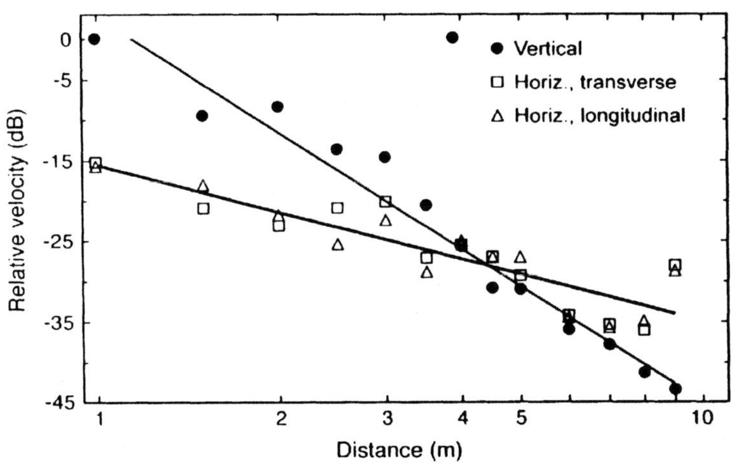

Fig. 4. Horizontal and vertical components of a Rayleigh wave in the sandy soil habitat of the Cape mole-rat, *Georychus*. Each *point* represents the maximum peak-to-peak velocity of the average of 50 simulated mole-rat thumps. An array of three orthogonally mounted geophones was positioned at various distances from the thumper and oriented as follows: horizontally and transverse to the axis of propagation (*open squares*), horizontally and longitudinal (parallel) to the axis of propagation (*open triangles*), and vertically (*closed circles*)

3.4 *Crotalus viridus* - the Western Rattlesnake

It has long been known that snakes are extremely sensitive to ground vibrations, but this was not quantified until the classical work of Hartline (1971). In this study, he demonstrated that snakes possess two sensory systems which respond to both airborne sound and substrate vibration: one subserved by the eighth nerve and inner ear (auditory), and the other mediated by the spinal cord and cutaneous mechanoreceptors (somatic). Based on multiunit evoked responses, it was concluded that the auditory system of the western rattlesnake (*Crotalus viridus*) is not particularly sensitive to airborne sound; its airborne sound threshold between 200 and 400 Hz is about 20 dB higher than the human auditory system in that frequency range. Nonetheless, it showed remarkable sensitivity to seismic vibrations: a 1-Å peak-to-peak vibration amplitude was suprathreshold (Hartline 1971). Evoked potential recordings failed to demonstrate that either the somatic or auditory system can distinguish between vibrations from the air and from the substrate. It was postulated that intensity information from the auditory and the somatic system could be compared centrally to determine the relative amplitudes of airborne and substrate-borne vibration in an unknown natural stimulus. However, although it is presumed that vibration detection may subserve prey detection in snakes, the function of their remarkable vibration sensitivity has not been confirmed behaviorally.

3.5 *Scincus scincus* - the Sandfish Lizard

The sandfish lizard (*Scincus scincus*, family Scincidae) spends most of its time buried in the Sahara Desert. Experiments with this animal in captivity demonstrate that *Scincus* is quite adept at detecting insects on or below the surface of the sand, and it appears to do so by reception of vibrations produced by the insects' movements (Hetherington 1989). When live potential prey were placed in a sand tank on one side of a partition composed of two sheets of Plexiglas separated by 1 cm, the buried lizard on the other side of the partition did not respond to the motion of either mealworms or crickets. It was also shown in this study that the location of the lizard relative to the sand surface was not critical for this behavior. Lizards buried in the sand generally responded to mealworms and crickets walking on the surface at distances of up to 15 cm.

The buried lizard's almost complete indifference to dead insects at the same distance supports the hypothesis that *Scincus* responds to signals produced by moving insects rather relying strictly on olfactory cues. Moreover, Hetherington (1989) measured the angle of lizard's head orientation upon its emergence from the sand following prey detection and demonstrated that, far from being random, the average emergence angle was not significantly different from the expected angle (that angle corresponding to the prey's actual location). Thus, remarkably, the sandfish lizard appears to be able to localize the vibratory cues while buried, and thereby gains an additional advantage over his prey.

3.6 *Eremitalpa granti namibensis* - the Namib Desert Golden Mole

The 18 species of golden moles, placed in the family Chrysochloridae, comprise seven genera that are found exclusively in sub-Saharan Africa, south of a line from Cameroon to Uganda. They are image-blind (DuBost 1968), and for the most part nocturnal, surface-foraging mammals, some genera possessing massively hyper-trophied mallei which presumably confer low-frequency, substrate-vibration sensitivity through inertial bone conduction (Mason 1998). (Inertial bone conduction results from the vibration of the skull which results in relative movement between the stapes footplate and the oval window due to ossicular inertia.)

Unlike many other subterranean mammals, the Namib Desert golden mole, *Eremitalpa granti namibensis*, lacks a permanent burrow system and can forage five or more km per night (Holm 1969) for its prey on the dune surface (Fielden et al. 1992). Foraging animals typically move between miniature sand dunes or mounds topped with dune grass (*Stipagrostis*) which contain the large majority of living biomass in the Namib Desert (Fielden et al. 1990). Recently, two hypotheses were directly tested regarding the foraging strategies adopted by *E. g. namibensis*: that foraging is random in these animals, that is, encounters with food resources are purely stochastic events, and that the mole's ability to detect prey is effective over short distances only (Narins et al. 1997).

To test the first hypothesis, a simple statistical test was performed to determine the probability that the golden mole would visit (for example), seven food patches in succession if it were indeed foraging randomly. For every foraging trail tested, the calculated cumulative probabilities of randomly encountering seven particular mounds in succession were so small that the hypothesis of random foraging could safely be abandoned. Consistent with this result, foraging trail segments between visited mounds appear remarkably straight, showing no evidence of a random search pattern. Rather, the data suggest an alternative working hypothesis, namely that the golden mole exhibits sensory-guided foraging behavior.

Foraging trails consist of footprints interspersed with characteristic sand disturbances in which the animal either head dips or "swims" under the sand. The function of these behaviors is not known but we suspected that they may be used to obtain a seismic "fix" on the next mound to be visited. To test this, the local seismic vibrations both on the top of a mound and in the flat sand at a distance of at least 20 m from the mound under test were measured. Resulting representative velocity spectra are shown in Fig. 5A. The spectrum recorded on the flat shows a relatively low-amplitude peak at about 120 Hz, whereas the spectral peak recorded from the mound is nearly 17 dB greater in amplitude and centered at 310 Hz. The difference spectrum (Fig. 5B) exhibits a peak frequency at 300 Hz with a relative magnitude above background noise of nearly 30 dB.

It was suggested that wind blowing the dune grass on the mounds could set up vibrations in the mounds that would result in the emission of a relatively strong, low-frequency seismic signal. Mounds would then act as seismic beacons for the golden moles that would be detectable from distances corresponding to typical intermound distances of 20-25 m. More data are needed to confirm or refute this idea.

What is clear is that the mallei of the golden moles in the genera *Chrysochloris* and *Eremitalpa* are massively hypertrophied (Cooper 1928; Findlay 1944; Mayer et al. 1995). In fact, out of the 117 species for which data are available, the golden moles *Chrysospalax, Eremitalpa* and *Chrysochloris* have the greatest ossicular mass relative to body size (Mason, pers. comm.). Functionally, the chrysochlorids appear to be low-frequency specialists, and it is likely that golden moles hear through substrate conduction (Narins et al. 1997).

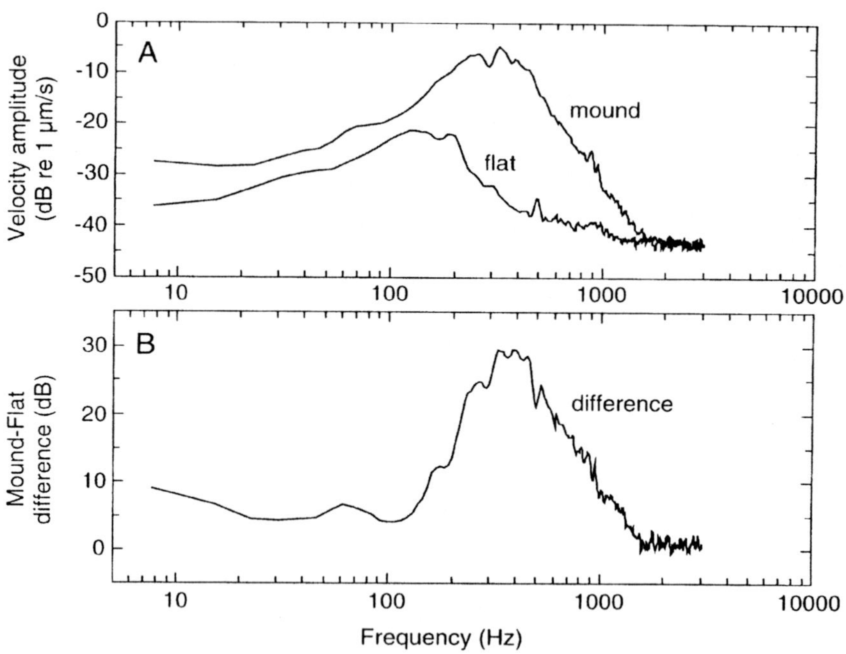

Fig. 5. A Comparison of the representative velocity amplitude spectra of geophone recordings made from the top of a medium-sized mound and from the flat sand. Recordings were made within 10 min of each other under identical conditions. **B** Difference between the two spectra shown in **A**. (Narins et al. 1997)

3.7 *Scaphiopus* spp. - the Spadefoot Toad and Others

Seismic cues may be used to signal environmental changes that are critical for an organism's reproductive behavior. For example, the emergence of burrowing species of spadefoot toads (*Scaphiopus*) in response to rain has been shown to be triggered by the seismic vibrations produced by the rain drops (Dimmit and Ruibal 1980). It is likely that other burrowing amphibian species respond to seismic vibrations of rainfall in a similar fashion. Likely candidates for this behavior are individuals of the relatively unknown genus *Breviceps*. These are entirely terrestrial and fossorial frogs endemic to sub-Sarahan Africa, that occupy a wide range of habitats including forest, montane grassland, open and closed woodland, and vegetated coastal dunes (Minter 1998).

For example, during the breeding season in Pietersburg, Republic of South Africa, the response to rain by populations of *Breviceps adspersus pentheri* (Microhylidae) is very rapid; the first calls were heard 15-30 min after rain began to fall (Minter 1998). Although it is possible that water seeping down to the buried frogs signaled the onset of the breeding season, the rapid response times suggest that other cues, perhaps vibrational, in fact trigger the emergence of these interesting animals. Future field experiments will undoubtedly be designed to distinguish between these (and other) possibilities.

3.8 *Dipodomys spectabilis* - the Banner-Tailed Kangaroo Rat

It has long been known that banner-tailed kangaroo rats, *Dipodomys spectabilis*, engage in foot-drumming behavior as a territorial signal (Randall 1984). In addition, there is some evidence that they may recognize foot-drum patterns and use them to distinguish neighbors from strangers (Randall 1984). The question of whether the airborne component or the seismic component is important in signaling territoriality and individual recognition was experimentally tested in an ingenious experiment in which an artificial thumper was used to simulate the foot-drumming pattern; the timing of the responses of individual kangaroo rats was then compared to the stimulus (Randall and Lewis 1997). In one set of experiments, kangaroo rats in their burrows were given a choice of two stimuli: (1) simultaneous seismic foot-drums from the mechanical thumper buried in the ground plus high-level broadband noise broadcast through a loudspeaker to mask any airborne sound generated by the buried thumper, or (2) simultaneous airborne foot-drumming sounds lacking any seismic components plus high-level broadband noise.

When the number of foot-drumming sequences initiated during the 7 s of seismic playback stimuli with airborne masking noise was compared to the number initiated in the 7 s before and after the playback, it was shown that the kangaroo rats initiated significantly fewer drumming sequences (suggesting they detected the seismic stimulus) during the seismic playback than either before or after the playback. In contrast, similar numbers of drumming sequences were initiated before, during and after the stimuli played back through the air. Moreover, by measuring the sound pressure inside a simulated kangaroo rat's chamber in response to a mixed seismic-plus-airborne stimulus such as an artificial thump, it was determined that in all cases tested, more than 98% of the peak acoustic power in the thump sound recorded in the chamber was transmitted directly through the ground from the site of the thump, and then radiated into the chamber.

Thus, it was concluded that the acoustic power transmitted through the ground was radiated by the chamber wall as airborne sound. It is of interest that sound enters the burrow chamber via the seismic channel 40 dB louder than sound entering from the air, down the chamber entrance. Thus on windless nights these animals are able to communicate out of the burrow to distant neighbors via airborne sound, and on windy nights inside the burrow to close neighbors (Randall and Lewis 1997). Remarkably, the banner-tailed kangaroo rat has evidently evolved an efficient communication system that utilizes seismic signals as an intermediate "carrier" of information ultimately broadcast (by the chamber walls) and detected in air.

3.9 *Chamaeleo calyptratus* - the Veiled Chameleon

Veiled chameleons are aggressive, brightly colored animals with a very high (nearly 5 cm) laterally compressed parietal crest (casque). They inhabit Yemen's high, dry plateau and may grow to 38 cm (Martin 1992). They spend the majority of their time in trees in which they forage for food, mate, and carry out other life functions. When handling chameleons, it is common in some species to feel a buzzing or vibrating sensation (D. Lawson, pers. comm.). Vibration production in response to touch has been reported in several species of chameleons in the genus *Brookesia* (Raxworthy 1991; Glaw and Vences 1994) and *Rhampholeon* (Raxworthy 1991). Recently, accelerometer measurements taken from a plant on which a male *Chamaeleo calyptratus* was perched revealed a conspicuous vibratory signal emanating from the body just anterior to the front legs (Barnett et al. 1999). When a female was placed on a plant with an adult male, he produced a variety of distinct vibratory signals. Signals were recorded in bouts of 1-14 notes, typically with a series of higher-pitched, shorter signals followed by one longer, low-pitched signal. These signals were accompanied by either head tilting or shaking.

The fact that these signals were produced only in the presence of a receptive female or when individuals were startled by direct touch suggests that they serve a communicative function. This is the first example reported of plant-borne vibration communication by a reptile, and again emphasizes the often cryptic nature of this signaling modality.

3.10 *Polypedates leucomystax* - the Common Malaysian Treefrog

Frogs in the family Rhacophoridae, the Old World treefrogs, comprise 186 species in 10 genera (Duellman and Trueb 1986). One of these genera, *Polypedates*, contains 11 species found in Japan, eastern China, and throughout tropical Asia. In order to characterize the evoked calling responses of one species in this genus, a series of acoustic playback experiments was carried out with the common Malaysian treefrog, *Polypedates leucomystax*, from Peninsular Malaysia, (Narins et al. 1998).

Relatively few rhacophorid frogs have been so studied, and *P. leucomystax* has the potential for providing an excellent neuroethological model of vertebrate acoustic behavior, because it is widely distributed, highly vocal, relatively accessible in its natural habitat, and of sufficient head size to facilitate electrophysiological investigations of natural sound encoding by the auditory system (Narins et al. 1998). These workers found that one color morph of the Malaysian treefrog, *Polypedates leucomystax*, exhibits a rather unusual behavior. Females living in dense mats of floating vegetation perch conspicuously on a reed or blade of grass and tap their rear toes rhythmically on the vegetation. This activity, which persists for several minutes, occurs in the dark and was only occasionally accompanied by vocalizations. Thus, it is likely that the tapping functions as a vibrational signal indicating the female's presence to neigh-boring males. This is the first known example of a vertebrate using vibratory signaling via a substrate other than the earth.

3.11 *Loxodonta africana* and *Elephas maximus* - the African and Asiatic Elephants

It has been recently shown that elephants produce significant low-frequency, substrate-borne vibrations during periods of vocalizations and movement (O'Connell et al. 1997). The substrate vibrations were characterized as vertically polarized (Rayleigh) waves with a velocity of about 250 ms^{-1}. These workers make the plausible suggestion that seismic vibrations produced concomitantly with airborne vocalizations could mediate long distance signaling in these animals. Whether these seismic signals are detected using a form of bone conduction involving the relative motion of the massive ossicles with respect to the skull (Reuter et al. 1998), or via sensory mechanoreceptors in the toes or feet (O'Connell et al. 1999), is an open question for future research.

4 Concluding Remarks

It is becoming clear that seismic communication and sensitivity to whole-body vibrations are more ubiquitous among the vertebrates than had been previously imagined. Vibration signaling provides a means for animals to communicate with each other without the effort of overcoming the often formidable ambient acoustic noise levels produced by other calling individuals. It is possible that various substrates in diverse habitats offer a relatively quiet channel waiting to be exploited by seismically active animals. Thus the vibration sensitivity of substrate-dwelling animals is likely to be widespread and may serve as a neuroethological model for understanding the evolution of vibrational communication systems in environments characterized by high levels of airborne noise.

Acknowledgments. I thank F. G. Barth for his kind invitation to participate in the Sensory Ecology meeting in Vienna, his warm hospitality during the meeting, and his personal understanding during the production of the resulting manuscript. Thanks are due as well to the Austrian Academy of Science for their generous support. I am most grateful to R. Cocroft for the loan of the *C. calyptratus* recording and to W. Yamada for providing the bimodal second-order kernel. Many thanks are due to E. R. Lewis for helpful discussions and to M. Mason for his helpful comments on the manuscript. Work supported by the UCLA Academic Senate and NIDCD grant no. DC00222.

References

Aicher B, Tautz J (1990) Vibrational communication in the fiddler crab, *Uca pugilator* I. Signal transmission through the substratum. J Comp Physiol 166: 345-353

Barnett KE, Cocroft RB, Fleishman LJ (1999) Possible communication by substrate vibration in a chameleon. Copeia 1999: 225-228

Benedix JH Jr, Pedemonte M, Velluti R, Narins PM (1994) Temperature dependence of two-tone rate suppression in the northern leopard frog *Rana pipiens pipiens*. J Acoust Soc Am 96: 2738-2745

de Boer E, de Jongh, HR (1978) On cochlear encoding: potentialities and limitations of the reverse-correlation technique. J Acoust Soc Am 63: 115-135

Brownell P (1977) Compressional and surface waves in sand: used by desert scorpions to locate prey. Science 197: 479-482

Brownell P, Farley RD (1979a) Detection of vibrations in sand by tarsal sense organs of the nocturnal scorpion, *Paruroctonus mesaensis*. J Comp Physiol 131: 23-30

Brownell P, Farley RD (1979b) Orientation to vibrations in sand by the nocturnal scorpion, *Paruroctonus mesaensis*. J Comp Physiol 131: 31-38

Capranica RR (1976) Auditory system: morphology and physiology of the auditory system. In: R Llinas, W Precht (eds) Frog Neurobiology. Berlin, Springer-Verlag, pp 551-575

Capranica RR, Moffat, AJM (1980) Nonlinear properties of the peripheral auditory system of anurans. In: AN Popper, RR Fay (eds) Comparative Studies of Hearing in Vertebrates, New York, Springer-Verlag, pp 139-165

Christensen-Dalsgaard J, Jørgensen MB (1996) One-tone suppression in the frog auditory nerve. J Acoust Soc Am 100: 451-457

Christensen-Dalsgaard, J, Narins, PM (1993) Sound and vibration sensitivity of VIIIth nerve fibers in the frogs *Leptodactylus albilabris* and *Rana pipiens pipiens*. J Comp Physiol A 172: 653-662

Cooper FC (1928) On the ear region of certain of the Chrysochloridae. Phil Trans R Soc [Biol] 216: 265-283

van Dijk P, Wit HP, Segenhout JM (1997a) Dissecting the frog inner ear with Gaussian noise I Application of high-order Wiener kernel analysis. Hearing Res 114: 229-242

van Dijk P, Wit HP, Segenhout JM (1997b) Dissecting the frog inner ear with Gaussian noise II Temperature dependence of inner-ear function. Hearing Res 114: 243-251

van Dijk P, Wit HP, Segenhout JM, Tubis A (1994) Wiener kernel analysis of inner ear function in the American bullfrog. J Acoust Soc Am 95: 904-919

Dimmit MA, Ruibal R (1980) Environmental correlates of emergences in spade-foot toads (*Scaphiopus*). J Herpetol 14: 21-29

DuBost G (1968) Les mammifères souterrains. 1. L'adaptation morphologique à la vie souterrain. Rev Écol Biol Sol 5: 99-133

Dunia R, Narins PM (1989) Temporal integration in an anuran auditory nerve. Hearing Res 39: 287-298

DuToit JT, Jarvis JUM, Louw GN (1985) Nutrition and burrowing energetics of the Cape mole-rat *Georychus capensis*. Oecologia 66: 81-87

Evans EF (1977) Frequency selectivity at high signal levels of single units in cochlear nerve and cochlear nucleus. In: Evans EF, Wilson JP (eds) Psychophysics and Physiology of Hearing. London, Academic Press, pp 185-192

Feng AS (1980) Directional characteristics of the acoustic receiver of the leopard frog (*Rana pipiens*): A study of eighth nerve auditory responses. J Acoust Soc Am 68: 1107-1114

Feng AS (1982) Quantitative analysis of intensity-rate and intensity-latency functions in peripheral auditory nerve fibers of northern leopard frogs (*Rana pipiens*). Hearing Res 6: 241-246

Feng AS, Narins PM, Capranica RR (1975) Three populations of primary auditory fibers in the bullfrog (*Rana catesbeiana*): their peripheral origins and frequency selectivities. J Comp Physiol 100: 221-229

Fielden LJ, Hickman GC, Perrin MR (1992) Locomotory activity in the Namib Desert golden mole, *Eremitalpa granti namibensis* (Chrysochloridae). J Zool Lond 226: 329-344

Fielden LJ, Perrin MR, Hickman GC (1990) Feeding ecology and foraging behaviour of the Namib Desert golden mole, *Eremitalpa granti namibensis* (Chrysochloridae). J Zool Lond 220: 367-389

Findlay GH (1944) The development of the auditory ossicles in the elephant shrew, the tenrec and the golden mole. Proc Zool Soc Lond 114: 91-99

Frishkopf LS, Capranica RR, Goldstein MH Jr (1968) Neural coding in the bullfrog's auditory system- a teleological approach. Proc IEEE 56: 969-980

Glaw F, Vences M (1994) A Field Guide to the Amphibians and Reptiles of Madagascar. Moos Druck, Leverkusen and FARBO, Köln, Germany

Hartline PH (1971) Physiological basis for detection of sound and vibration in snakes. J Exp Biol 54: 349-371

Heffner RS, Heffner HE (1992) Hearing and sound localization in blind mole rats (*Spalax ehrenbergi*). Hearing Res 62: 206-216

Hetherington TE (1989) The use of vibratory cues for detection of insect prey by the sandswimming lizard *Scincus scincus*. Anim Behav 37: 290-297

Holm E (1969) Contribution to the knowledge of biology of the Namib Desert golden mole, *Eremitalpa granti namibensis* Bauer and Niethammer 1959. Sci Pap Namib Des Res 41: 37-42

Lee YW, Schetzen M (1965) Measurement of the Wiener kernels of a nonlinear system by cross-correlation. Int J Control 2: 237-254

Lewis ER, Narins PM (1985) Do frogs communicate with seismic signals? Science 227: 187-189

Lewis ER, Narins PM (1999) The acoustic periphery of amphibians: anatomy and physiology. In: RR Fay, AN Popper (eds) Comparative Hearing: Fish and Amphibians, New York, Springer-Verlag, pp 101-154

Liff HJ, Goldstein MH Jr (1970) Peripheral inhibition in auditory fibers in the frog. J Acoust Soc Am 47: 1538-1547

Marmarelis PZ, Marmarelis VZ (1978) Analysis of Physiological Systems. The white-noise Approach. Plenum Press, New York, London

Martin J (1992) Chameleons-Nature's Masters of Disguise. Blandford, London. Mason M (1998) Functional anatomy of the middle ear of insectivores. J Acoust Soc Am 103: 2827

Mayer Av, O'Brien G, Sarmiento EE (1995) Functional and systematic implications of the ear in golden moles (Chrysochloridae). J Zool London 236: 417-430

Megela AL (1984) Diversity of adaptation patterns in responses of eighth nerve fibers in the bullfrog, *Rana catesbeiana.* J Acoust Soc Am 75: 1155-1162

Megela AL, Capranica RR (1981) Response patterns to tone bursts in peripheral auditory systems of anurans. J Neurophysiol 46: 465-478

Minter LR (1998) Aspects of the reproductive biology of *Breviceps*. Doctoral thesis, Johannesburg, Republic of South Africa, University of the Witwatersrand, 281 p

Narins PM (1987) Coding of signals in noise by amphibian auditory nerve fibers. Hearing Res 26: 145-154

Narins PM (1990) Seismic communication in anuran amphibians. Bioscience 40: 268-274

Narins PM, Feng AS, Yong H-S, Christensen-Dalsgaard J (1998) Morphological, behavioral, and genetic divergence of sympatric morphotypes of the treefrog *Polypedates leucomystax* in Peninsular Malaysia. Herpetologica 54: 129-142

Narins PM, Lewis ER, Jarvis JUM, O'Riain J (1997) The use of seismic signals by fossorial southern African mammals: a neuroethological gold mine. Brain Res Bull 44: 641-646

Narins PM, Lewis ER, Lopez PT, Moore SW (1985) Acoustic behavior of a vibration-sensitive frog. J Acoust Soc Am 77: Suppl no 1, S52

Nevo E (1961) Observations on Israeli populations of the mole rat *Spalax E. ehrenbergi* Nehring 1898. Mammalia 25: 127-144

Nevo E, Heth G, Pratt H (1991) Seismic communication in a blind subterranean mammal: a major somatosensory mechanism in adaptive evolution underground. Proc Natl Acad Sci USA 88: 1256-1260

O'Connell CE, Arnason BT., Hart LA (1997) Seismic transmission of elephant vocalizations and movement. J Acoust Soc Am 102: 3124

O'Connell CE, Hart LA, Arnason BT (1999). Comments on "Elephant hearing" [Reuter T, J Acoust Soc Am 104, 1122-1123 (1998)]. J Acoust Soc Am 105: 2051-2052

Rado R, Himelfarb M, Arensburg B, Terkel J, Wollberg Z (1989) Are seismic communication signals transmitted by bone conduction in the blind mole rat? Hearing Res 41: 23-30

Rado R, Levi N, Hauser H, Witcher J, Adler N, Intrator N, Wollberg Z, Terkel J (1987) Seismic signalling as a means of communication in a subterranean mammal. Anim Behav 35: 1249-1251

Rado R, Terkel J, Wollberg Z (1998) Seismic communication signals in the blind mole rat (*Spalax ehrenbergi*): electrophysiological and behavioral evidence for their processing by the auditory system. J Comp Physiol A 183: 503-511

Randall JA (1984) Territorial defense and advertisement by footdrumming in bannertail kangaroo rats (*Dipodomys spectabilis*) at high and low population densities. Behav Ecol Sociobiol 16: 11-20

Randall JA, Lewis ER (1997) Seismic communication between the burrows of kangaroo rats, *Dipodomys spectabilis*. J Comp Physiol A 181: 525-531

Raxworthy CJ (1991) Field observations on some dwarf chameleons (*Brookesia* spp.) from rainforest areas of Madagascar, with the description of a new species. J. Zool Lond 224: 11-25

Reuter T, Nummela S, Hemilä S (1998) Elephant hearing. J Acoust Soc Am 104: 1122-1123

Wiener N (1958) Nonlinear Problems in Random Theory. MIT Press, Cambridge, MA

Wolodkin G, Yamada WM, Lewis ER, Henry KR (1997) Spike rate models for auditory fibers. In: ER Lewis, GR Long, RF Lyon, PM Narins, CR Steele, E

Hecht-Poinar (eds) Diversity in auditory mechanics. Singapore, World Scientific Press, pp 104-110

Yamada WM (1997) Second-order Wiener Kernel Analysis of Auditory Afferent Axons of the North American Bullfrog and Mongolian Gerbil Responding to Noise. Doctoral dissertation, Graduate Group in Neurobiology, University of California, Berkeley

Yamada WM, Lewis ER (1999) Predicting the temporal responses of non-phase-locking bullfrog auditory units to complex acoustic waveforms. Hearing Res 130: 155-170

Yamada WM, Lewis ER (2000) Demonstrating the Wiener kernel description of tuning and suppression in an auditory afferent fiber: Predicting the AC and DC responses to a complex novel stimulus. In: H Wada, T Takasaka (eds) Symposium on Recent Developments in Auditory Mechanics, Singapore, World Scientific Publishers, In press

III.3 Lateral Line Research: the Importance of Using Natural Stimuli in Studies of Sensory Systems

Horst Bleckmann, Joachim Mogdans and Guido Dehnhardt

Institut für Zoologie, Universität Bonn, Poppelsdorfer Schloß, 53115 Bonn, Germany

Abstract
The mechanosensory lateral line is a hydrodynamic receptor system which allows fishes and some aquatic amphibians to detect minute water motions generated by conspecifics, predators, or prey. The sensory units of the lateral line, the neuromasts, can occur free-standing on the surface of the skin or embedded in lateral line canals. The morphological design of the peripheral lateral line varies across species and has long been thought to represent an adaptation to the hydrodynamic conditions that prevail in the habitat of a given species. This chapter argues that in order to fully comprehend lateral line information processing it is imperative to take into account the ecology of fishes, meaning that natural stimulus and noise conditions have to be considered. In all previous studies on lateral line function sinusoidal water motions were applied to a stationary fish in a low-noise environment, a highly unnatural situation. Thus, these studies have not revealed specialized form-function relationships. However, data from our laboratory indicate that morphological and physiological spezializations of the lateral line system can be correlated when experiments are performed in which the natural situation of fishes is considered, for instance when the lateral line is tested with hydrodynamic stimuli applied under different background noise conditions.

Key words Lateral line, mechanoreception, fish, ecology, sensory physiology

1 Introduction

Extant fishes constitute about half the total number of all living vertebrates. Most vertebrates - including fish - have a well-developed olfactory, auditory, and visual system. Some fish lack a visual system, others, like Gymnotidae and Siluridae, have an additional electrosensory system. All extant fishes and most aquatic amphibians possess a hydrodynamic sensory system, the mechanosensory lateral line. Thus the lateral line is one of the most important sensory systems of aquatic anamniotes.

In contrast to the electrosensory system of fishes (review Bullock and Heiligenberg 1986) and the visual, auditory and somatosensory (mechanosensory) systems of terrestrial vertebrates and invertebrates (review Delcomyn 1997), the mechanosensory lateral line is still poorly investigated. In consequence, our knowledge regarding the behavioral relevance and physiology of the lateral line is far from complete. This may have several reasons. First, compared to visual, acoustic, and tactile stimuli, it is difficult to generate, manipulate, quantify, and measure water motions. Second, humans lack a hydrodynamic sense and thus cannot easily grasp the biological importance and role of hydrodynamic sensory systems. This, no doubt, makes it difficult to ask biologically relevant questions and consequently to design lateral line experiments.

This chapter will not review the lateral line in detail. Instead, it will focus on the information necessary to follow and understand our argument that we will fully comprehend lateral line information processing only if we take into account the ecology of fishes. This means that we have to consider the hydrodynamic conditions that are prevailing in the habitat of a given species and determine biologically relevant stimuli and background noise conditions. Using this information, experiments have to be defined in which more natural stimulus conditions can be created, for instance experiments in which hydrodynamic signals can be applied in background noise. Physiological data from such experiments can then be correlated to the various morphological specializations of the lateral line system that were already pointed out by Dijkgraaf (1952). For more general information on lateral line systems the reader is referred to Coombs et al. (1988), Coombs et al. (1992), Bleckmann (1994), Montgomery et al. (1995a), and Schellart and Wubbels (1998). Because of limitations in available space we will, whenever possible, cite only recent review articles.

2 Morphology of the Periphery

The smallest sensory unit of the lateral line is the neuromast which occurs freestanding on the skin, in pits, on pedestals raised above the skin, and - if we exclude amphibians - in fluid-filled canals. The sensory receptor cells within a neuromast are hair cells. Hair cells carry a hair bundle at the apical surface which is comprised of stereovilli and a single eccentrically placed kinocilium. The hair bundle protrudes into a gelatinous cupula which connects the bundle with the water surrounding the fish.

The design and number of lateral line canals as well as the design and number of superficial and canal neuromasts can be very different in even closely related species. Superficial neuromasts are smaller, contain fewer hair cells, and are innervated by fewer nerve fibers than canal neuromasts (Münz 1989). Within a given animal, but also between species, cupula length and shape may vary (Bleckmann 1994). The variability of lateral line canals includes their number, placement and branching pattern, canal width, canal compartmentalization and the number, size and placement of canal pores (e.g., Coombs et al. 1988).

3 Central Pathways

3.1 Ascending Pathways

Lateral line information reaches the brain via lateral line nerves. Upon entering the brain, individual nerve fibers bifurcate into an ascending and a descending branch which terminate throughout the entire medial octavolateralis nucleus (MON) (e.g. Alexandre and Ghysen 1999). The terminal pattern implies that activity in any afferent fiber reaches all the various parts of the medullary lateral line column. Cells of the MON project bilaterally to a secondary brainstem nucleus, to the torus semicircularis (lateral mesencephalic nucleus in elasmobranchs) and to the contralateral MON (e.g. Boord and Montgomery 1989; McCormick and Hernandez 1996). The torus semicircularis projects to the diencephalon which relays information to the telencephalon (Wullimann 1998). In both cartilaginous fishes (Bleckmann et al. 1989) and teleosts (Prechtl et al. 1998) there is a distinct region in the pallium that responds to quasinatural water movement stimuli.

3.2 Descending Recurrent Pathways

The MON receives descending input from a secondary brainstem nucleus, which receives input from the MON and the torus semicircularis (Montgomery et al. 1995). Discussed functions of descending recurrent inputs include gain control and the establishment of central representations of expected sensory input (Montgomery et al. 1995b).

Lateral line neuromasts receive efferent innervation from a rhombencephalic (e.g., Song and Northcutt 1991) and, in some species, also from a diencephalic nucleus (Roberts and Meredith 1989). In amphibians (*Xenopus laevis*) the activity of efferent lateral line fibers is influenced by centers in or rostral to the diencephalon (Münz and Claas 1991). The efferent lateral line system can inhibit and/or excite afferent activity (Roberts and Meredith 1989).

4 Physiology

4.1 Physiology of the Periphery

Since the work of Harris and van Bergeijk (1962), researchers have stimulated the lateral line with constant volume vibrating spheres. To keep experimental conditions defined, fishes are fixed to a stationary holder. Recordings are not begun before the water in the tank is quiet. A minishaker drives the sphere with constant frequency and amplitude.

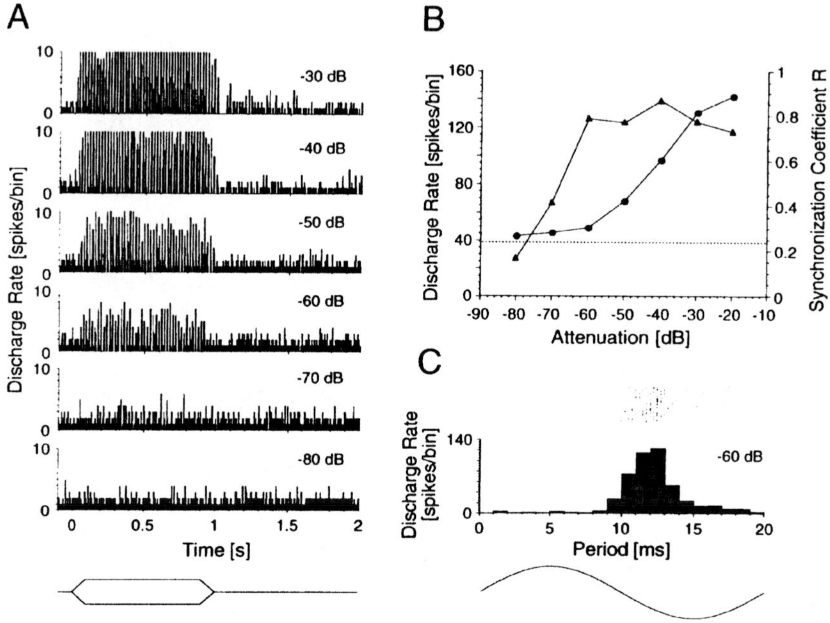

Fig. 1 A-C. Characterization of the responses of a goldfish posterior lateral line nerve fiber (PLLN) to different levels of a 50 Hz constant-amplitude sine wave stimulus generated by a vibrating sphere of 8 mm diameter. **A** PST histograms (binwidth 2 ms) of the responses to ten stimulus repetitions. *Top trace* original recording; *bottom trace* stimulus. Vibration amplitudes were 144, 38, 15, and 4 μm for -30 to -60 dB attenuation, and < 4 μm for attenuations greater than -60 dB. At levels of -40 dB and stronger, the unit responded with at least one highly phase-locked spike per cycle. For this reason, many bins in the PST histogram are of identical height. **B** Input-output function of the fiber. Discharge rates (line connecting circles, left-hand axis) and synchronization coefficients R (lines connecting triangles, right-hand axis) are plotted as a function of level (in rel. dB). An attenuation of -20 dB corresponds to a vibration amplitude of 425 μm. **C** Raster diagram of the distribution of spikes within each cycle of the 50 Hz stimulus and the corresponding period histogram. Graph derived from the data shown in A (-60 dB attenuation)

A dipole (vibrating-sphere) stimulus, applied in a low-noise environment, has many advantages. It is well-defined, easy to generate and manipulate, and causes robust sustained responses from primary lateral line afferents (Bleckmann 1994).

Using a dipole stimulus one can learn, for instance, that the working range of lateral line neuromasts covers frequencies from less than 1 Hz up to about 150 Hz (Kalmijn 1989), that primary lateral line afferents have small receptive fields, that displacement thresholds are as low as 0.01 mm, that superficial neuromasts rank between velocity and acceleration detectors, and that the lateral line canal system functions as an acceleration detector (e.g., Schellart and Wubbels 1998; Coombs and Janssen 1989). In addition we can learn (1) that primary afferents phase lock to sinusoidal stimuli, (2) that their dynamic amplitude range covers up to 90 dB, (3) that their sustained responses encode stimulus duration (Fig. 1) and (4) that the responses of primary afferents change cosine-like with stimulus direction and quickly diminish with source distance (Bleckmann 1994).

4.2 Central Physiology

Compared with primary afferents, the physiological properties of central lateral line units are different in many respects. Ongoing activity is low or absent and responses to constant-amplitude dipole stimuli are usually rapidly adapting (e.g. Bleckmann and Bullock 1989). Receptive fields are generally larger but their size may also be similar to receptive fields of primary afferents (Schellart and Wubbels 1998; Fig. 2). If stimulated with a vibrating sphere many central units are remarkably insensitive, 26-30 % of all MON units (Kröther, unpubl.; Mogdans and Goenechea 1999) and 10% of all units recorded in the torus semicircularis (Plachta et al. 1999) do not respond to a dipole stimulus. Central units which do respond to a dipole stimulus show no or only a weak phase locking (Bleckmann and Bullock 1989; Plachta et al. 1999). Their frequency response usually is similar to that of primary afferents and barely changes along the neuraxis. Sharply tuned toral lateral line units have been found, however (Müller et al. 1996). Threshold sensitivity and dynamic amplitude range of central lateral line units decreases from medulla to telencephalon (Bleckmann and Bullock 1989) but the amount of data is not sufficient to make this a general statement.

Experiments with dipole stimuli, no doubt, have uncovered many important physiological properties of the lateral line. This approach has failed, however, to find a species-specific form-function relationship of the peripheral lateral line (Coombs et al. 1988). Dipole stimuli did reveal some specific response properties of central units as, for instance, in the degree of phase coupling, dynamic amplitude range, or receptive field organization (e.g., Bleckmann and Bullock 1989; Coombs et al. 1998). However, the shape of threshold curves obtained from central units recorded at different levels of the neuraxis (Bleckmann and Bullock 1989; Bleckmann and Münz 1990) is similar to the uniform shape of primary afferent threshold curves. Thus, using vibrating-sphere stimuli, neither specialized peripheral form-function relationships nor the emergence of highly specific response properties of units recorded along the neuraxis have been found.

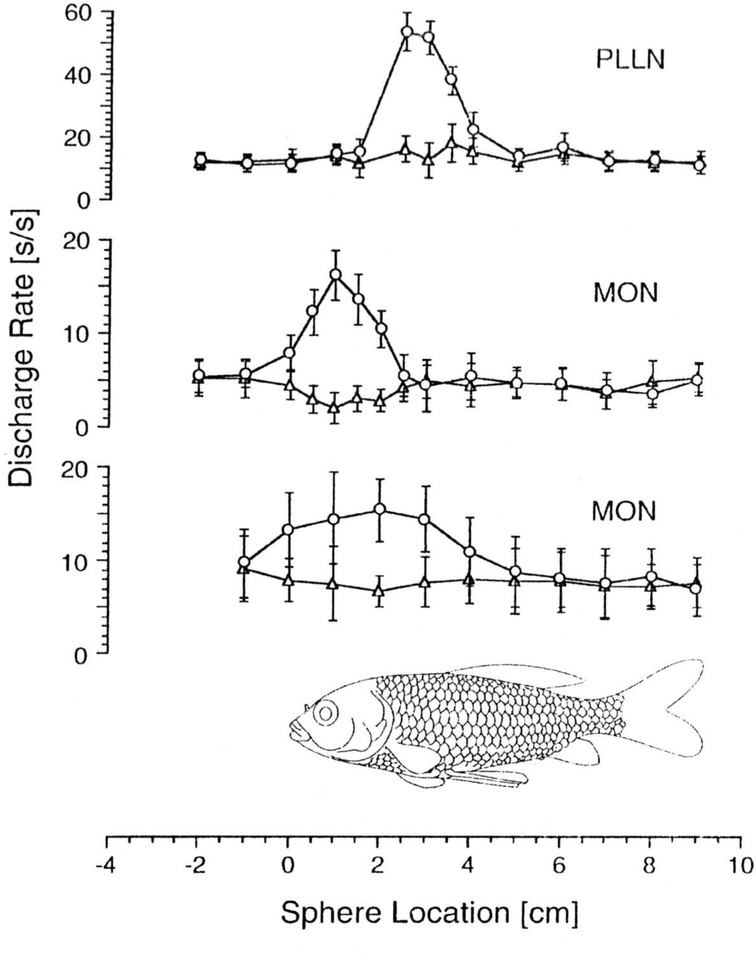

Fig. 2. Receptive fields of a goldfish PLLN fiber (top) and two MON units (*center* and *bottom*). Circles represent average unit responses to 20 repetitions of a 50 Hz constant-amplitude sine wave stimulus (duration 1 s) presented at different locations along the side of the fish. *Triangles* represent average unit activity during a period of 1 s before stimulus presentation. *Vertical* bars represent standard deviations. Vibration amplitudes were 80 μm (PLLN) and 100 μm (MON). Note the differences in discharge rate between PLLN and MON units (different scalings of the Y-axes). *From top to bottom* the width of the receptive fields measured at 50% of the maximum evoked rate were 1.55, 1.73, and 3.86 cm. The *fish drawing* is scaled to match the size of the fish in the experiment in which data for the *top* and *center* graph were recorded (tip of snout at 0 cm, caudal peduncle at 7.4 cm). The fish used in the experiment in which the *bottom trace* was obtained was 4 mm smaller. (Fish drawing kindly provided by A.Grotefeld. Receptive field data provided by S.Kröther)

5 What Can We Learn from Other Sensory Systems?

If we look for commonalities in the processing by other sensory systems we repeatedly find (1) a separation of sensory channels, (2) a systematic representation of stimulus parameters along central maps and (3) the emergence of more and more selective responses to specific, biologically relevant stimuli as one ascends to higher brain levels. Examples for the separation of sensory channels include the amplitude and phase channels in auditory (e.g., Konishi 1986) and electrosensory systems (e.g., Heiligenberg 1991), the parvocellular and magnocellular channels in the primate visual system (e.g., DeYoe and van Essen 1988), the cold and warmth channels in the somatosensory system (Hensel 1974) and the pheromonal and ordinary smell channels in the brains of insects and vertebrates (Hildebrand 1995). Sensory maps often show a relation to the spatial order of the receptors. We know topographic projections from the somatosensory (Catania and Kaas 1995), electrosensory (Bodznick and Schmidt 1984; Heiligenberg 1991), visual (Vanegas et al. 1984), and auditory (Heil and Scheich 1992) system. In chemotopic maps of invertebrates and vertebrates sensory cells expressing a certain receptor protein project to the same glomerulus (Hildebrand 1995; Friedrich and Korsching 1998). In this way, identity and concentration of odorants are encoded by glomerular activity patterns.

Examples of highly selective units include sign-selective neurons in the diencephalon of weakly electric fish (Keller 1988), song-selective neurons in the telencephalon of birds (Whaling et al. 1997) and face-selective units in the temporal cortex of primates (Rolls 1984). Central maps may be constructed and ordered by computations (Heiligenberg 1991). Examples include auditory space maps in owls (Konishi 1986), echo-delay maps in bats (Suga 1990), visual orientation-preference maps in primates (Chapman et al. 1996), and auditory frequency-modulation maps in birds (Heil et al. 1992).

6 Peripheral and Central Sensory Channels

Up to now parallel central lateral line channels have not been found. However, the peripheral lateral line consists of at least two channels. One processes superficial, the other canal neuromast input. Each neuromast contains two sets of hair cells which are oriented in opposite directions (Coombs et al. 1988). Due to the directional sensitivity of hair cells, each neuromast gives rise to two populations of afferent fibers with opposite directional sensitivity. Thus peripheral sensory channels separate information about direction of water motion with respect to the long axis of a neuromast. Using vibrating-sphere stimuli, researchers failed to demonstrate that the above sensory channels continue into the CNS. With one exception (see below) they also failed to uncover central lateral line maps in physiological experiments and the emergence of highly selective response properties at higher brain levels.

Anatomical studies (Alexandre and Ghysen 1999) and studies in which water surface waves have been used as a stimulus (Zittlau et al. 1986) clearly show that there are at least some central lateral line maps. In all bony fishes and in amphibians primary lateral line afferents distribute such in the medulla that fibers from the anterior lateral line nerve are represented ventromedially and fibers from the posterior lateral line nerve dorsolaterally (e.g., McCormick 1989; Song and Northcutt 1991). The projections from the anterior and posterior lines run parallel but do not mix. Moreover, the position of neuromasts is represented in the MON in that the projection of anterior neuromast lies ventrolateral to the projections of more posterior neuromast. The same applies to the projections of the posterior lateral line (Alexandre and Ghysen 1999). Thus, the primary projections of the lateral line of fishes are doubly somatotopic. There may also be a crude somatotopy with the dorsolateral and ventrolateral surfaces of the trunk represented ventrally and dorsally, respectively, in the MON of some fishes (Song and Northcutt 1991). There are no indications that superficial and canal neuromasts map differentially in the MON or that vertically and horizontally oriented neuromasts are mapped separately (Song and Northcutt 1991). Whether the two hair cell populations in a neuromast map differentially is also not clear (Fritzsch 1989).

What are the reasons why lateral line research does not match the progress made in the analysis of information processing in other sensory systems? One possibility is, of course, that many types of information processing known from other sensory systems (e.g. central maps, parallel pathways, highly selective central units) are rare or even do not exist in the fish lateral line system. This is unlikely, however. For instance, in amphibians a central computed lateral line map has been demonstrated. Tectal lateral line units of the clawed frog, *Xenopus laevis*, and the axolotl, *Ambystoma mexicanum*, encode the direction of propagation of quasi natural surface wave stimuli (Zittlau et al. 1986; Bartels et al. 1990). Clearly, with dipole stimuli tectal directional maps would not have been discovered.

7 Natural Hydrodynamic Stimuli

When we evaluate lateral line research we should keep in mind that background noise was always kept low and that the applied stimuli were usually quite artificial. In rivers and creeks, as well as along the ocean shoreline, the water constantly moves and even in seemingly quiet lakes some water movements may be present. Furthermore, even if the water does not move, the fish may move, or both. Thus, in natural environments the lateral line rarely faces stillwater conditions. In addition, animate sources of natural hydrodynamic stimuli rarely vibrate with constant frequency and amplitude.

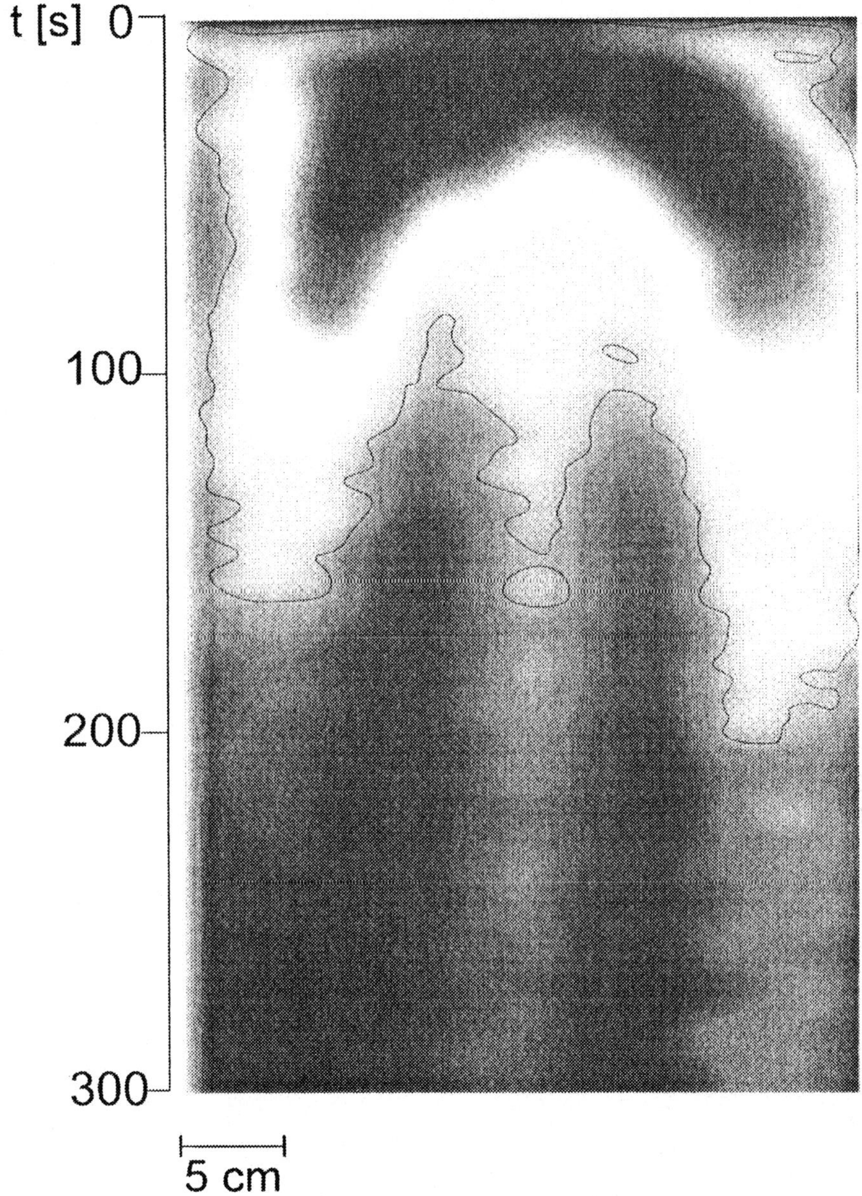

Fig. 3. Grey-scale coded water velocity behind a swimming fish averaged over the columns of the camera field. The axis of ordinates shows the width of the camera field, the abscissa shows the time, which has passed since the fish entered the camera field. The black line indicates a water velocity of 0.2 mms^{-1}. (For further information see Hanke et al. 2000)

In order to create more natural, complex lateral line stimuli, researchers (e.g., Wojtenek et al. 1998) have stimulated fishes with small moving objects. Such objects cause short transient water motions followed by an ill-defined long-lasting wake. If primary lateral line afferents are stimulated with water motions caused by a moving object, they respond with excitation followed by inhibition or vice versa. A given response pattern inverses when object motion direction is reversed. About 70% of all primary afferents discharge many bursts of spikes after an object has passed the fish (Fig. 4A). These bursts are probably caused by the wake of the moving object. About 27% of posterior lateral line nerve fibers do not continue to respond after the object has passed the fish (Mogdans and Bleckmann 1998). Units of the first type most likely receive input from superficial, those of the second type from canal neuromasts.

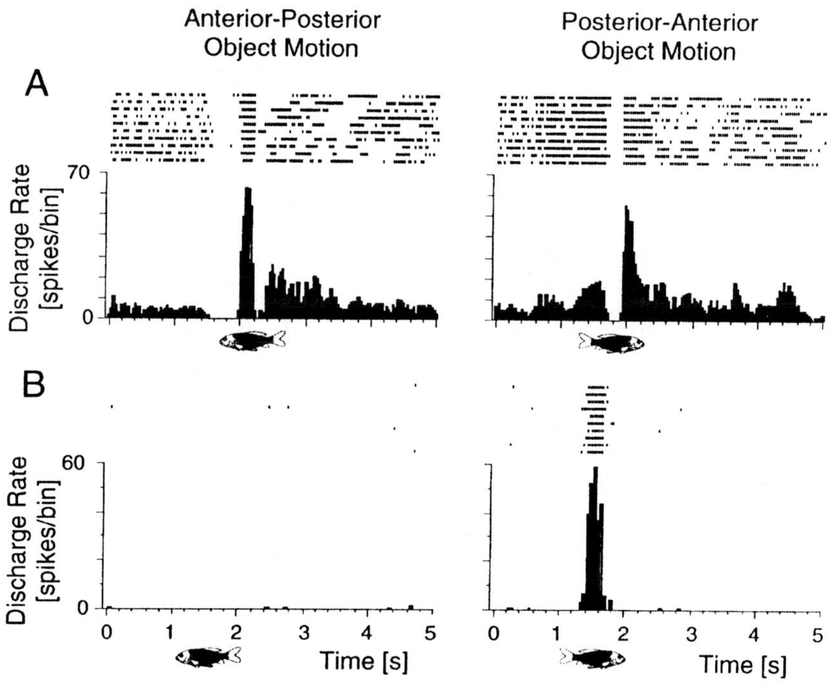

Fig. 4 A, B. Responses of goldfish lateral line units to ten repetitions of an object passing the fish laterally from anterior to posterior (*left*) or from posterior to anterior (*right*). Raster diagrams and the corresponding PST histograms are shown. **A** Responses of a PLLN fiber. Note that a triphasic discharge pattern consisting of reproducible excitatory and inhibitory periods was followed by multiple unpredictable bursts of spikes (object speed 15 cms^{-1}, PSTH binwidth 20 ms). **B** Responses of a unit in the torus semicircularis. Note that the unit responded only when the object passed the fish from posterior to anterior (object speed 12.5 cms^{-1}, PSTH binwidth 50 ms). The *fish drawings* represent location, size and orientation of the fish relative to the path of the moving object

Using moving object stimuli one immediately learns that some higher-order lateral line neurons show very specific response properties. Many units respond to a moving object with only a few spikes (Müller et al. 1996; Wojtenek et al. 1998), some units are inhibited while the object moves by (Mogdans et al. 1997). In terms of spike number, medullary units are usually not directionally sensitive. However, toral units often respond only if the object moves in a certain direction (Fig. 4B). With a dipole stimulus units of this type would not have been discovered. Recent physiological experiments with moving objects indicate that there is a lateral line map in the torus semicircularis of fish. Units in the anterior lateral torus respond while the object moves along the anterior part of the fish. Units at increasingly more posterior locations respond while the object moves along increasingly more caudal parts of the fish (Bleckmann and Zelick 1993, Plachta unpubl.).

8 Behavior

Animals increase their biological success by reacting to subtile aspects of the environment (Krebs and Davis 1991). Besides visual, acoustic, and olfactory cues, lateral line input is the basis for many behavioral decisions. Hydrodynamic input is used, for instance, for prey detection, predator avoidance, intraspecific communication, schooling behavior, object discrimination, and wave source localization (Bleckmann 1994).

Some fish use the dispersion and damping properties of the water surface to determine the distance to a surface wave source, thus indicating that they exploit the complicated physical properties of the water surface for prey localization (Bleckmann et al. 1989). However, in most experiments the behavioral tasks the fish had to fulfill were quite simple, i.e., most behavioral experiments were not designed to elucidate the full behavioral capabilities of fish to discriminate complex water motions under still and running water conditions.

9 Background Noise

In order to fully understand the capabilities of the lateral line we need more behavioral and physiological experiments in flow tanks. Water currents are a dominant and pervasive feature of many aquatic environments. Montgomery et al. (1997) nicely showed that the superficial lateral line system of fishes controls rheotaxis at low current velocities. Engelmann (unpubl.) recorded the responses of primary lateral line afferents of goldfish to a dipole stimulus under both still-water and running water conditions. Under stillwater conditions all units responded highly sensitively and phase-locked to sinusoidal water motions. Without a frequency analysis responses from superficial and canal neuromasts could hardly be distinguished. However, under running water conditions (background water velocity 10

cms^{-1}) two types of units immediately became apparent. Depending on flow direction (anterior-to-posterior vs. posterior-to-anterior), type I units responded to water flow with an increase or decrease of ongoing activity. If the direction of flow was such that ongoing activity was increased, the responses to a vibrating sphere were masked (Fig. 5). The ongoing activity of type II units barely changed in response to water flow. Consequently responses of type II units to a vibrating sphere stimulus were not masked (Fig. 5). There are many indications that type I units innervate superficial and type II units canal neuromasts.

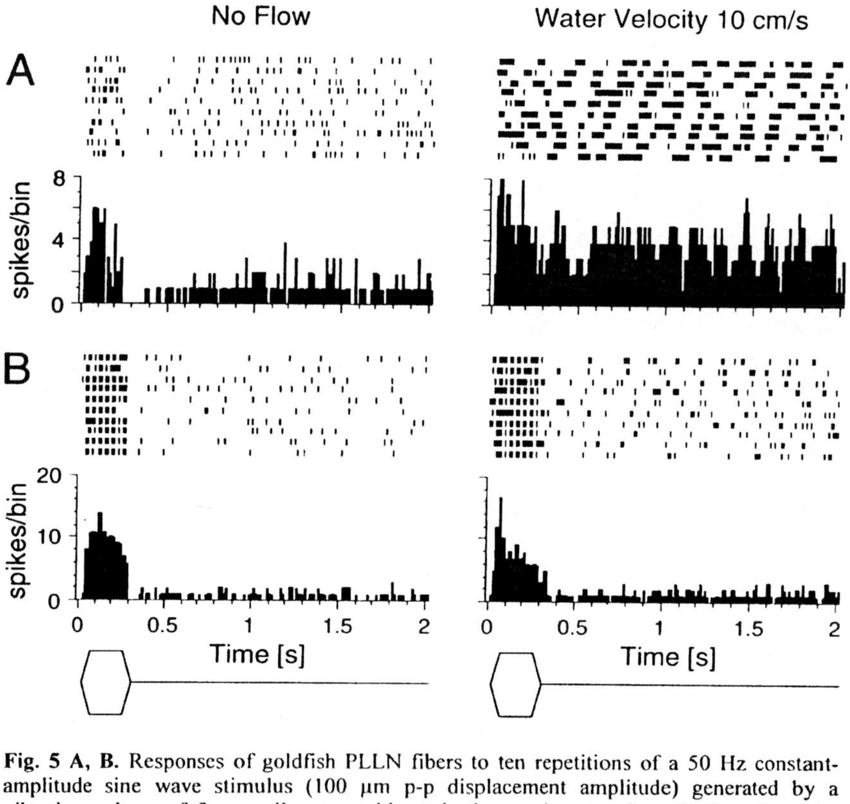

Fig. 5 A, B. Responses of goldfish PLLN fibers to ten repetitions of a 50 Hz constant-amplitude sine wave stimulus (100 μm p-p displacement amplitude) generated by a vibrating sphere of 8 mm diameter without background water flow (*left*) and in the presence of a 10 cms^{-1} DC background flow (*right*). Flow direction was from anterior to posterior. Raster diagrams and the corresponding PST histograms (binwidth 5 ms) are shown. **A** Responses of a type I unit. **B** Responses of a type II unit. Note that the background flow masks the responses of type I units but not those of type II units. (Data kindly provided by J. Engelmann)

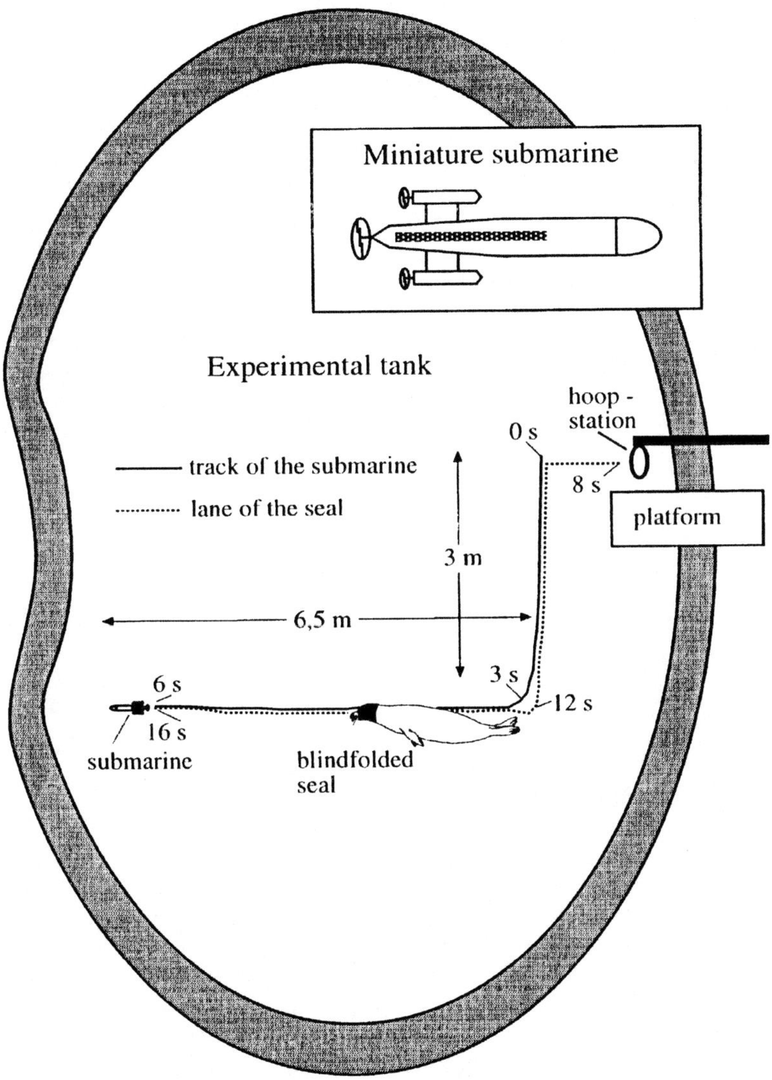

Fig. 6. Experimental design for the study of hydrodynamic trail following in Harbor seals. The *order of numbers* along the trail of the submarine and the lane of the seal reflect the course of a trial. The *inset* shows the miniature submarine driven by one main propeller. The experimental pool was filled with natural seawater that did not allow remote control of the submarine. For this reason two small additional propellers served to change the course

of the submarine. The main propeller of the submarine was started under water via a solenoid switch. Due to the momentum of the submarine, a mercury-inclination switch started one of the small propellers, thus inducing a change of swimming direction. It was unpredictable at what time which of the two small propellers was started during the course of a trial. For further explanations see text

The above experiments clearly show that background flow in combination with dipole stimuli is a powerful tool to analyze lateral line function. Flow tank experiments immediately reveal why running water fish have only a small number of superficial neuromasts (Coombs et al. 1992). In running water DC flow immediately drives superficial neuromasts into saturation and therefore renders them useless. We believe that we will uncover a clear form-function relationship of the various lateral lines only if we study the respective system under hydrodynamic conditions which come close to the hydrodynamic conditions a fish faces in real life.

10 Detection of Fish-Borne Wakes

The idea that hydrodynamic sensory systems like the lateral line are useful for the detection of complex water motions like fish wakes (cf. Fig. 3) is supported by recent experiments with Harbor seals. By means of their facial vibrissae, Harbor seals detect minute water movements (Dehnhardt et al. 1998). Blind seals successfully hunt in the open ocean (Newby et al. 1970). Thus seals can sense, locate, and pursue fish without vision. One potential cue which might guide seals to their prey are the hydrodynamic trails generated by swimming fish (see above).

Recent experiments (Dehnhardt and Mauck, unpubl.) suggest that blindfolded seals can track fish wakes. Blindfolded seals were trained to follow the wake of a miniature submarine (Fig. 6, inset). In order to exclude acoustic cues the seal was stationed with its head in a hoop so that it was outside the water up to its foreflippers. While in the hoop station the seal additionally wore head phones playing white noise. When the seal was stationed the submarine started from a platform in an arbitrary direction and unpredictably changed its direction while moving across the experimental pool (Fig. 6). After 5 s the motor of the submarine turned off and the seal was allowed to start its search 2 seconds after the submarine surfaced (Fig. 6). In 78% of the trials (n = 326) the blindfolded seal successfully tracked the wake. From its hoop station the diving seal always started its search by swimming straight into the pool. As soon as it came across the point where the submarine had started, it immediately changed its swimming direction and exactly followed the hydrodynamic trail. In trials in which the seal missed the submarine's trail, it started its search deeper or shallower than the submarine's starting position. However, when it found the trail, the seal never had a problem to find the submarine. Using vibrating spheres as a stimulus we would have totally underestimated the hydrodynamic reception capabilities of seals. Under natural conditions the detection and decoding of complex historical hydrodynamic signals in a world full of noise may be crucial for the survival of many aquatic animals.

This short review hopefully has illustrated that to understand hydrodynamic reception requires detailed investigations of the physical structure of the hydrodynamic environment in which an animal lives. It is now time to supplement studies with stationary vibrating spheres in a low-noise environment with studies in which dipole stimuli are applied in a noisy environment. In addition, more natural, complex water motions, applied both in low- and high-noise environments, should be introduced in lateral line research. We believe that only if we ask more subtle questions will we have a chance to uncover the full capabilities and adaptations of the lateral line and other hydrodynamic sensory systems.

Acknowledgments. We thank TH Bullock, P Görner and FG Barth for commenting on the MS. Support for research reported here came from the German Science Foundation (Bl 242 7-1; Mo 718/2-1; De 538/5-1) and from the BMBF to JM.

References

Alexandre D, Ghysen A (1999) Somatotopy of the lateral line projection in larval zebrafish. Proc Natl Acad Sci 96:7558-7562

Bartels M, Münz H, Claas B (1990) Representation of lateral line and electrosensory systems in the midbrain of the axolotl, *Ambystoma mexicanum*. J Comp Physiol A 167: 347-356

Bleckmann H, Bullock TH (1989) Central nervous physiology of the lateral line system, with special reference to cartilaginous fishes. In: Coombs S, Görner P, Münz H (eds) The Mechanosensory Lateral Line. Neurobiology and Evolution. New York: Springer, pp. 387-408

Bleckmann H, Tittel G, Blübaum-Gronau E (1989) The lateral line system of surface-feeding fish: Anatomy, physiology, and Behavior. In: Coombs S, Görner P, Münz H (eds) The Mechanosensory Lateral Line. Neurobiology and Evolution. New York, Springer, pp. 501-526

Bleckmann H, Münz H (1990) Physiology of lateral-line mechanoreceptors in a teleost with highly branched, multiple lateral lines. Brain Behav Evol 35: 240-250

Bleckmann H (1994) Reception of hydrodynamic stimuli in aquatic and semi-aquatic animals. In: Rathmayer W (ed) Progress in Zoology, Vol 41. Stuttgart, Gustav Fischer, pp. 1-115

Bleckmann H, Zelick R (1993) The responses of peripheral and central mechanosensory lateral line units of weakly electric fish to moving objects. J Comp Physiol A 172: 115-128

Bodznick D, Schmidt AW (1984) Somatotopy within the medullary electrosensory nucleus of the little skate, *Raja erinacea*. J Comp Neurol 225:581-590

Boord RL, Montgomery JC (1989) Central mechanosensory lateral line centers and pathways among the elasmobranchs. In: Coombs S, Görner P, Münz H (eds) The Mechanosensory Lateral Line. Neurobiology and Evolution. New York, Springer, pp. 323 -340

Bullock TH, Heiligenberg W (1986) Electroreception. New York: John Wiley and Sons, pp. 1-722

Catania KC, Kaas JH (1995) Organization of the somatosensory cortex of the star-nosed mole. J Comp Neurol 351:549-567

Chapman B, Stryker MP, Bonhoeffer T (1996) Development of orientation preference maps in ferret primary visual cortex. J Neurosci 16:6443-6453

Coombs S, Janssen J, Webb JF (1988) Diversity of lateral line systems. Evolutionary and functional considerations. In: Atema J, Fay RR, Popper AN, Tavolga WN (eds) Sensory Biology of Aquatic Animals. New York, Springer, pp. 553-593

Coombs S, Janssen J (1989) Peripheral processing by the lateral line system of the mottled sculpin (*Cottus bairdi*). In: Coombs S, Görner P, Münz H (eds) The Mechanosensory Lateral Line. Neurobiology and Evolution. New York, Springer, pp. 299-319

Coombs S, Janssen J, Montgomery J (1992) Functional and evolutionary implications of peripheral diversity in lateral line systems. In: Webster DB, Fay RR, Popper AN (eds) The Evolutionary Biology of Hearing. New York, Springer, pp. 267-294

Coombs S, Mogdans J, Halstead M, Montgomery J (1998) Transformation of peripheral inputs by the first-order lateral line brainstem nucleus. J Comp Physiol A 182:609-626

Dehnhardt G, Mauck B, Bleckmann H (1998) Seal whiskers detect water movements. Nature 394:235-236

Delcomyn F (1997) Foundations of Neurobiology. New York: Freeman and Company, pp. 1-648

DeYoe EA, van Essen DC (1988) Concurrent processing streams in monkey visual cortex. Trends in Neurosci 11:219-226

Dijkgraaf S (1952) Bau und Funktionen der Seitenorgane und des Ohrlabyrinths bei Fischen. Experientia 8: 205-215

Friedrich RW, Korsching SI (1998) Combinatorial and chemotopic odorant coding in the zebrafish olfactory bulb - visualized by optical imaging. Neuron 18:737-752

Fritzsch, B (1989) Diversity and regression in the amphibian lateral line and electrosensory system. In: Coombs S, Görner P, Münz H (eds) The Mechanosensory Lateral Line. Neurobiology and Evolution. New York, Springer, pp. 99-114

Hanke W, Brücker C, Bleckmann H (2000) The ageing of the low-frequency water disturbances caused by swimming goldfish and its possible relevance to prey detection. J Exp Biol 203: 1193-1200

Harris GG, Bergeijk WA van (1962) Evidence that the lateral line organ responds to near-field displacements of sound sources in water. J Acoust Soc Am 34:1831-1841

Heil P, Langner G, Scheich H (1992) Processing of frequency-modulated stimuli in the chick auditory cortex analogue: Evidence for topographic representations and possible mechanisms of rate and directional sensitivity. J Comp Physiol A 171:583-600

Heil P, Scheich H (1992) Spatial representation of frequency-modulated signals in the tonotopically organized auditory cortex analogue of the chick. J Comp Neurol 322:548-565

Heiligenberg W (1991) The neural basis of behavior: A neuroethological view. Annu Rev Neurosci 14:247-267

Hensel H (1974) Cutaneous thermoreceptors. In: Autrum H, Jung R, Loewenstein WR, MacKay DM (eds) Electroreceptors and Other Specialized Receptors in Lower Vertebrates. New York, Springer, pp. 79-110

Hildebrand JG (1995) Analysis of chemical signals by nervous systems. Proc Acad Nat Sci USA 92:67-74

Kalmijn AJ (1989) Functional evolution of lateral line and inner ear sensory systems. In: Coombs S, Görner P, Münz H (eds) The Mechanosensory Lateral Line. Neurobiology and Evolution. New York, Springer, pp. 187-216

Keller CH (1988) Stimulus discrimination in the diencephalon of *Eigenmannia*: the emergence and sharpening of a sensory filter. J Comp Physiol A 162:747-757

Konishi M (1986) Centrally synthesized maps of sensory space. Trends in Neurosci 100:163-168

Krebs JR, Davies NB (1991) Behavioural Ecology. London, Edinburgh, Boston, Blackwell scientific publications, pp. 1-482

McCormick CA (1989) Central lateral line mechanosensory pathways in bony fish. In: Coombs S, Görner P, Münz H (eds) The Mechanosensory Lateral Line. Neurobiology and Evolution. New York, Springer, pp. 341-364

McCormick CA, Hernandez DV (1996) Connections of the octaval and lateral line nuclei of the medulla in the goldfish, including the cytoarchitecture of the secondary octaval population in goldfish and catfish. Brain Behav Evol 47:113-138

Mogdans J, Bleckmann H, Menger N (1997) Sensitivity of central units in the goldfish, *Carassius auratus*, to transient hydrodynamic stimuli. Brain Behav Evol 50:261-283

Mogdans J, Bleckmann H (1998) Responses of the goldfish trunk lateral line to moving object. J Comp Physiol A 182:659-676

Mogdans J, Goenechea L (1999) Responses of medullary lateral line units in the goldfish, *Carassius auratus*, to sinusoidal and complex wave stimuli. Zoology (in press)

Montgomery J, Coombs S, Halstead M (1995a) Biology of the mechanosensory lateral line in fishes. Rev Fish Bioland Fisheries 5:399-416

Montgomery JC, Coombs S, Conley RA, Bodznick D (1995b) Hindbrain sensory processing in lateral line, electrosensory, and auditory systems: a comparative overview of anatomical and functional similarities. Auditory Neuroscience 1:207-231

Montgomery JC, Baker CF, Carton AG (1997) The lateral line can mediate rheotaxis in fish. Nature 389:960-963

Müller HM, Fleck A, Bleckmann H (1996) The responses of central octavolateralis cells to moving sources. J Comp Physiol A 179:455-471

Münz H (1989) Functional organization of the lateral line periphery. In: Coombs S, Görner P, Münz H (eds) The Mechanosensory Lateral Line. Neurobiology and Evolution. New York, Springer, pp. 285-298

Münz H, Claas B (1991) Activity of lateral line efferents in the axolotl (*Ambystoma mexicanum*). J Comp Physiol A 169:461-469

Newby TC, Hart FM, Arnold RA (1970) Weight and blindness of Harbour seals. J Mammal 51:152

Plachta D, Mogdans J, Bleckmann H (1999) The responses of midbrain lateral line units to amplitude modulated hydrodynamic stimuli. J Comp Physiol A 185: 405-417

Prechtl JC, Emde G von der, Wolfart J, Karamürsel S, Akoev GN, Andrianov YN, Bullock TH (1998) Sensory processing in the pallium of a mormyrid fish. J Neurosci 18: 7381-7393

Roberts BL, Meredith GE (1989) The efferent system. In: Coombs S, Görner P, Münz H (eds) The Mechanosensory Lateral Line. Neurobiology and Evolution. New York, Springer, pp. 445-459

Rolls ET (1984) Neurons in the cortex of the temporal lobe and in the amygdala of the monkey with responses selective for faces. Hum Neurobiol 3:209-222

Schellart NAM, Wubbels RJ (1998) The auditory and mechanosensory lateral line system. In: Evans DH (ed) The Physiology of Fishes. New York, CRC Press, pp. 245-282

Song J, Northcutt RG (1991) The primary projections of the lateral-line nerves of the Florida gar, *Lepisosteus platyrhincus*. Brain Behav Evol 37:38-63

Suga N (1990) Biosonar and neural computation in bats. Sci Amer 34-41

Vanegas H, Williams B, Essayag E (1984) Electrophysiological and behavioral aspects of the teleostean optic tectum. In: Vanegas H (ed) Comparative Neurology of the Optic Tectum. New York, Plenum Press, pp. 121-161

Whaling CS, Solis MM, Doupe AJ, Soha JA, Marler P (1997) Acoustic and neural bases for innate recognition of song. Proc Natl Acad Sci USA 94:12694-12698

Wojtenek W, Mogdans J, Bleckmann H (1998) The responses of midbrain lateral line units of the goldfish *Carassius auratus* to moving objects. Zoology 101:69-82

Wullimann MF (1998) The central nervous system. In: Evans DH (ed) The Physiology of Fishes. New York, CRC Press, pp. 245-282

Zittlau KE, Claas B, Münz H (1986) Directional sensitivity of lateral line units in the clawed toad *Xenopus laevis* Daudin. J Comp Physiol A 158:469-477

IV.1 The Metabolic Cost of Information- a Fundamental Factor in Visual Ecology

Simon B. Laughlin

Department of Zoology, University of Cambridge, Downing St, Cambridge CB2 3EJ, UK

Abstract
To better understand the adaptations of form to function that constitute much of contemporary visual ecology, we examine the costs and the benefits associated with the first stages of vision. Taking the blowfly as a model, the major metabolic costs are the metabolic energy used to support the mass of the eye during locomotion and the metabolic energy used to generate electrical signals. These account for 10% of total metabolic rate at rest, and 3% in flight. This significant contribution to total metabolic rate helps to explain the numerous adaptations that maximize the efficiency with which the components of sensory systems capture and encode information. To further understand these adaptations, metabolic costs are related to sensory benefit by modelling basic optical and biophysical constraints, and by making measurements. The appropriate measure of benefit is the quantity of information coded, in bits. The cost is the consumption of ATP. Cost-benefit relationships follow the law of diminishing returns and this increases the pressure to reduce sense organs to the essential minimum. Ultimately the costs and the uses made of information by individuals should be related to fitness.

Key words Vision, allometry, information theory, cost-benefit, energy

1 Introduction

Sensory ecology is the study of the contributions made by sense organs to ecology and behavior (Lythgoe 1979). These contributions are of crucial importance because sensory information enables animals to make the adaptive physiological and behavioral responses that permit survival and reproduction. It follows that much of sensory ecology concerns adaptations of sense organs and sensory processes that, by extracting and processing the appropriate sensory information, enable a particular species to pursue a particular life-style in a particular habitat. The extent and variety of these adaptations, together with the subtle differences observed between closely related species, suggest that sensory function has a large effect on fitness. Significant costs are being balanced by considerable benefits. We will consider the relationship between cost and benefit in sensory ecology, concentrating on the uptake and coding of sensory signals by eyes. Here the benefits can be assessed by optical measures, resolution and sensitivity, and by the quantity of information that is coded. The costs are assessed in terms of the consumption of metabolic energy. Energy consumption is determined by the size of a sense organ, and by the electrical current used in receptors and neurons to encode and transmit information.

2 Adaptations Suggest Costs and Benefits

Adaptations of form to function are particularly prominent in well-developed sensory systems. For example, the pinnae of bats' ears come in many shapes and sizes (Altringham 1996). The pinnae contribute to the sensitivity, directionality and frequency tuning of hearing. Many interspecific variations in the frequency content and temporal structure of bats' calls are adaptations that improve echolocation in particular habitats (Neuweiler 1989). Thus the variety of pinnae ultimately relate to sensory ecology. The benefits of larger and more highly sculpted pinnae, higher sensitivity and better resolution, are balanced by costs, such as the drag induced in flight. Costs will tend to reduce the size of pinnae.

Eyes vary widely in design, shape and size (Walls 1942). Anatomical resolving power, defined as the reciprocal of the angular separation of photoreceptors, increases with the length or radius of the eye. Kirschfeld (1976) examined the relationship between anatomical resolving power and body height. Taking 18 species with simple or compound eyes, ranging in size from *Drosophila* to human, the anatomical resolving power increases by 3 orders of magnitude, roughly in proportion to body height. A resolution proportional to body height provides a constant subjective resolution because a given object on the ground subtends the same number of photoreceptors. Deviations from this scaling relationship, such as the enlarged eyes of jumping spiders and falcons and the reduced eyes of bats, relate directly to sensory ecology (Kirschfeld 1976).

The scaling of eye to body height for constant subjective resolution is not seen in better-defined taxonomic groups. Across the vertebrates, including birds, fish, reptiles and mammals from shrew to whale, the ratio between eye length and body length decreases as size increases (Hughes 1977).

The largest vertebrate eyes have axial lengths of 35 mm to 50 mm but occur in animals of very different size, such as the ostrich, the eagle, the zebra, the giraffe, the elephant and the baleen whale. Hughes (1977) observes "given the nature of the elements common to its construction and the physical world to which it is adapted, a globe from 35 to 50 mm diameter suffices to meet the most stringent requirements of species over a 10^4 – fold range of body size".

Could Leuckart's Law, swifter-moving species have larger eyes, determine the scaling of eye to body (Brooke et al. 1999)? Faster moving animals need to resolve objects further ahead and if resolution increased in proportion to speed, objects would be resolved at a constant time to contact. On average, larger animals move fewer body lengths per second and have eyes that are a smaller fraction of body length. However, the hypothesis that anatomical resolving power rises in proportion to speed does not hold in birds (Brooke et al. 1999).

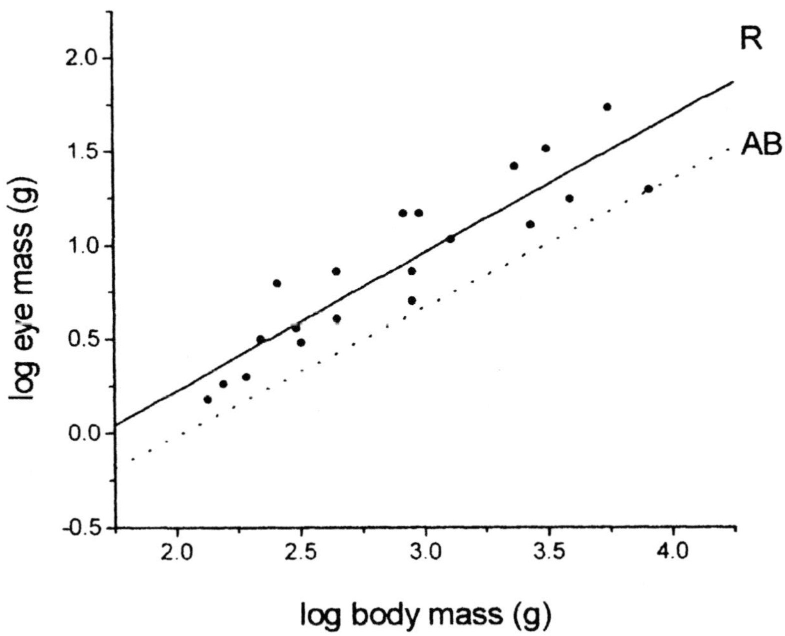

Fig. 1 A. Scaling relationships in eyes. The allometric relationship for eye mass on body mass for raptors, *R*, compared to the average relationship for birds, *AB*. The line for raptors, *R*, was fitted to the data points, which are from 20 species. The average bird line, *AB*, was fitted to data from 104 families (data points not shown). The graph is replotted from Brooke et al. (1999).

The allometric relationship between eye mass (the cube of eye length) and body mass exhibits an exponent of 0.68 (Fig. 1). Given the scaling relationship between flight speed and body mass, an exponent of 0.5 is required to scale resolution to speed. The scatter of points on allometric plots (Fig. 1) shows that species of similar weight differ in eye size. These differences suggest species-specific compromises between investment in the visual system and investment in other body parts. Some of the reasons for increasing the investment in eyes are obvious. Raptors, with their need for high acuity, and owls, with their need for high sensitivity, have eyes that are 1.4 (Fig. 1) and 2.2 times larger than the average for birds of the same body weight. Why are many eyes smaller than average? What factors have promoted reductions in eye size and can these factors be quantified?

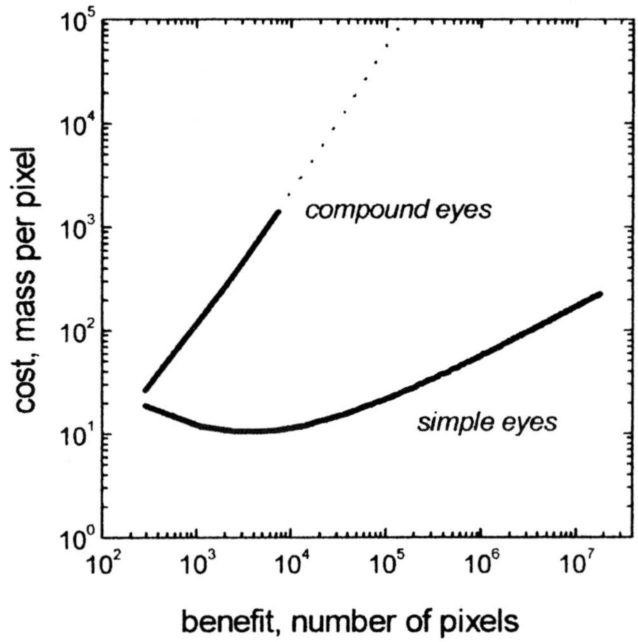

Fig. 1 B. Scaling relationships in eyes. The cost of pictorial information, measured as eye mass per pixel, rises with the benefit of the eye, taken as the total number of pixels covering a unit solid angle. Note that simple eyes provide a given benefit at much lower cost than apposition compound eyes. Further details are given in the text, and in Laughlin (1998), from which this graph has been adapted

The consumption of metabolic energy is an obvious candidate (Land 1981) and it is thought to promote the reduction of vision and visual centres in subterranean animals, such as the mole rat (Nevo 1998). The metabolic cost of sensation includes the energy required to build, maintain and carry a sensory system, and as we shall see, the energy need to power its receptors and neurons. The costs of building and maintenance increase in proportion to eye mass and, when eyes contribute relatively little to air or water resistance, so does the cost of carrying the eye. The cost of carriage approximates the total rate of consumption of metabolic energy required for locomotion multiplied by the fraction of body mass that is eye. Given that the costs of building, maintaining and carrying the eye are proportional to eye mass, we can examine relationships between costs and benefits.

3 Costs and Benefits in Image Formation

Take a simple (single-lens) eye. Define its cost as its mass, and its benefit as the quantity of information that it captures (Laughlin 1998). Each photoreceptor is one pixel on the retinal image. The information captured depends upon the number of photoreceptors (pixels) in the retina and the signal-to-noise ratio (SNR) in each pixel. Now make two simplifying assumptions. (1) All eyes have the same F-ratio (focal length/lens aperture), irrespective of size. (2) The number of receptors increases in proportion to focal length squared, to maintain retinal coverage, but photoreceptor length and diameter are held constant. Because these assumptions tend to maintain the SNR in photoreceptors, independent of eye size, the information gained increases in proportion to the number of photoreceptors on the retina, as $(\text{eye length})^2$. Metabolic cost increases with mass, as $(\text{eye length})^3$. Therefore the unit cost of information, the cost per pixel of the image, rises with eye size, and hence with the total number of pixels coded per unit area (Fig. 1).

The shallow optimum, a minimum cost per pixel, is for small eyes containing a few hundred pixels and is due to a tradeoff between investment in image quality (focal length and lens aperture) and investment in photoreceptors. We conclude that basic optical principles dictate that the unit cost of information (the cost per pixel) increases with eye size and hence with the total amount of information coded (the total number of pixels). In other words, this optical cost-benefit relationship follows the law of diminishing returns. A similar analysis has been applied to apposition compound eyes (Laughlin 1998), and its result draws attention to the hopeless inefficiency of this widespread optical mechanism (Fig. 1). The law of diminishing returns increases the pressure to minimize investment because the unit cost is rising with total expenditure.

4 Efficiency and Cost

Costly mechanisms should be adapted for efficient usage. Indeed, the assumption that improvements in efficiency promote mechanisms that perform better underpins adaptationist explanations of sensory structure and function. Adaptations that maximize the effectiveness of structures are common-place in vision (Walls 1942). Such excellent engineering is made possible because biological systems are highly adaptable. The variety and variability of components (molecules, cells, and cell assemblies), combine with the specificity and spatial precision of developmental mechanisms. Together they enable organisms to generate subtle and ingenious solutions to the practical problems of collecting physical and chemical energy, transducing this energy, and gaining information from it.

Engineering for high performance is illustrated by two examples from optics. The nocturnal spider *Dinopis* depends upon night vision to locate its prey. *Dinopis* maximizes the sensitivity of its photoreceptors by using an exceptionally powerful lens, with F-ratio 0.58, to concentrate the image on its retina. The severe spherical aberrations of the lens are reduced by constructing a refractive index gradient within its bulk (Blest and Land 1977). Similar gradients reduce aberrations in mammalian lenses (Hughes 1977). The second example is the deep pit found in the retinas of raptors, chameleons and jumping spiders. Each of these animals requires high spatial resolution to locate prey. To maximize the anatomical resolving power of photoreceptors, the eye is made as large as possible and spans the relevant body cavity. The pit has a similar profile in each animal because it is constructed to form a diverging lens (Williams and McIntyre 1980). This lens increases spatial resolution beyond the limit set by eye length, by converting the eye into a telephoto system.

Faced with numerous examples of this type (Walls 1942; Hughes 1977; Land 1981; Martin 1983; Nilsson 1989), we must agree with Charles Darwin that many eyes are "organs of extreme perfection and complication" (Darwin 1859). Perfection points to the selective advantage of good vision, and suggests that natural selection is favouring organisms that operate efficiently within physical, biophysical and developmental constraints, to make the best use of their resources. Perfection is not limited to optics, it extends to neural coding.

5 The Optimization of Neural Coding

Neural coding in the retina provides a compelling example of the efficient use of a limited resource. A given retina contains a certain number of photoreceptors and neurons, each with a limited ability to transmit information. A cell's information capacity is ultimately limited by the number of different messages that it can transmit in a given time, as determined by two factors, the number of discriminable signal levels, and the rate at which the cell can change these levels. In electrical signalling both of these factors are constrained by elementary biophysics (Laughlin 1989). The maximum number of discriminable levels is the cell's response range divided by the amplitude of the noise. Noise is always present because it is generated by the elementary molecular events that produce and process cell signals (e.g. photon absorption, synaptic vesicle release, binding and dissociation of transmitter, channel opening). The rate at which receptors and neurons can change signal level is limited by the membrane time constant.

Barlow (1961) proposed that the goal of early visual processing is to fill the limited information capacity of retinal neurons by coding signals efficiently. Its neat packaging of information requires the removal of redundancy from the incoming signals. Redundant components convey no new information because they repeat, or can be predicted from, other parts of the data. Because natural objects are not constructed at random, they have predictable parts and this introduces redundancy into natural signals. Lateral inhibition and transient responses, common processes in sensory systems, remove redundancy by emphasizing the differences between neighbouring signals (Barlow 1961).

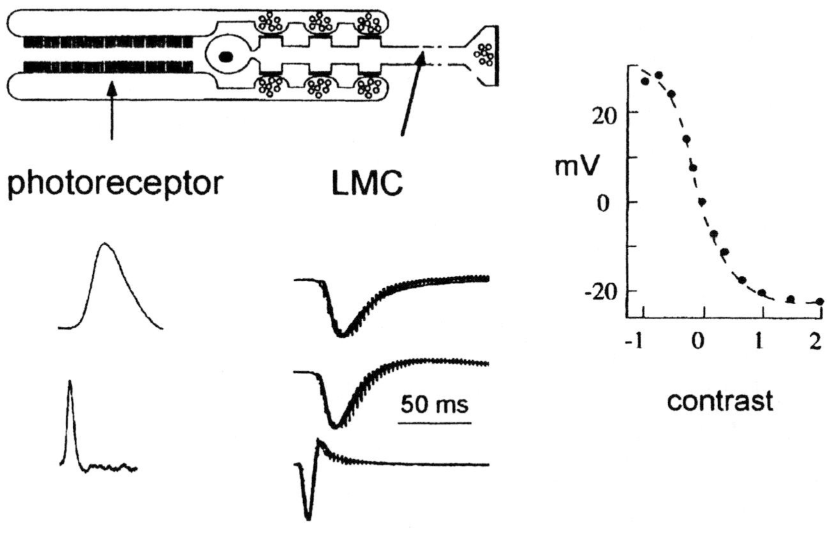

Fig. 2. The optimization of neural coding in the blowfly compound eye. Intracellular responses to flashes of light recorded from a photoreceptor and a large monopolar cell (LMC) when dark-adapted (*top trace*), fully light-adapted (*bottom trace*) and weakly light-adapted (*middle trace*, shown for LMC only). For the LMC waveforms the *continuous lines* are the responses of the optimum encoder, predicted from information theory, and the *vertical bars* show the mean and S.D. of responses recorded from cells (van Hateren 1992a). The graph plots the relationship between mV response amplitude and stimulus contrast in an LMC. The response (*data points*) follows the optimum encoder (*dashed curve*) predicted from information theory (Laughlin 1981). (After Laughlin 1998)

Information theory (Shannon 1949) has been used to determine the efficiency with which natural signals are coded (Laughlin 1981; Srinivasan et al. 1982; Atick 1992; van Hateren 1992b; Rieke et al. 1997). Given a signal of a certain shape (e.g. waveform or spectrum), and reliability (signal-to-noise ratio) and a set of neural constraints on coding (e.g. the number of signal levels, neural noise), information theory predicts the optimum encoder, defined as the set of operations that transfers the greatest quantity of information from image to neuron. The shapes of signals and the form of redundancy are measured in terms of the statistical properties of natural images (probability distributions, correlation functions, power spectra). Signal-to-noise ratios are derived from calibrations of photon flux, and by direct recordings from photoreceptors and neurons. This data on signal and noise is fed into the theory to derive the optimum encoder. The relevant properties of the neurons are measured and compared with the optimum.

The compound eye of the blowfly demonstrates the high degree of coding efficiency achieved in eyes. Each facet of the compound eye defines a single pixel in the retinal image. A precisely defined group of photoreceptors and interneurons is devoted to coding the relative light intensity, the contrast, at each pixel. This neural module consists of six identical photoreceptors, R1-6, and two postsynaptic interneurons, the large monopolar cells (LMCs) L1 and L2 (Fig. 2). Contrast is coded by the amplitudes of the analogue voltage signals in photoreceptors and LMCs. The relationship between response amplitude and contrast (Fig. 2) is matched to the statistical distribution of contrast in natural scenes (Laughlin 1981) to ensure that all of the LMCs response levels are used equally often. This matching maximizes the information transmitted.

Both the spatial and the temporal filtering of signals are optimized. Van Hateren (1992a,b) defined the statistics of natural input signals encountered at one pixel of a blowfly compound eye, using the power spectrum of natural images, the range of velocities at which images move, the fly's optics and the rate at which photoreceptors absorb photons. He then used information theory to predict the response of the optimum encoder. This is a filter that manipulates the natural inputs so as to maximize the information coded by an LMC, within the limitations imposed by the two neural constraints, response range and neural noise. The correspondence between the responses recorded from LMCs and the response of the optimum filter is remarkable (Fig. 2). To remain optimal, the LMC response waveform changes with the background light level. At low light levels the response is slow and monophasic, to average out photon noise (random fluctuations that result from the Poisson statistics of photon absorption). At the highest light levels the optimum response is fast and biphasic, to pick out rapid changes and reject redundancy in time, just as lateral inhibition removes redundancy in space. Between these two extremes, the LMC response changes gradually from one form to the other, maintaining the optimum over an intensity range of 10^5. This virtuoso performance in optimum tuning is a concerted effort, brought about by an increase in the speed of photoreceptor response with light adaptation, and adjustments to the transfer function of the photoreceptor – LMC synapses (Laughlin 1998). The human visual system behaves in a similar manner, to optimize information uptake within the constraints of natural image statistics and neural information capacity (Atick 1992).

This precise optimization of response strongly supports Barlow's proposal that the goal of retinal processing is to maximize the information coded by neurons; but is this goal relevant? Information theory treats all bits of information as units of equal value. This disregard of context and meaning apparently contradicts the purpose of sensory processing, namely the assignment of biological significance to incoming data. However, one must use a context-independent measure, such as the bit, when one is assessing the overall quality of a large data set from which many different meanings will be extracted. The visual system interrogates the retinal image to discern shape, depth, motion, color, and polarization, under a variety of lighting conditions, and with objects at different angles and distances.

Note that efficient coding involves a careful match between sensitivity and signal. Systems that code efficiently are adhering to one of the oldest principles in sensory ecology: they place their receptors and neural machinery where they expect the signals to be (Laughlin 1983). For example, bats and electric fish tune their receptors to the signals that they use for communication and active sense. Moreover, many sensory systems take account of redundant and predictable components; e.g. efference copy exploits the predictable outcomes of releasing energy into the environment (Bell et al. 1993). These examples reinforce the proposition that, because the resources available for sensory processing are limited, Natural Selection favours the individuals who use their resources efficiently so as to improve the ratio between benefit and cost. We can quantify one cost-benefit relationship, the metabolic energy required to code information.

6 The Metabolic Cost of Neural Information

Measurements from the blowfly compound eye suggest fundamental relationships between energy and information, of the type seen in telecommunications and computation, and first suggested by Szilard's analysis of Maxwell's demon (Laughlin et al. 1998). The analogue electrical responses of photoreceptors and LMCs are analyzed to determine the rates at which they transmit information about the stimulus contrast at their pixel. From the biophysical properties of the photoreceptor and LMC one calculates the current required to generate the electrical responses (Fig. 3), and this gives the metabolic cost.

The information transmission rates were measured by driving photoreceptors and LMCs with random stimuli of natural contrast and daylight intensity (de Ruyter van Steveninck and Laughlin 1996). Signal and noise were extracted from responses and converted into their respective power spectra, *S(f)* and *N(f)*, where f is frequency in Hz. Because signal and noise are Gaussian, the rate at which information transmitted, I, in bits s^{-1}, is given by

$$I = \int_0^{b} \log_2[1 + S(f) / N(f)] df$$

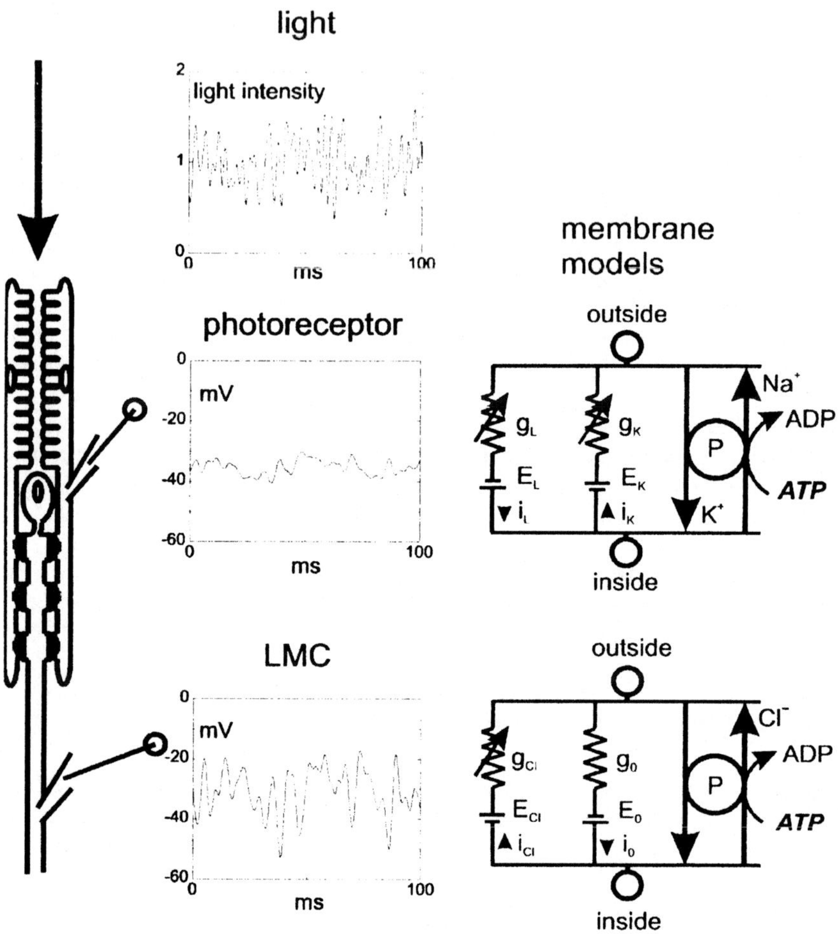

Fig. 3. The signals and circuits that define the metabolic cost of information in the blowfly compound eye (Laughlin et al. 1998). The *traces* show a brief segment of the random modulation of the light stimulus, and the responses recorded from a photoreceptor and an LMC. These responses were analyzed to give the information transmission rate in bits s^{-1}. The membrane models were constructed and solved using biophysical data to give the metabolic cost, as the rate of hydrolysis of ATP to ADP by a Na^+/K^+ or Cl^- pump. *Symbols: i* current; *E* reversal potential of ionic current; *g* ionic conductance; *P* pump. *Subscripts: L* light-gated; *K* potassium; *Cl* chloride; *0* ionic species with zero reversal potential. (After Laughlin et al. 1998)

This classic formula (Shannon 1949) shows how the information transmission rate increases in proportion to bandwidth *b*, (the frequency range from zero to the highest frequency transmitted, *b*) and as the log of the frequency dependent SNR, $S(f)/N(f)$. (For a clear introduction to information theory, in the context of neural coding, consult Rieke et al. 1997). Applying this formula to signal and noise power spectra, a photoreceptor transmits 1000 bits s^{-1} and the postsynaptic LMC 1600 bits s^{-1}. The LMC transmits at a higher rate because the convergence of signals from six photoreceptors has increased the SNR. Serial EM reconstruction shows that each photoreceptor drives an LMC through 220 identical chemical synapses (Nicol and Meinertzhagen 1982). Transmission across this massive parallel array of synapses improves the SNR because the synaptic noise (noise generated independently in each synapse) is reduced by averaging over many synapses. From the bit rates in pre- and postsynaptic cells, and the numbers of synapses involved, each synapse transmits 55 bits s^{-1}.

The metabolic cost of the information transmitted by photoreceptors, LMCs and synapses is calculated from the electrical current used to produce voltage responses (Laughlin et al. 1998). To estimate currents one constructs membrane models from biophysical data. These electrical circuits (Fig. 3) incorporate the ionic batteries that drive current, the conductances through which ionic currents flow, and the pumps that keep the batteries charged by maintaining the ionic concentration gradients. The circuits (Fig. 3) are solved, using appropriate measurements of total membrane conductance and membrane potential to give the flux of ions across the membrane and the rates at which pumps must hydrolyze ATP to restore these ions and maintain the concentration gradients that drive current.

Dividing the ATP hydrolysis rate by the bit rate gives a cost of 7×10^6 ATP molecules per bit in a photoreceptor and approximately 2×10^6 ATP/bit in an LMC. The LMC value is lower because efficient coding cuts costs. Redundant components are removed when the signal is transferred from photoreceptor to LMC. These components cost ATP to generate, but carry no information. It costs 50 times less, about 4×10^4 ATP molecules, to transmit a bit across a single photoreceptor-LMC synapse. The reasons for this lower cost are instructive. Because it is generated by 1320 identical synapses (220 from each of six photoreceptors), the cost of the LMC signal is 1320 times the cost at a single synapse. However, the rate at which information is transmitted by the LMC is only 29 times the rate at a single synapse. The massive shortfall in LMC bit rate results from massive redundancy – each of the 1320 synapses is carrying the same signal. This redundancy is the inevitable consequence of using parallel synapses to reduce the effects of synaptic noise. Each synapse severely contaminates the signal with noise, hence its low rate of 55 bits s^{-1} (de Ruyter van Steveninck and Laughlin 1996). To rid the system of this synaptic noise, the signal is transmitted through parallel synapses. This improves the SNR by $\sqrt{1320}$, and raises the information transmission rate to 1600 bits s^{-1}. Essentially, the total cost of transmission is increasing in proportion to the number of synapses, but the rate of transmission of information is increasing as the log of SNR. Thus when parallel channels (in this case synapses) are used to increase the bit rate, the cost per bit also increases.

The array of photoreceptor-LMC synapses typifies any neural system that has to improve its information rate by increasing the number of unreliable components (e.g. transmitter molecules, ion channels, synapses) used to transmit the signal (Laughlin et al. 1998). As commonly observed in communications engineering, increasing the rate at which information is transmitted also increases the cost per bit. Here is another law of diminishing returns that heightens the selective pressure to reduce the unit cost by reducing performance to the lowest acceptable level. However, the metabolic costs of capturing and coding information will only influence the form and function of sense organs and neural codes when these costs make a significant contribution to an animal's total energy budget. Let us examine this contribution.

7 The Contribution of Retinal Coding to Metabolic Rate

Multiplying the rate of ATP hydrolysis per photoreceptor by the total number of photoreceptors in the blowfly eye gives the total photoreceptor consumption. Allowing for mitochondrial proton leak (Rolfe and Brown 1997), approximately five ATP molecules are generated for each oxygen molecule consumed. The ATP consumed by photoreceptors in daylight converts to 11% of the total oxygen used by a resting blowfly. Adding the photoreceptor-LMC synapses, 13% of resting energy consumption is devoted to electrical signalling in the two compound eyes. This large retinal electricity bill is a good reason to maximize the information gained from electrical signals by optimizing all components of coding and processing. The high level of consumption by photoreceptors has also led to energy saving adaptations. The study of these adaptations takes visual ecology into the time domain.

8 Dynamics, Cost and a Visual Ecology in Time

Insect compound eyes exhibit differences in response dynamics that economize photoreceptor energy consumption. In the majority of retinas, invertebrate and vertebrate, phototransduction accelerates with lightadaptation. In blowfly this acceleration is carefully adjusted to optimize performance, by converting the response from slow to brisk (Fig. 2). However, looking across the Diptera, light adaptation accelerates the responses of some species more than others. The photoreceptors of tipulids (craneflies) scarcely speed up at all (Laughlin and Weckström 1993). Response dynamics are matched to visual ecology because, as previously observed (Autrum 1958), it is the slower-flying insects that have slower responses.

Fast responses are metabolically expensive because they require a membrane with a short time constant. The membranes of fast photoreceptors contain delayed rectifier potassium conductances that adapt the membrane for a fast response by lowering the membrane resistance, and hence time constant. This low resistance draws large currents. The membranes of slow photoreceptors are specialized to reduce current (Weckström andLaughlin 1995). Unlike the delayed rectifiers of fast cells, the slow cell potassium conductance responds transiently because it inactivates. Inactivation reduces current flow and, by increasing resistance, increases the membrane time constant. The longer time constant smooths out rapid fluctuations in membrane voltage, but unlike fast cells, where this smoothing would destroy fast signals, in slow cells this smoothing is advantageous (Laughlin 1996). In an insect that moves slowly and views the world through poor optics, a rapid change in receptor response is almost certainly noise.

Fast and slow photoreceptors exemplify visual ecology in the time domain. The dynamics of vision are matched to the dynamics of natural signals, as dictated by life style and habitat, to strike appropriate balances between costs and benefits. Outside of insects, comparative data is conspicuously absent, but the matching of dynamics to life style might be widespread. Birds have higher flicker fusion frequencies than other vertebrates (pigeon 140 Hz; chicken 100 Hz – cited in Laughlin and Weckström 1993) and there are suggestive differences among fish (Lythgoe 1979). Matching occurs at higher levels of visual processing. The motion sensitive neurons of hovering insects (e.g. hawkmoths and hoverflies), respond better at lower velocities than those of insects whose flight is less well stabilised (e.g. butterflies, blowflies and bees). Sensitivity to lower velocities is achieved with longer time delays in motion detectors (O'Carroll et al. 1997). Warming the retina could match response dynamics to visual ecology (Block 1986). Fast moving fish (swordfish, tuna) heat their retinas and fast moving insects (dragonflies, honeybees) heat their heads (Heinrich 1993). Heating improves the temporal resolving power of photoreceptors (Tatler et al. 2000).

The significance of economizing on photoreceptor energy usage is emphasized by the degree to which photoreceptor response dynamics are fine-tuned. Adaptations of response dynamics occur among photoreceptors of the same spectral and morphological class, within one species. There are examples of a sexual dimorphism of photoreceptor voltage response. In bibionids and in the housefly *Musca*, the male has an eye region that is enlarged to detect, follow, intercept and hopefully mate with females (Land 1997). The photoreceptors in this male region have faster responses which, when combined with their better optics, improves the resolution of small and rapidly moving targets (Laughlin and Weckström 1993; Hornstein et al. 2000). In the retina of the male blowfly the photoreceptors at the front of the eye have the fastest response and this is associated with tracking (Burton et al. 2000). These variations suggest that, to economize on energy usage, response dynamics are matched to the spatial and temporal qualities of images and to behavioral requirements.

9 Comparing Retinal Costs

As we have seen (Sect. 7), electrical coding by photoreceptors and LMCs accounts for 13% of the blowfly's resting metabolic rate. When the blowfly takes off the total metabolic rate increases 30-fold and the electrical cost is now <0.4%. However, in flight the cost of carrying the eye is approximately 3% of total metabolic rate (the eye is 3% of body weight). Thus signalling costs dominate at rest and carriage dominates in flight (Laughlin 1995; O'Carroll 1998). Because signalling and carriage contribute significantly to total metabolic rate, it is advantageous to maximize efficiency and minimize mass. Dragonflies have the largest apposition compound eyes and they reduce the mass of the retina with large air-filled tracheae. In addition to catering for energy supply, these tracheae are saving energy.

10 Energy, Information and Fitness

The subject of sensory ecology, the contributions made by sense organs to ecology and behavior (Lythgoe 1979), exists because these contributions have promoted survival by helping species to compete for, and occupy, favourable niches. As forcefully argued by Chittka (this Vol.) and by evolutionary physiologists (Garland and Carter 1994), the ultimate measure of the benefit of a set of physiological mechanisms is their contribution to fitness (Laughlin 1998). To establish the relationship between sensation and fitness we must move towards experiments in which we manipulate sensory function and observe the effects on lifetime reproductive success. The approach set out in this chapter, the analysis of costs and benefits in terms of the metabolic energy required to achieve a certain level of performance, is a useful step. When studying the effects of sensation on fitness one should manipulate systems in which sensory information is costly, is sensitive to metabolic investment, and contributes directly to reproductive success.

References

Altringham JD (1996) Bats, Biology and Behaviour. Oxford University Press, Oxford

Atick JJ (1992) Could information-theory provide an ecological theory of sensory processing. Network-Computation In Neural Systems 3: 213-251

Autrum H (1958) Electrophysiological analysis of the visual systems of insects. Exp Cell Res Suppl 5: 426-439

Barlow HB (1961) Possible principles underlying the transformation of sensory messages. In: Rosenblith WA (ed) Sensory Communication. MIT Press, Cambridge, Mass., pp 217-234

Bell CC, Hopkins CD, Grant K (1993) Contributions of electrosensory systems to neurobiology and neuroethology - Proceedings of a conference in honor of the scientific career of Thomas Szabo. J Comp Physiol A 173: 657-763

Blest AD, Land MF (1977) The physiological optics of *Dinopis subrufus* L. Koch: a fish-lens in a spider. Proc R Soc Lond B 196: 197-222

Block BA (1986) Structure of the brain and eye heater tissue in marlins, sailfish, and spearfishes. J Morphol 190: 169-189

Brooke MD, Hanley S, Laughlin SB (1999) The scaling of eye size with body mass in birds. Proc R Soc Lond B 266: 405-412

Burton B, Tatler B, Laughlin SB (2000) Photoreceptor response dynamics vary with retinal position. Neuron (submitted)

Darwin C (1859) The Origin of Species by Means of Natural Selection. Murray, London

de Ruyter van Steveninck RR, Laughlin SB (1996) The rate of information-transfer at graded-potential synapses. Nature 379: 642-645

Garland T, Carter PA (1994) Evolutionary physiology. Ann Rev Physiol 56: 579-621

Heinrich B (1993) The Hot-Blooded Insects. Springer, Berlin, Heidelberg, NewYork

Hornstein EP, O'Carroll DC, Anderson JC, Laughlin SB (2000) Sexual dimorphism matches photoreceptor performance to behavioral requirements. Proc R Soc B Lond B (submitted)

Hughes A (1977) The topography of vision in mammals of contrasting life style: comparative optics and retinal organisation. In: Crescitelli F (ed) Handbook of Sensory Physiology, Vol. VII/5. Springer, Berlin, Heidelberg, New York, pp 613-756

Kirschfeld K (1976) The resolution of lens and compound eyes. In: Zettler F, Weiler R (eds) Neural Principles in Vision. Springer, Berlin, Heidelberg, New York, pp 354 - 370

Land MF (1981) Optics and vision in invertebrates. In: Autrum H (ed) Comparative Physiology and Evolution of Vision in Invertebrates. B: Invertebrate Visual Centers and Behaviour I. Springer, Berlin, Heidelberg, New York, pp 471-592

Land MF (1997) Visual acuity in insects. Annual Rev of Entomol 42: 147-177

Laughlin SB (1981) A simple coding procedure enhances a neuron's information capacity. Z Naturforsch 36c: 910 - 912

Laughlin SB (1983) Matching coding to scenes to enhance efficiency. In: Braddick OJ, Sleigh AC (eds) Physical and Biological Processing of Images. Springer, Berlin, Heidelberg, NewYork, pp 42-52

Laughlin SB (1989) The reliability of single neurons and circuit design: a case study. In: Durbin R, Miall C, Mitchison G (eds) The Computing Neuron. Addison-Wesley, New York, pp 322 - 336
Laughlin SB (1995). Towards the cost of seeing. Nervous Systems and Behaviour, 4th International Congress of Neuroethology, Georg Thieme, Cambridge
Laughlin SB (1996) Matched filtering by a photoreceptor membrane. Vision Res 36: 1529-1541
Laughlin SB (1998) Observing design with compound eyes. In: Weibel ER, Taylor CR, Bolis L (eds) Principles of Animal Design. Cambridge University Press, Cambridge, pp 278-287
Laughlin SB, de Ruyter van Steveninck RR, Anderson JC (1998) The metabolic cost of neural information. Nature Neurosci 1: 36-41
Laughlin SB, Weckström M (1993) Fast and slow photoreceptors - a comparative study of the functional diversity of coding and conductances in the Diptera. J Comp Physiol A 172: 593-609
Lythgoe JN (1979) The Ecology of Vision. Clarendon Press, Oxford
Martin GR (1983) Schematic eye models in vertebrates. Prog Sensory Physiol 4: 43-81
Neuweiler G (1989) Foraging ecology and audition in echolocating bats. Trends in Ecol Evol 4: 160-166
Nevo E (1998) Evolution of a visual system for life without light: optimization versus tinkering in blind mole rats. In: Weibel ER, Taylor RC, Bolis L (eds) Principles of Animal Design. Cambridge University Press, Cambridge, pp 288-298
Nicol D, Meinertzhagen IA (1982) An analysis of the number and composition of the synaptic populations formed by photoreceptors of the fly. J Comp Neurol 207: 29-44
Nilsson DE (1989) Vision, optics and evolution - Nature's engineering has produced astonishing diversity in eye design. Bioscience 39: 298-307
O'Carroll DC (1998) The cost of seeing. In: Elsner N, Wehner R (eds) New Neuroethology on the Move. Proceedings of the 26th Goettingen Neurobiology Conference. Georg Thieme, Stuttgart, p 150
O'Carroll DC, Laughlin SB, Bidwell NJ, Harris RA (1997) Spatio-temporal properties of motion detectors matched to low image velocities in hovering insects. Vision Res 37: 3427-3439
Rieke F, Warland D, de Ruyter van Steveninck RD, Bialek W (1997) Spikes - exploring the neural code. MIT Press, Cambridge, Mass.
Rolfe DFS, Brown GC (1997) Cellular energy utilization and molecular origin of standard metabolic rate in mammals. Physiol Rev 77: 731-758
Shannon CE (1949) Communication in the presence of noise. Proc Inst Radio Eng 37: 10-21
Srinivasan MV, Laughlin SB, Dubs A (1982) Predictive coding - a fresh view of inhibition in the retina. Proc R Soc Lond B 216: 427-459

Tatler B, O'Carroll DC, Laughlin SB (2000) Temperature and the temporal resolving power of fly photoreceptors. J Comp Physiol A 186: 399-407

van Hateren JH (1992a) Real and optimal neural images in early vision. Nature 360: 68-70

van Hateren JH (1992b) Theoretical predictions of spatiotemporal receptive-fields of fly LMCs, and experimental validation. J Comp Physiol A 171: 157-170

Walls GL (1942) The Vertebrate Eye and its Adaptive Radiation. Cranbrook Institute of Science, Bloomfield Hills

Weckström M, Laughlin SB (1995) Visual ecology and voltage-gated ion channels in insect photoreceptors. Trends Neurosci 18: 17-21

Williams DS, McIntyre P (1980) The principal eyes of a jumping spider have a telephoto component. Nature 288: 578-580

IV.2 The Design of Compound Eyes and the Illumination of Natural Habitats

Eric J. Warrant

Department of Zoology, University of Lund, Helgonavägen 3, S-22362 Lund, Sweden

Abstract
Compound eyes are the most abundant eye design in the animal kingdom, and probably its most adaptable. In this chapter I describe the ways in which compound eyes have evolved in response to the intensity and direction of natural illumination. Species active in bright light generally have no need for compound eyes of high sensitivity. Instead, their eyes frequently possess acute zones, regions where spatial resolution is greatly enhanced, and where the spatial layout of ommatidia is often matched to the spatial layout of the habitat. Such matched filtering is not confined to habitat structure. Acute zones may also behave as matched filters for an animal's motion flow-field or for localizing other animals. In dim light, sensitivity becomes the overriding priority, and at night, or in the depths of sea, compound eyes of immense sensitivity have evolved. Those from increasing depths in the sea also tend to have an increasingly dorsal bias that reflects the dominance of dorsal (down-welling) illumination. These and other aspects of compound eye design are illustrated using a broad range of aquatic and terrestrial invertebrates as examples.

Key words Compound eye, light intensity, matched filter, resolution, sensitivity

1 Introduction

The daylight seen by a butterfly flitting across a sunny meadow is 15 orders of magnitude brighter than that seen by a crustacean swimming 1 km below the surface of the sea. Both of these animals have compound eyes optimized for vision at these extremes, and both require their eyes to survive. Of course, no single eye is able to see over this entire range. Instead, butterfly eyes are designed for vision in bright daylight, and for a range of intensities a few orders of magnitude lower than this, in an extended world of contrast features. Deep sea crustaceans, on the other hand, have compound eyes designed for seeing in a world where dim blue daylight comes from a small angle directly above, and where the darkness in other directions can be pierced by bright pin-points of bioluminescence. How have these two extreme habitats - and indeed all others between them - shaped the design of compound eyes?

The most obvious and dramatic effect has been the evolution of two broad optical classes of compound eyes. One class - apposition eyes (Fig. 1A) - is adapted best to bright light, the other class - *superposition* eyes (Fig. 1B) - to dim light. In apposition eyes, the abundance of light has allowed spatial resolution to be maximized, and foveal eye regions of higher acuity are very common. The opposite is generally true in superposition eyes: spatial resolution is usually sacrificed in the drive to capture as much light as possible.

Remarkably, there are many exceptions to this rule. Ecological pressures, including predation or competition for a limited food source, have forced many species to occupy habitats with light intensities that are not optimal for their eye designs. Superposition eyes can be found in many species active in bright light, and apposition eyes in many that are active in dim light. When one looks more carefully at these exceptional species, one finds surprising optical and neural adaptations that have re-optimized the eye for the new habitat. This remarkable richness and adaptability of compound eye design to the intensity and direction of natural illumination is the subject of this chapter.

2 The Design of Compound Eyes

2.1 Basic Types

Compound eyes, by far the most widespread eye design in the animal kingdom, are possessed by insects, most crustaceans, myriapods and even some clams and polychaetes. An excellent review of the optics and design of compound eyes can be found in Nilsson (1989).

Table 1. Resolution ($\Delta\phi$ & $\Delta\rho$), sensitivity (S) and habitat in compound eyes

Species	Name	Habitat	Eye	S	$\Delta\phi$	$\Delta\rho$
Ciulfina[1]*	praying mantis	D	App	0.025	0.8	0.85
Apis[2]	honeybee	D	App	0.22	1.9	2.6
Ocypode[3]*	ghost crab	D	App	0.05	0.26	0.5
Cirolana[4]	isopod	S	App	5091	20	45
Anopheles[5]	mosquito	N	App	8.1	8	37
Onitis[6]	dung beetle	N	Sup	56	2	4
Deilephila[7]	hawkmoth	N	Sup	101	1	2.5
Oplophorus[8]	shrimp	S	Sup	3300	8	15
Macroglossum[9]*	hawkmoth	D	Sup	8.5	1	1.3
Ocybadistes[10]	skipper butterfly	D	Sup	11.3	2	2.2

$\Delta\rho$ is the acceptance angle [the half-width of the spatial receptive field (deg)], $\Delta\phi$ is the interommatidial angle (deg) and S is the optical sensitivity to an extended white light source ($\mu m^2 sr$). *App* apposition eye; *Sup* superposition eye; *D* diurnal; *S* deep sea; *N* nocturnal; * values quoted for acute zone. Sources: 1 Horridge and Duelli 1979; 2 Warrant et al. 1996; 3 Zeil et al. 1986; 4 Nilsson and Nilsson 1981; 5 Land et al. 1999; 6 McIntyre and Caveney 1998; 7 Warrant unpubl.; 8 Land 1976, 1981; 9 Warrant et al. 1999; 10 Land 1984

All compound eyes are composed of tube-like optical units called ommatidia. Each ommatidium consists of a lens element (the cornea and crystalline cone) that focuses light incident from a narrow region of space onto an underlying bundle of photoreceptors (the rhabdom). A compound eye may contain as few as 6 ommatidia, as in some ants, or as many as 30 000 as in large dragonflies. The number of ommatidia is usually a good indicator of the spatial resolution of a particular eye. The greater the number of ommatidia, and the denser their packing, the finer the resolution of the eye. In fact, the angle between neighbouring ommatidia - the interommatidial angle, $\Delta\phi$ - sets the anatomical resolution of the eye (Snyder 1979): $1/(\Delta\phi\sqrt{3})$. An eye with an interommatidial angle of 0.24° (as in the dorsal eyes of some dragonflies: Sherk 1978) has greater anatomical resolution than an eye with an interommatidial angle of 4° (as might be found in a ground beetle). To put this into perspective, the angle between the cones of our fovea is around 0.01°, so our anatomical resolution is significantly greater! The interommatidial angles of various compound eyes are given in Table 1.

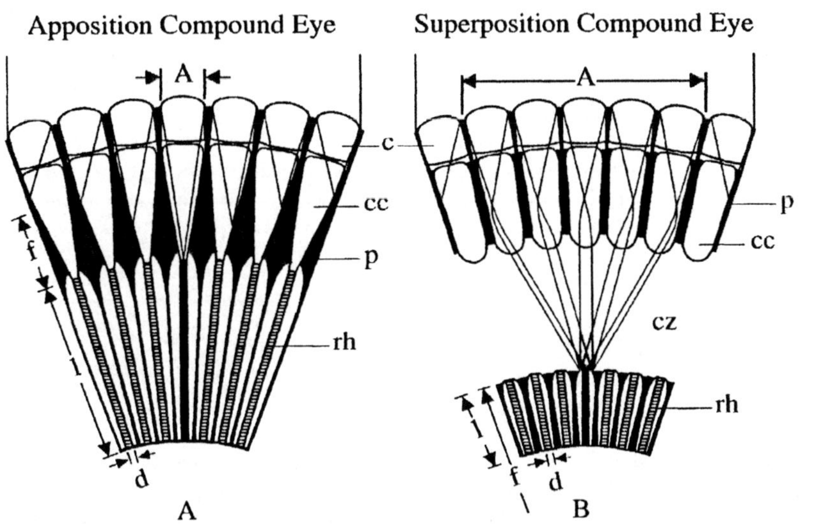

Fig. 1 A,B. Schematic diagrams of the two main compound eye designs. **A** Apposition eye. **B** Superposition eye (of the refracting type, relying on crystalline cones with internal gradients of refractive index). The paths and fates of parallel light rays, incident on the external eye surface, are indicated in each (*shaded area*). For each design the target rhabdom is shaded *black*. *A* Diameter of the aperture; *f* focal length (which, in superposition eyes, is measured from the eye's centre of curvature, not indicated); *c* corneal facet lens; *cc* crystalline cones; *p* screening pigment; *rh* rhabdom; *cz* clear zone; *l* rhabdom length; *d* rhabdom diameter

The ultimate indicator of resolution in a compound eye is the size of the photoreceptor's spatial receptive field (often referred to as its angular-sensitivity function). This can be measured optically or electrophysiologically (see Warrant and McIntyre 1993), and its half-width - the acceptance angle, $\Delta\rho$ - is an excellent guide to how well resolved a compound eye is. Smaller values of $\Delta\rho$ indicate better resolution (Table 1). Very often (but not always), eyes which strive for sensitivity sacrifice resolution and have wider acceptance angles (Warrant and McIntyre 1992).

In apposition eyes (Fig. 1A) each ommatidium is isolated from its neighbours by a sleeve of light-absorbing screening pigment, thus preventing light reaching the rhabdom from all but its own small corneal lens. It is this small lens aperture (only 15 μm wide in some species) which severely limits the amount of light that can reach each rhabdom. Apposition eyes are therefore typical of bright habitats. This limitation is slightly offset in higher flies which possess a special variation of the apposition design known as neural superposition (Kirschfeld 1967). In this design rhabdomeres in seven neighbouring ommatidia view the same direction in space, and send their axons to the same cartridge in the first optic neuropil, the lamina. This improves sensitivity and contrast discrimination since each visual channel receives light through seven lenses.

In superposition eyes (Fig. 1B), the pigment sleeve is withdrawn and a wide optically transparent area (the clear zone) is interposed between the lenses and the retina. This clear zone - and specially modified crystalline cones - allow light from a narrow region of space to be collected by a large number of ommatidia (comprising the superposition aperture) for focusing onto a single rhabdom. Unlike the crystalline cones of apposition eyes, those of some superposition eyes have evolved refractive index gradients (Exner 1891), those of others reflective surfaces (Vogt 1975, Land 1976), and in still others a combination of both (Nilsson 1988). The optics of such cones allows as many as 2000 ommatidia to collect light for a single rhabdom (as in some large nocturnal moths: Warrant, unpubl. data). Not surprisingly, superposition eyes are typical of animals active in dim habitats.

2.2 Design Principles, Constraints and Trade-Offs

2.1.1 Acute Zones and Matched Filtering

Despite their incredible adaptability, compound eyes are rarely more than a few millimetres wide, and this is a serious drawback. Light reaches the rhabdoms through tiny corneal lenses that are frequently no more than a few tens of micrometres in diameter (A in Fig. 1A). This is not only a problem for photon collection in dim light. Even in bright light such small lenses cause diffraction, and thus blurring, of the incoming light rays, thereby limiting spatial resolution. A solution to both problems is to make the lenses larger: diffraction is reduced and light capture improves. Unfortunately, for the same eye radius R, larger lenses mean a wider interommatidial angle: $\Delta\phi = A/R$ radians. This, in turn, means poorer anatomical resolution. A simple way out of this dilemma is to scale up the size of the eye (i.e. make R larger), and allow larger lenses without sacrificing $\Delta\phi$: but there are limits to this, too. Animals with compound eyes are generally rather small and there is a limit to how big an eye they can carry. Apart from anything else, the metabolic cost of having a larger eye is quite enormous and sets a serious limit to how big a particular eye can be (see

Laughlin, this Vol.). Instead, a wonderful compromise has evolved that maintains eye size. Rather than scaling up the whole eye, some animals have a small part of their eye, and thus a small part of their visual field, scaled up. A local increase in eye radius may not only allow an increase in lens size (which reduces diffraction and improves light capture) but if eye radius is increased sufficiently, it might even allow a simultaneous reduction in $\Delta\phi$ (thus sharpening anatomical resolution). Unfortunately, any benefits gained in one eye region necessarily come at the cost of other regions. Despite this, it turns out that these specialized eye regions, known as acute zones, are quite common, especially in apposition eyes from bright habitats (Horridge 1978). Until very recently, they were thought impossible in superposition eyes due to the optics of this type of eye, but at least one diurnal hawkmoth now appears to possess them (Sect. 3.2).

The presence of an acute zone implies that one region of the visual world is more important to an animal than others. Exactly which region depends on several factors, many reflecting transient needs, such as the need to find a mate or prey, or the need to detect the advance of a predator. Others reflect more permanent needs, not the least of which is the structure of the habitats where animals live. In his terrain theory of eye design, Hughes (1977) proposed that the spatial layout of sampling stations in an eye is matched to the spatial layout of the habitat. As beautifully explained by Rüdiger Wehner (1987), such matched filtering greatly simplifies processing, "freeing the brain from the need to perform more intricate computations to extract the information finally needed for fulfilling a particular task". Another aspect of habitat structure, the way in which scene details flow through an animal's visual field when it moves, is also processed with the help of matched filtering. These three aspects of an animal's habitat - its structure, its motion flow field and the locations of other animals - have been identified by Land (1989a, 1999) as the main influences on the evolution of acute zones. For a more comprehensive review, the interested reader is referred to Land's beautiful accounts.

2.1.2 The Struggle to Capture Light

In dim habitats the luxury of an acute zone is rarely permitted and the capture of light by the entire eye becomes the overriding priority. This does not mean to say that matched filtering is absent. Dim habitats still have a structure, present a motion flow field and contain mates and prey. Rather, the matched filtering found in animals from dim habitats is not as obviously manifested in regional variations of eye structure. As we shall soon see, matched filtering also occurs in the neural circuitry which processes visual information, and this applies equally well to animals from dim habitats as to those from bright ones.

As light levels fall, so too does the reliability of vision. There are several reasons for this, not the least of which is the random nature of photon arrivals on the retina. A photoreceptor that absorbs N photons during one integration time will experience an uncertainty, or shot noise, of $\sqrt{N}$ photons associated with this sample, that is, $N \pm \sqrt{N}$ photons (Land 1981; Warrant and McIntyre 1993). This noise reduces the reliability of intensity measurements, and thereby discrimination of the contrasts which define spatial details in a scene. The relative proportion of noise ($\sqrt{N}/N$, that is, $1/\sqrt{N}$), and thus the unreliability of vision, will increase with decreasing light intensities (i.e. lower N). In other words, as light intensities fall, increasing noise levels prevent animals from seeing finer contrasts, thus restricting their vision to coarser details of higher contrast (which is analogous to a CCD video camera providing noisier and less distinct images in dimmer light). Another reason why vision becomes unreliable at low light levels is that the biochemical pathways responsible for transduction can occasionally be activated in the absence of light (Barlow 1956). These activations are thermal in origin, produce electrical impulses indistinguishable from those produced by real photons, and are more frequent at higher retinal temperatures. At very low light levels this dark noise can significantly contaminate visual signals, especially in vertebrates whose measured rates of dark noise tend to be high (see Aho et al. 1988). In invertebrates the rates of dark noise measured to date have been quite low (see Warrant 1999). Finally, even if a photon is unambiguously absorbed, the small voltage response it produces in the photoreceptor (called a quantum bump) is not identical from one photon to the next. Bumps vary in amplitude, latency and duration, and this inconsistency is also a form of noise, called transducer noise, that adds to the uncertainty of vision (Lillywhite and Laughlin 1979).

Irrespective of their source, all types of noise have the same consequence for vision: less reliable detection of contrasts in the natural world. In dim light, the only solution to the problem is to have an eye design that maximises the capture of light, even at the expense of other visual capacities. The way this has been achieved is best seen in Land's extremely useful equation for the optical sensitivity S (Kirschfeld 1974, Land 1981), adjusted for white light illumination (Warrant and Nilsson 1998):

$$S = \left(\frac{\pi}{4}\right)^2 A^2 \left(\frac{d}{f}\right)^2 \left(\frac{kl}{2.3 + kl}\right), \tag{1}$$

where A is the diameter of the aperture (pupil), d and l are the diameter and length of the rhabdoms and k is their absorption coefficient (units of μm^{-1}). The optical sensitivity is the ratio of the number of photons absorbed by a rhabdom to the number emitted per unit area of an extended white source. If all lengths are in μm, then S has units of μm^2steradian. Compound eyes constructed to see in dimmer light tend to have higher optical sensitivities (Table 1). This is why superposition eyes are usually so superior in dim light. Their large superposition apertures (proportional to A^2: Fig. 1B) deliver a tremendous amount of light to each rhabdom. And compared to apposition eyes, the rhabdoms also tend to be wide (d in Fig. 1) with larger spatial receptive fields (proportional to $(d/f)^2$) that capture a larger fraction of incident light (ie. have greater kl). Compound eyes also frequently possess a reflective tapetum that lines the back of the eye. The tapetum reflects incoming light back through the retina for a second chance of absorption, thereby effectively up to doubling the length of the rhabdoms [i.e. $l \rightarrow 2l$ in Eq. (1)].

As mentioned above, some superposition eyes may employ as many as 2000 ommatidia to focus light onto a single rhabdom. Because of aberrations and optical misalignments, it is not a trivial task to direct 2000 individual beams of light so that they all meet at one point on the retina. In larger superposition eyes it might not actually be possible, with a blurry image being the unfortunate result. Even if all the light is perfectly coincident, it may still be impossible to trap the very wide cone of incident light rays within the target rhabdom by total internal reflection (Warrant and McIntyre 1990a, 1991). High sensitivity may come only at the cost of spatial resolution. Nevertheless, recent work has shown that image quality need not be all that blurry: many diurnal superposition eyes are actually diffraction limited (Land 1984, Warrant et al. 1999), and aberrations in many others are not as big a problem as previously thought (Nilsson et al. 2000). In addition, many superposition eyes have tackled the problem of preserving the image in the retina by isolating their rhabdoms in reflective tapetal sheaths, ensuring total internal reflection for any angle of incidence (Land 1984; Warrant and McIntyre 1991).

3 Compound Eyes for Bright Habitats

I mentioned above that the abundance of light in bright habitats has allowed compound eyes to develop acute zones, regions where the visual channels are more tightly packed and sample the world with greater spatial resolution. These acute zones invariably reveal remarkable matched filters - for habitat structure, motion and the location of mates and prey - and it is these to which we turn our attention next.

3.1 Apposition Eyes

3.1.1 Matched Filters for Flat Habitats

Matched filtering is clearly seen in the apposition eyes of animals adapted for life in a flat habitat: a desert ant that runs across a salt pan (Wehner 1987), a water strider that stands on a pond (Dahmen 1991), a backswimmer that looks upwards through the water surface in search of stranded insects (Schwind 1980), or a fiddler crab that scans a flat intertidal beach for mates (Zeil et al. 1986). All experience a bright world dominated by the horizon, and all possess eyes having an elongated horizontal region of enhanced resolution known as a visual streak (Fig. 2). Because most objects of interest for these animals occur either at or very near the horizon, visual streaks concentrate the majority of the eye's sampling stations - and thereby visual capacity - at the same locations. Take the fiddler crab (Fig. 2A). Its apposition eyes are vertically elongated, almost cylindrical, and located on long stalks above the carapace. The shape of the eye means that the radius of curvature in the vertical eye plane is much greater than in the horizontal eye plane, resulting in the vertical interommatidial angles ($\Delta\phi_v$) being much smaller (by up to a factor of 4) than the horizontal ones. $\Delta\phi_v$ steeply narrows towards the horizon (from both above and below) becoming smallest (0.3°) just along the eye's equator. The terrain of the fiddler crab's habitat not only shapes the design of its eye, it also shapes its behaviour. Fiddler crabs categorize

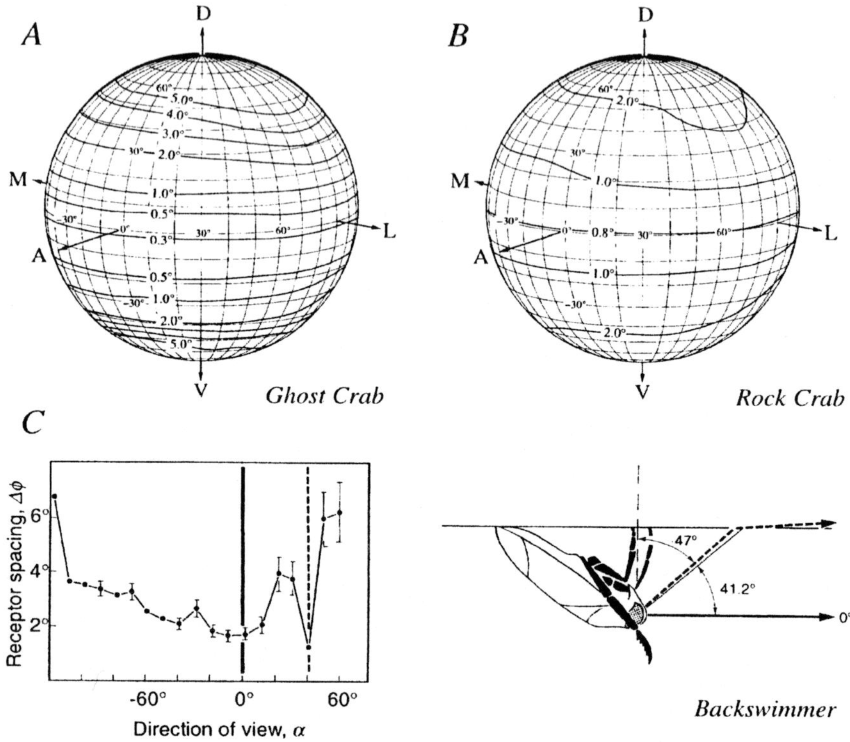

Fig. 2A-C. Apposition eyes for a flat world. **A, B** The distribution of interommatidial angles ($\Delta\phi$: shown as isolines) in the visual fields of the left eyes of two crab species: the ghost crab *Ocypode ceratophthalmus* (**A**) and the rock crab *Leptograpsus variegatus* (**B**). Visual fields are projected onto a sphere, where A = anterior, M = medial, L = lateral, V = ventral and D = dorsal. Note the very steep decline in $\Delta\phi$ at the ghost crab's horizon compared to the very much shallower decline in the rock crab. Data taken from Zeil et al. 1986. **C.** The vertical distribution of interommatidial angles in the backswimmer *Notonecta glauca* (*left*), and its normal position hanging from the underside of water surfaces (*right*). Plus and minus directions of view (α) refer to the dorsal and ventral parts of the eye looking upwards and downwards, respectively. Note that $\Delta\phi$ is smallest in the parts of the eye looking along the edge of Snell's window (*dashed lines*) and directly ahead (*solid lines*). (Original data from Schwind 1980; diagram Wehner 1987)

moving objects as mates or predators based on their position relative to the horizon (Layne et al. 1997). An excellent demonstration of Hughes' terrain theory is seen by comparing the fiddler crab eye with that of a rock crab living in a complicated rocky world where the horizon does not dominate (Fig. 2B). It is clear that the visual streak of the rock crab is much shallower and less developed than in its shore-living relative.

Visual streaks are also found in the eyes of an insect that hangs suspended from the underside of water surfaces: the backswimmer *Notonecta*. The fact that water has a higher refractive index than air means that the entire 180° dome of the sky is compressed to a 97° cone of light underwater. Within this cone of light, called Snell's window, all the features of the terrestrial world above can be found, including the flat water surface, which is located at the edge of the cone (Walls 1942, p. 378). By looking upwards along the edge of the cone, a suspended backswimmer is able to have a periscopic view of the outside water surface and see anything, including prey, which might be trapped on it. The water surface is an important horizon for the backswimmer, and the ventral part of the eye possesses a well-developed visual streak (Fig. 2C) that watches the surface along the edge of Snell's window (Schwind 1980). This matched filtering does not stop at the optics of the eye: in the optic lobe there are cells which have their visual fields coincident with the visual streak, and which respond maximally to prey-sized objects on the water surface (Schwind 1978). But the water surface is not the backswimmer's only horizon. Frontwards, the backswimmer can also see the environment of the pond and any item of interest that might be located there. There is a second visual streak which views this direction as well! (Schwind 1980)

3.1.2 Matched Filters for Detecting Mates and Prey Against the Sky

The need to see other animals, whether they be potential mates, possible food items or advancing predators, is so important in the lives of animals that it has driven the evolution of a wide range of visual specializations. In compound eyes this has resulted in a fantastic variety of acute zones (Land 1989a). In even more extreme cases entire extra compound eyes have evolved, to create divided eyes, like those of male bibionid flies which are specialized for the dorsal detection of females (Zeil 1983). We will encounter divided eyes again when we discuss the superposition eyes of krill.

In many insects the apposition eyes show a strong sexual dimorphism. Female brachyceran flies, like blowflies and hoverflies, have their eyes spaced well apart, but in males the eyes are joined at the top of the head. This extra area of eye constitutes a frontal-dorsal acute zone used by the males to keep sight of females during high-speed pursuits (Land and Eckert 1985). Amusingly, these acute zones are often referred to as love spots. They are clearly seen in the male hoverfly *Volucella pellucens*, which has large love spots located frontally, 20° above the equator (Fig. 3B). The interommatidial angle ($\Delta\phi$) here falls to just 0.7° (Warrant and Olsson, unpubl.). The size of the acute zone (the eye region where, say, $\Delta\phi < 1.1°$) occupies 2230 deg^2 of the visual field (shaded area in Fig. 3B). In females there is also an acute zone, directed exactly frontally (Fig. 3A). As we will see, this is a useful adaptation for matching the flow field during flight. In females, $\Delta\phi$ only falls to 0.9°, and the area of the acute zone ($\Delta\phi < 1.1°$) is a mere 23% as large as that of males (510 deg^2: shaded area in Fig. 3A).

The acute zones of male flies are not restricted to the eye surface. Below the eye there is an intricate neural pathway that is specific to males. First, the connections of photoreceptor axons to the lamina are quite different in the acute zone compared to both the female's eye and the rest of the male's eye (as found in houseflies: Franceschini et al. 1981). Higher up the brain, in the lobula, Strausfeld and colleagues (Strausfeld 1991; Gilbert and Strausfeld 1991; Gronenberg and Strausfeld 1991)

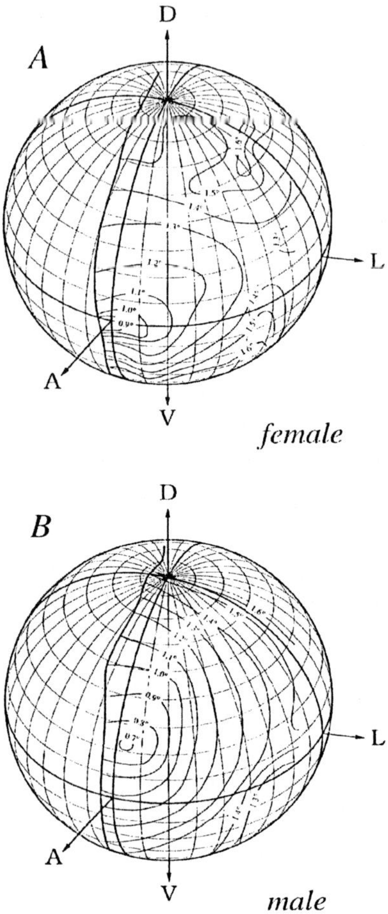

Fig. 3A,B Sexual dimorphism in the apposition eyes of the hoverfly *Volucella pellucens.* **A** Female. **B** Male. The visual fields of the left eyes of the two sexes, and interommatidial angles shown by isolines, are projected onto spheres. Terminology and conventions as in Fig. 2. Both sexes possess acute zones directed frontally, an adaptation for processing optical flow during forward flight (see Sect. 3.1.3). Compared to the female however, the male has a much larger acute zone directed 20°-30° dorsally (shaded regions, where $\Delta\phi < 1.1°$). The male's acute zone is used for fixating females during sexual pursuit. (Warrant and Olsson, unpubl. data)

identified large male-specific visual cells that respond maximally to small dark objects that move across a bright background in the frontal-dorsal visual field. This seems to be an excellent matched filter for silhouetted females flying against the sky!

Similar cells are also found in the dragonfly brain, not for spotting females, but for spotting prey. These insects have possibly the most advanced dorsal acute zones of all apposition eyes (Fig. 4), with values of $\Delta\phi$ falling to an incredibly low 0.24° in *Anax junius* (Sherk 1978). The facet diameters here are huge (62 μm) and this, together with a yellow screening pigment between the ommatidia (which is transparent to the light which reconverts rhodopsin: Labhart and Nilsson 1995), helps to ensure high sensitivity. The acute zone, clearly visible even to our own naked eye (Fig. 4A), has its region of highest resolution distributed in an elongated dorsal strip (Fig. 4B). Just as in male flies, this region eventually feeds its high resolution and sensitivity to large specialized cells in the brain, which behave as matched filters for fly-sized objects against a bright background (Fig. 4C; Olberg 1981, 1986).

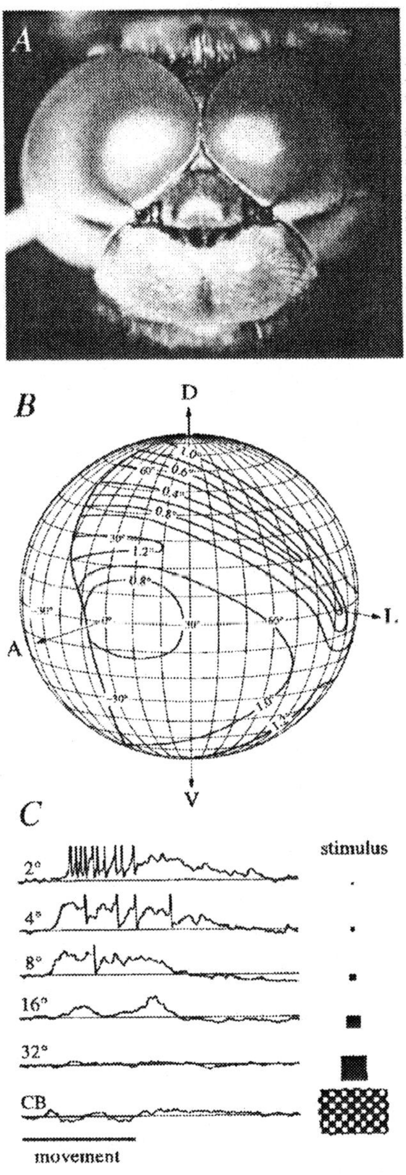

Fig. 4A-C. The dorsal apposition eyes of dragonflies: a matched filter for detecting prey. **A** In the dorsal regions of many dragonfly eyes the eye radius is longer, the facets larger and the interommatidial angles finer. Picture courtesy of D.-E. Nilsson. **B** The visual field of the eye of *Anax junius*, and interommatidial angles shown by isolines, are projected onto spheres. Terminology and conventions as in Fig. 2. *Anax* has two acute zones: (1) a shallow frontal acute zone (an adaptation for processing optical flow during forward flight: see Sect. 3.1.3), and (2) an intense strip-like acute zone directed about 60° above the horizon, where $\Delta\phi$ falls to just 0.24°. This acute zone is used for fixating prey during pursuit. Data from Sherk (1978) modified from Land (1989a). **C** Electrophysiological recordings of action potentials in a descending interneuron (DIT1) from the ventral nerve cord of the dragonfly *Aeschna umbrosa*. This cell, which has a visual field coincident with the visual field of the acute zone, is most responsive to small dark prey-sized targets (in this case 2° in size) moving against a bright background. The movements of large targets, or of wide checkerboard (*CB*) patterns, elicits no response from the cell. (After Olberg 1986)

3.1.3 Matched Filters for a Moving World

Animals very rarely view a stationary world. Because of their own movements, or the movements of objects in their environment, the scenes which animals look at are constantly moving. This very basic feature of the visual world has led to a wonderful array of matched filters, and the best-known examples come from the eyes of animals living in bright habitats.

As an animal walks or flies forwards, or translates, the features of the world flow by in the opposite direction. Features directly ahead appear almost stationary, whereas things at the side appear to pass by rapidly. Features nearby appear to pass by faster than those (like the horizon) which are further away. This flow field of features around a moving animal (Gibson 1950) has expressed itself in the design of compound eyes.

Because of the velocity distribution within the flow field, it becomes increasingly

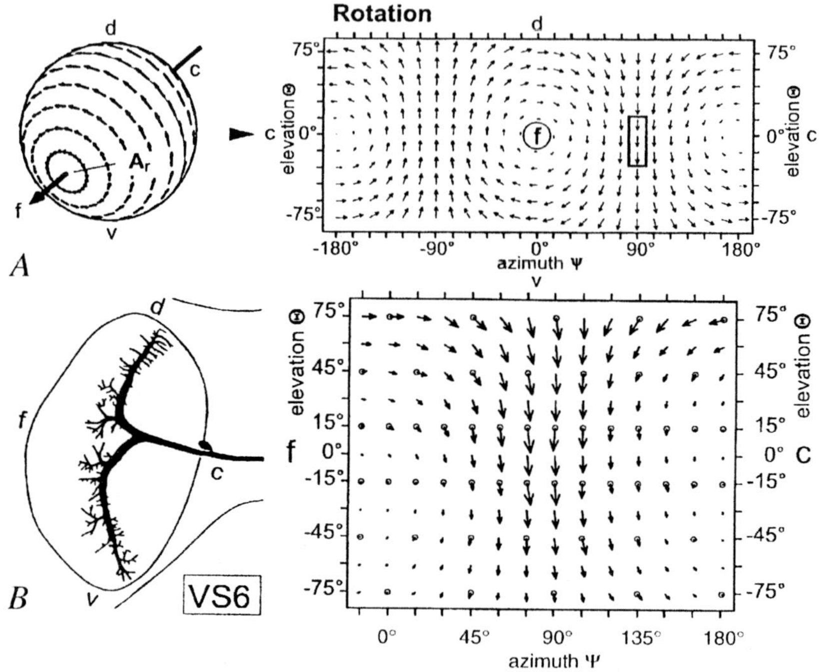

Fig. 5A,B. Optomotor cells in the lobula plate of the blowfly *Calliphora erythrocephala* act as matched filters for optic flow during flight. **A** Local motion flow-field vectors at various azimuths Θ and elevations Ψ experienced during rotation about an axis of rotation A , in this case the body axis. At a lateral position with $\Psi = 90°$, the local vectors are ventfally (v) oriented. At the opposite position ($\Psi = -90°$), the local vectors are dorsally (d) oriented. Frontally at the pole ($\Psi = 0°$, $\Theta = 0°$: circled f), the local vector is zero. Between these extremes the local vectors swirl in a clockwise direction around the frontal pole, and anticlockwise around the caudal (c) pole. **B** The local directional preferences of optomotor cell VS6 for small moving black spots at different azimuths Θ and elevations Ψ. Spots were presented to the cell at various Θ and Ψ, and the preferred direction of movement measured electrophysiologically. The local directional preferences show a striking similarity to the motion vectors experienced by the fly during rotation (**A**), and this cell would respond maximally to such a flow field. In a similar manner, other optomotor cells respond maximally to translation and pitch. (After Krapp and Hengstenberg 1996)

difficult for a translating animal to see features moving at more and more lateral positions. If the photoreceptors have a fixed integration time Δt, the motion of flow field images from front to back across the eye will cause blurring. An object moving past the side of the eye (with velocity v) will appear as a horizontal spatial smear whose angular size will be approximately $v\Delta t$ degrees. This effectively widens the local optical acceptance angle ($\Delta\rho$) to a new value of $\sqrt{(\Delta\rho^2 + (v\Delta t)^2)}$ (Srinivasan and Bernard 1975; Snyder 1977). The extent of this widening is worse at the side of eye (higher v) than at the front (lower v). In order to maintain an optimum sampling ratio of $\Delta\rho / \Delta\phi$ (Snyder 1977, 1979) the equatorial increase in $\Delta\rho$ posteriorly should be matched by an increase in $\Delta\phi$. This indeed seems to be the case in the apposition eyes of many flying animals (Land 1999), including the hoverfly *Volucella* (Fig. 3): $\Delta\phi$ increases progressively from front to back.

Of course, flow-field information continues to be processed in the brain. In the humble blowfly this occurs, in my opinion, with one of the most remarkable matched filters of all. For more than two decades, the lobula plate of the blowfly has been known to contain cells, known as horizontal (H) and vertical (V) cells, which respond to wide field motion. Some cells apparently prefer upward or downward motion, others leftward or rightward. Recently, Krapp and Hengstenberg (1996) examined very small regions of the visual fields of these cells, and discovered that the local direction preference was usually very different to the global preference. In fact they found that the local preferred direction changed in a predictable manner from one region to the next, building up an orderly map of directions across the entire visual field of the cell (Fig. 5B). The most remarkable feature of these maps is that they are astonishingly good matches to the maps of motion vectors that describe flow fields (Fig. 5A). Cell VS6 (Fig. 5B) has a map of preferred motion directions that is extremely similar to the map of motion vectors describing roll, the field of rotating features that results from a turn around the body axis (rotation: Fig. 5A). This means that if a fly makes a roll turn, VS6 will be maximally stimulated. Signals from these cells can then be used to activate compensatory flight manoeuvres that correct the fly's course. Similar matching can be found in other cells, including VS1 which matches pitch, the field of upwardly moving features resulting from a nose dive. The cell VS8 matches a field consisting of both pitch and roll. In other words, taken as a group, these cells respond vigorously whenever the fly experiences optic flow, an impressive matched filter for the motion of the world.

3.2 Superposition Eyes

Finally, a discussion of eye designs in bright habitats would be incomplete without mentioning the odd case of diurnal superposition eyes. The superposition design, the favourite of dim habitats, is surprisingly common in diurnal insects, especially among butterflies, moths and beetles. These superposition eyes generally have reduced apertures (see below) but if designed properly, the extra light afforded by this design can actually improve vision even in bright light. The aerobatic hummingbird hawkmoth has such an eye. The eye even has local acute zones in the retina, and a superposition aperture that is largest fronto-ventrally, in a region of the eye used for fixating flowers during feeding (Warrant et al. 1999). Such features were previously thought impossible for superposition eyes. Its acute zones - a horizontal visual streak for horizon detection and a highly acute frontal zone for flower fixation - are achieved by packing more rhabdoms into the retina than there are facets in the overlying cornea, with up to four rhabdoms for every facet (Fig. 6). This results in a remarkable superposition eye specialized for fixating flowers during feeding in bright light: the brightest and sharpest image of the flower is formed on the most acute part of the retina. Meanwhile, the visual streak aligns with the horizon to ensure stable hovering.

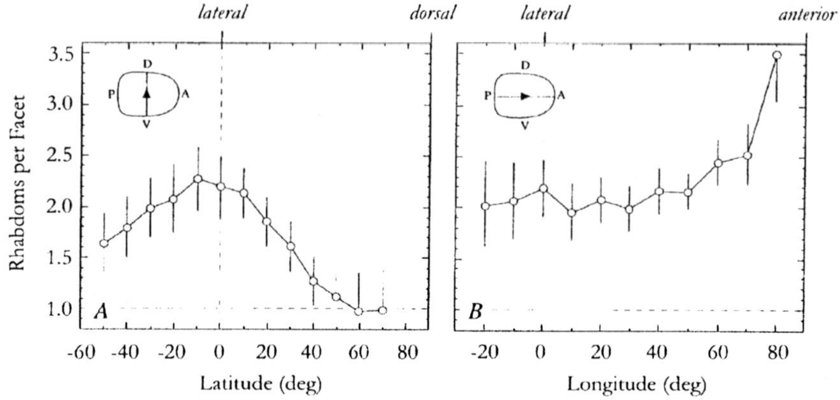

Fig. 6A,B Local acute zones in the retina of the hummingbird hawkmoth *Macroglossum stellatarum*. These acute zones, not visible from outside the eye, are built by abandoning the usual ratio of one rhabdom per facet, and instead packing more rhabdoms into the retina. The number of rhabdoms per facet is shown (**A**) along the dorso-ventral meridian (longitude 0°), and (**B**) along the posterior-anterior equator (latitude 0°). Except for the extreme dorsal (and possibly the extreme ventral) part of the eye, the number of rhabdoms always exceeds the number of facets. Note that rhabdom density increases at the equator to form a horizontal visual streak (**A**), and that density also increases markedly along the equator towards the front of the eye to form a frontal acute zone (**B**). These increases in density are accompanied by decreases in the inter-receptor angle, which falls to 1.4° at the equator in **A**, and to around 1° at the front of the eye in **B**. *D* dorsal; *V* ventral; *A* anterior; *P* posterior. (From Warrant et al. 1999)

As in flies, the sampling grid dilutes from front to back along the streak, presumably to match the flow field during forward flight. All of these unusual adaptations produce an eye of exceptional resolution and sensitivity, with the narrowest photoreceptor receptive fields yet measured in a superposition eye (ca. 1.3°: Warrant, Bartsch and Günther, in prep.)

4 Compound Eyes for Dim Habitats

As night advances, or as one penetrates the depths of the sea, light intensities fall dramatically. A bright sunny day is more than 100 million times brighter than a moonless starry night. In the clearest oceans and lakes, light intensity is reduced by 1.5 orders of magnitude for every 100 m of depth. At 700 m there would be insufficient daylight to sustain human vision, and below 1000 m there would be insufficient daylight to sustain vision of any kind (Denton 1990). The eyes of those animals that do live at tremendous depths are not designed to see daylight at all, but rather the point-source bioluminescent flashes produced by other animals, an interesting topic beyond the scope of this chapter.

Despite the problems of noise and visual uncertainty that accompany dim light, a rich variety of animals are nocturnal or live in the deep sea. Those with compound eyes are particularly challenged because their eyes are small. They are generally forced to forsake the well-developed acute zones typical of bright habitats, in favour of eye designs that maximize light capture. Superposition eyes, of course, are a popular option, but there are even apposition eyes that have evolved with extraordinary powers of light collection. We have already discussed some of the possible optical strategies for collecting light by examining Eq. (1): larger apertures, longer photoreceptors and wider receptive fields all improve sensitivity. So, too, does a higher absorption coefficient k, and it is interesting to note that in deep-sea animals k is typically 5 times greater than in terrestrial animals (see Warrant and Nilsson 1998). Eyes from dimmer habitats also typically have much lower F-numbers, a parameter much used by photographers for comparing the light-gathering power of lenses. The F-number is simply the focal length divided by the aperture diameter (f/A). A lower F-number indicates a brighter extended image and a greater sensitivity [sensitivity is inversely proportional to F-number squared: Eq. (1)].

Even with the most sensitive optical construction possible, a compound eye may still not collect enough light to meet the visual needs of the animal. In this case, there is still one more strategy available. This strategy, which resides in the neural circuits processing the incoming visual signal, involves neural summation of light in space and time. When light gets dim, the visual system increases its exposure time, that is, the time during which a sample of photons is counted. The greater photon catch that results creates a much more reliable image. Longer exposure times can be created by slowing down the photoreceptors in dim light (the common method in nocturnal vertebrates), or by some higher mechanism. This temporal summation, however, only comes at a price: long exposure times drastically degrade the perception of fast-moving objects. Too much temporal summation could be disastrous for a fast-flying nocturnal animal that needs to rapidly judge the presence of approaching obstacles!

Eyes can also improve image quality by summing photons in space. Instead of each ommatidial channel collecting photons in isolation (as in bright light), the transition to dim light could activate specialized lateral neurons which couple the channels together into groups (lateral neurons are common in the lamina and medulla, but as yet it is unknown whether they mediate coupling). In this way each group - themselves now defining the channels - could collect vastly more photons over a much wider visual angle, that is, with a greatly enlarged receptive field. Unfortunately, this improvement in photon catch also comes at a price: this spatial summation results in a simultaneous and unavoidable loss of spatial resolution. Despite being much brighter, the image becomes necessarily coarser.

Even though summation compromises spatial and temporal resolution, the gains in photon catch are so enormous that vision in dim light can be greatly improved. In fact, the improvements are often so great that some animals are able to see reliably at light intensities 100 000 times dimmer than those in which they would normally become blind (Warrant 1999).

How do these strategies for increasing light capture, both optical and neural, manifest themselves in the compound eyes of animals living in dim habitats?

4.1 Terrestrial Habitats

4.1.1 Superposition Eyes

Superposition eyes, by virtue of their large apertures, wide receptive fields and low F-numbers, are the compound eyes most suited for vision in dim light [Eq. (1); Table 1]. This design is much more flexible than previously thought (*Macroglossum*'s eye being but one of many newly discovered variations: see Sect. 3.2), and superposition eyes can be adapted to a whole range of different habitat intensities. A lovely demonstration of this can be seen in the superposition eyes of dung beetles from the single genus *Onitis* (McIntyre and Caveney 1998: Fig. 7). Individual species fly in search of dung at different times of day, and the superposition apertures of nocturnal species (width A = 845 μm in *O. aygulus*) are considerably larger than those of crepuscular species (A = 655 μm in *O. alexis*). These in turn are more than twice as large as those of diurnal species (A = 309 μm in *O. belial*). Moreover, the nocturnal species *O. aygulus* has huge contiguous rhabdoms (13 x 86 μm) compared to the diurnal species *O. belial* where they are small (6.5 x 32 μm) and widely spaced. Unlike *O. aygulus,* diurnal species like *O. belial* also have sheaths of screening pigment around their rhabdoms, which cuts down light flux even more (Warrant and McIntyre 1991). These differences are reflected in the sensitivities (S) of their eyes: S = 129 μm^{2}sr in *O. aygulus* but only 3.5 μm^{2}sr in *O. belial*. Despite their high sensitivity, the eyes of the nocturnal species have quite decent resolution, with photoreceptor receptive fields only 4° wide (Warrant and McIntyre 1990b). In comparison, the diurnal blowfly apposition eye has

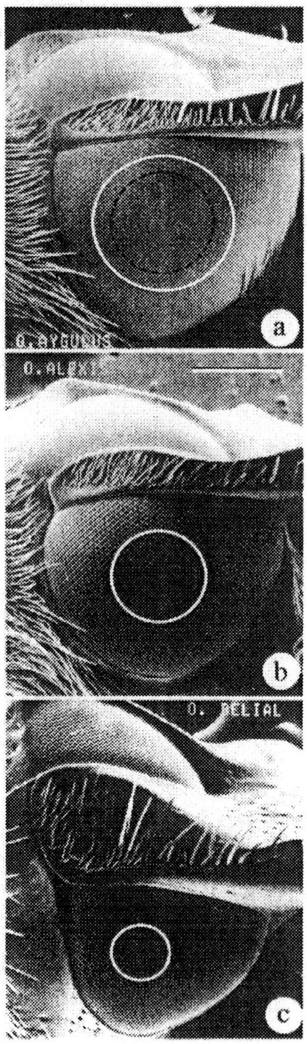

Fig. 7a-c. The size of the superposition aperture in three species of onitine dung beetles. **a** The nocturnal *Onitis aygulus*. **b** The crepuscular *Onitis alexis*. **c** The diurnal *Onitis belial*. The circular superposition aperture is indicated in white on the surface of each eye. The *dashed circles* refer to effective apertures, theoretically derived apertures in which each facet contributes light equally. In reality, facets near the edge of the aperture contribute less light than those near the centre. Note how the superposition aperture is smaller in beetles from brighter habitats. Bar 0.5 mm. (McIntyre and Caveney 1998)

receptive fields only half as wide (Hardie 1979).

An even better comparison of performance in apposition and superposition eyes can be seen in the motion-sensitive cells responsible for flow field analysis (which we discussed earlier in Sect. 3.1.3). These cells are located in the lobula plate of the optic lobe, and are possessed by many different flying insects, including the nocturnal elephant hawkmoth *Deilephila elpenor*. This insect has well developed superposition eyes that it uses for locating flowers at night. Once located, *Deilephila* hovers in front of the flower, locates the entrance of the nectar reservoir and then sucks it, a visually demanding task even in bright light. Lobula plate cells in both flies and moths respond vigorously to sinusoidal patterns of black-and-white stripes (called gratings) which move in a certain preferred direction. If the stripes are made finer and finer, there comes a point where the insect can no longer resolve the stripes and can thus no longer see them move. When this occurs the cell ceases to respond. David O'Carroll and I recently made recordings from such cells in *Deilephila* and also in the blowfly *Calliphora*, which has an apposition eye (Warrant and O'Carroll, in prep.). We chose two moving grating patterns, one with broad stripes and one with fine stripes, and by recording from lobula plate cells we tested the ability of the moth and the fly to see these gratings at different light intensities (early dusk, mid-dusk, moonlight and starlight: Fig. 8). Response (spike) histograms show that the hawkmoth is unable to see the fine grating, even at the brightest intensity. Flies, on the other hand, can

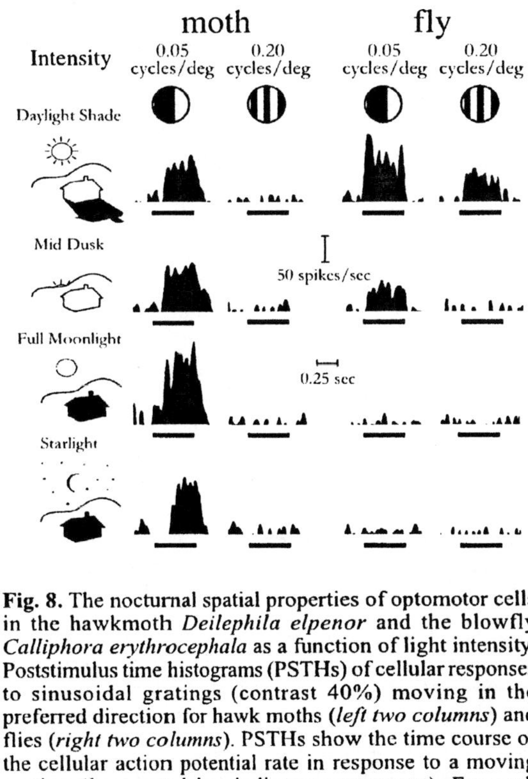

Fig. 8. The nocturnal spatial properties of optomotor cells in the hawkmoth *Deilephila elpenor* and the blowfly *Calliphora erythrocephala* as a function of light intensity. Poststimulus time histograms (PSTHs) of cellular responses to sinusoidal gratings (contrast 40%) moving in the preferred direction for hawk moths (*left two columns*) and flies (*right two columns*). PSTHs show the time course of the cellular action potential rate in response to a moving grating (*horizontal bar* indicates movement). For each insect, PSTHs are shown for four monitor intensities (from the *upper row*: early-dusk, mid-dusk, moonlight and starlight) for each of two gratings, one coarse (0.05 cycles deg^{-1}, *left column*), and one fine (0.2 cycles deg^{-1}, *right column*). Flies can respond to movements of the fine grating, but only at the brightest intensity, and fail to respond to either grating when intensity falls below mid-dusk levels. Whilst being unable to respond to movements of the fine grating at any intensity, moths maintain a strong response to the coarse grating at all light intensities. (Warrant and O'Carroll, in prep)

see it, which shows that the fly apposition eye is better resolved in bright light than the hawkmoth superposition eye. By mid-dusk, the fly apposition eye is unable to collect enough light to see the movements of the finer grating. By the time intensities have dropped to moonlight levels, it has also lost its ability to see the coarser grating. At some intensity between mid-dusk and moonlight the fly becomes blind. Even though the hawkmoth is unable to respond to movements of the fine grating at any intensity, they maintain a strong response to the coarse grating at all intensities, even starlight. These physiological results are in good accord with the moth's behaviour: with the aid of image intensification equipment, I have observed *Deilephila* flying and hovering confidently in starlight.

The superior light gathering capacity of superposition eyes - their great advantage for animals active in dim light - is clearly the reason why the hawkmoth outperforms the fly at low intensities. Compared to the fly's apposition eye (F-number = 2.6), the optics of *Deilephila*'s superposition eye (F-number = 0.6), deliver more than 100 times as much light to the retina. At a higher level, spatial and temporal summation appears to improve sensitivity even more (Warrant and O'Carroll, in prep.).

4.1.2 Apposition Eyes

As already mentioned, apposition eyes are not suited to dim light. Simply put, their small lenses have trouble collecting enough light for reliable vision. Despite this, various factors including predation, and competition for a limited food resource, forced many diurnal animals with apposition eyes to become nocturnal. In the tropics for instance, some species of bees and wasps, some with very rich visual behaviour, have become entirely nocturnal, despite their eye design (Warrant, Kelber and Wcislo, in prep.). Other nocturnal insects, such as crane flies and mosquitoes, seem to possess apposition eyes simply because their ancestors, who were likewise nocturnal, also possessed them (Nilsson and Ro 1994, Melzer et al. 1997, Land et al. 1999). These eyes, however, were not ordinary apposition eyes, but had special optical and neural adaptations for vision in dim light. Some of these adaptations are still present in nocturnal species even today, and it is to these which we turn our attention now.

A lovely example of how apposition eyes have adapted to dim light can be found among the mosquitoes (Land et al. 1997, 1999). These insects have conquered all light intensities, from brightest sunshine (e.g. the nectar-feeding *Toxorhynchites brevipalpis*) to starlight (e.g. the blood-sucking *Anopheles gambiae*). The apposition eyes of *Anopheles* have rhabdoms with huge receptive fields (d/f = 37°), and very wide interommatidial angles ($\Delta\phi$ = 8°). The large corneal lenses, which are almost hemispherical, have a diameter A = 28 µm and deliver a tremendously wide (60°) cone of light (and thereby a very bright image) to each rhabdom. Such a wide cone of light is potentially disastrous, because a normal rod-shaped rhabdom, with a refractive index only slightly higher than that of the surrounding cells, is unable to internally reflect a cone of light greater than about 20° (Warrant and McIntyre 1990a, 1993). *Anopheles* has overcome this problem with a wonderful adaptation: instead of having normal rod-shaped rhabdoms, they have conical ones, and these trap up to nine times as much light (Land et al. 1997: Fig. 9A). In contrast, the day-active

Toxorhynchites has an open rhabdom, with the individual rhabdomeres separated and rod-like. As in higher flies, it is likely that this mosquito possesses neural

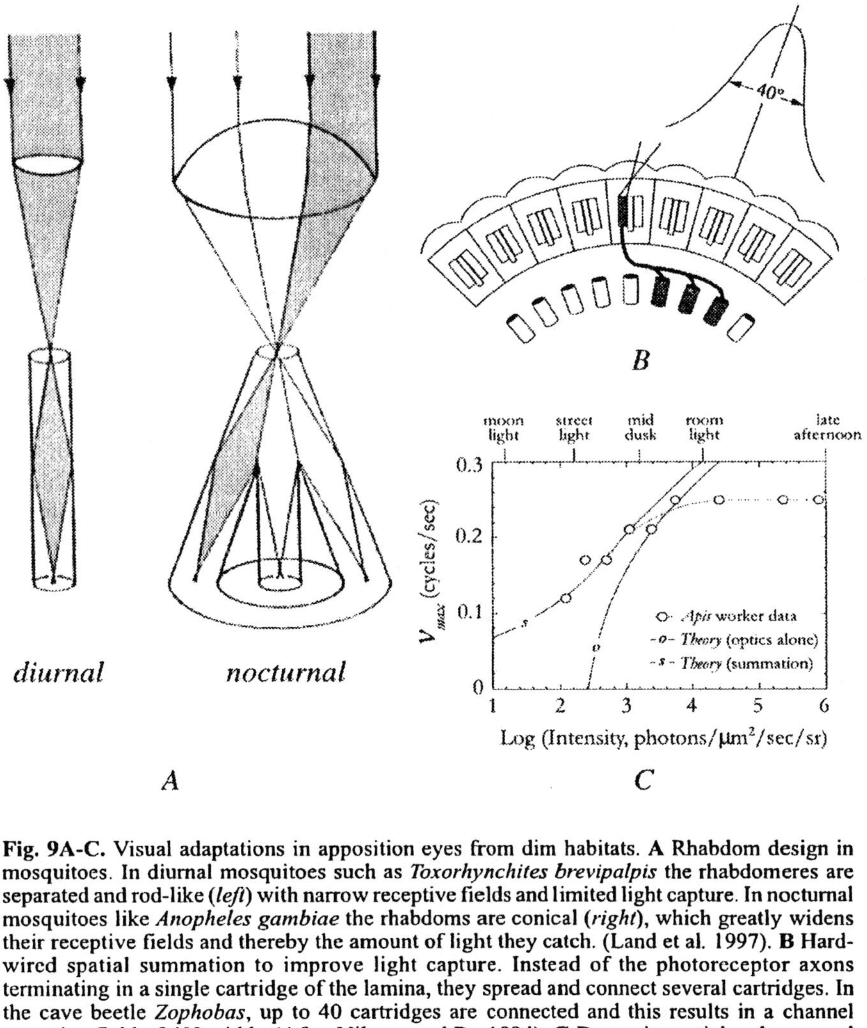

Fig. 9A-C. Visual adaptations in apposition eyes from dim habitats. **A** Rhabdom design in mosquitoes. In diurnal mosquitoes such as *Toxorhynchites brevipalpis* the rhabdomeres are separated and rod-like (*left*) with narrow receptive fields and limited light capture. In nocturnal mosquitoes like *Anopheles gambiae* the rhabdoms are conical (*right*), which greatly widens their receptive fields and thereby the amount of light they catch. (Land et al. 1997). **B** Hard-wired spatial summation to improve light capture. Instead of the photoreceptor axons terminating in a single cartridge of the lamina, they spread and connect several cartridges. In the cave beetle *Zophobas*, up to 40 cartridges are connected and this results in a channel receptive field of 40° width. (After Nilsson and Ro 1994). **C** Dynamic spatial and temporal summation to improve light capture. Spatial resolution, measured behaviourally as the finest spatial frequency detectable (v_{max}), is shown for the European honeybee as a function of light intensity (*open circles*). If light is collected by the optics of isolated ommatidia in the bee's apposition eye (no spatial or temporal summation: *-o-*), spatial resolution is predicted to decline with intensity faster than the data, with bees becoming blind (i.e. $v_{max} = 0$) by about mid-dusk. With optimum spatial and temporal summation (*-s-*), spatial resolution is predicted to decline less rapidly with intensity, a prediction that fits the data quite well below mid-dusk intensities. (Warrant et al. 1996).

superposition eyes. Compared to *Anopheles*, the corneal lenses of *Toxorhynchites* are smaller (A = 22 μm) and deliver much narrower (<30°) and dimmer cones of light to rhabdoms of much narrower receptive field (d/f = 3.2°). These features, together with a much finer interommatidial angle ($\Delta\phi$ = 2.8°), make the light-adapted diurnal apposition eye of *Toxorhynchites* an order of magnitude less sensitive than the dark-adapted nocturnal apposition eye of *Anopheles* (Land et al. 1999).

The eyes of mosquitoes show that clever optical specializations can turn an insensitive apposition eye into a decent light collector at night. Although it has not been shown, it is quite possible that nocturnal mosquitoes collect even more light by using hard-wired spatial summation. In the eyes of cave beetles, and several other insects with open rhabdoms like those of *Anopheles*, the photoreceptor axons do not simply terminate in the underlying lamina cartridge (as in most apposition eyes). Instead, they spread laterally within the lamina, coupling up to 40 cartridges (Warrant and Gislén, in prep.). This type of wiring has the potential to sum signals from large groups of ommatidia (Fig. 9B) and thereby greatly improve sensitivity in dim light (albeit at the cost of spatial resolution). This type of wiring, likely to be widespread within the primitive nocturnal Diptera, is thought to be the ancestral precursor to neural superposition, the wiring found in higher Diptera (Nilsson and Ro 1994).

The conical rhabdoms of mosquitoes and the hard-wired summation of cave beetles have most likely never developed among bees and wasps. These animals have typical diurnal apposition eyes, and their movement into the nocturnal niche is probably comparatively recent. An interesting example is the honeybee genus *Apis*, with four species including the well-known European honeybee *Apis mellifera carnica* and its Africanized sister race, *Apis mellifera scutellata*. The Africanized honeybee, and the southeast Asian giant honeybee *Apis dorsata*, will forage during dusk and dawn, and even throughout the night if a moon half-full or larger is present in the sky. This is despite the fact that the honeybee apposition eye should in theory be blind by mid-dusk (Warrant et al. 1996). Behavioural experiments show, however, that even the strictly day-active European honeybee is capable of seeing course habitat features, like large pale flowers, at moonlight intensities. This ability can be explained only if bees optimally sum photons over space and time (Fig. 9C: Warrant et al. 1996). According to theory, an Africanised bee could forage slowly in moonlight if it possessed a visual exposure time of around 120 ms and summed signals from groups consisting of no more than seven ommatidia, both requirements being rather modest and within the capacity of an insect visual system.

4.2 Aquatic Habitats

The deep sea is a very special visual environment. With increasing depth it is not only the intensity of light that changes dramatically. Its colour and direction change, too. Light becomes much bluer in deeper water, finally concentrating to a narrow wavelength band around 475 nm (Tyler and Smith 1970). In addition to becoming blue-shifted, almost all of the daylight available for vision comes increasingly from above. This dim downwelling blue light provides the backdrop against which aquatic animals can spot prey floating above, or against which they themselves can be seen from below. Not surprisingly, this peculiar light habitat has led to very specialized eye designs, especially among the compound eyes.

In the dim blue depths of the sea one can find compound eyes of tremendous sensitivity, a fact that applies not only to superposition eyes, but to apposition eyes as well. A very sensitive superposition eye is that of the deep sea shrimp *Oplophorus* (Land 1976, 1981), an animal which lives at a depth of about 500 m. With an F-number of 0.38, a reflective tapetum, and enormous rhabdoms (32 x 100 μm), these eyes attain a very high sensitivity of 3300 $\mu m^2 sr$, one of the largest values known in the animal kingdom. Unfortunately, they probably cannot distinguish point light sources closer than 15° apart (Land 1976), but presumably they do not need to either. Even greater sensitivity can be found, amazingly, in an apposition eye. The deep sea isopod *Cirolana* has apposition eyes possessing 60 ommatidia of giant proportions (Nilsson and Nilsson 1981). Huge corneal lenses (150 μm wide) of short focal length (100 μm), and gigantic rhabdoms (90 x 90 μm) sitting in cups of reflective pigment, endow *Cirolana* with an incredible sensitivity of 5091 $\mu m^2 sr$. As in *Oplophorus* this high sensitivity comes only at the cost of resolution: in *Cirolana*, the ommatidia have visual fields of about 45°, and are separated from one another by an angle of between 10° and 20°. But for the scavenging life style that *Cirolana* leads, sensitivity is probably of greater importance. Indeed, its close relative *Aega* has 300 ommatidia and is an active hunter, a life style that probably demands better visual acuity (Nilsson and Nilsson 1981).

The darkness of the deep sea has not been the only driving force in the evolution of aquatic vision. The fact that almost all available daylight comes directly from above has also been a major force. To catch as much of the downwelling light as possible, and to enhance the discrimination of any small object that might be silhouetted against it, many aquatic eyes have evolved narrow, dorsally directed visual fields. This is particularly obvious in compound eyes.

The large apposition eyes of hyperiid amphipods are an excellent example, as shown in the lovely study of Land (1989b). These highly mobile crustaceans can be found at all depths down to 1000 m, with different species being adapted for life at different depths (and thus different light intensities). Their eyes reflect this adaptation too. Near the surface the position of the sun determines the direction from which most light is incident, but due to underwater scattering, animals just beneath the surface see light from other directions as well. This all-round visual world is clearly reflected in the eyes of surface living species like *Thamneus*. This amphipod has eyes of quite uniform construction (Fig. 10A), with rather even (but somewhat low) anatomical resolution in all directions of the visual field (Fig. 10B). A species inhabiting a depth of around 200 m experiences a light environment quite different to that of *Thamneus*. Compared to the surface, the visual world is no longer all-around: daylight is 1000 times dimmer and incident almost entirely from above (ie. from within ca. ± 50° of vertical: Jerlov 1976). The eyes of amphipods living at these depths, like *Parapronoe* (Fig. 10C), differ accordingly. In *Parapronoe* the dorsal part of eye has become greatly enlarged, with enormous corneal facets adapted to the dim downwelling light (122 μm wide compared to 35 μm in *Thamneus*). In

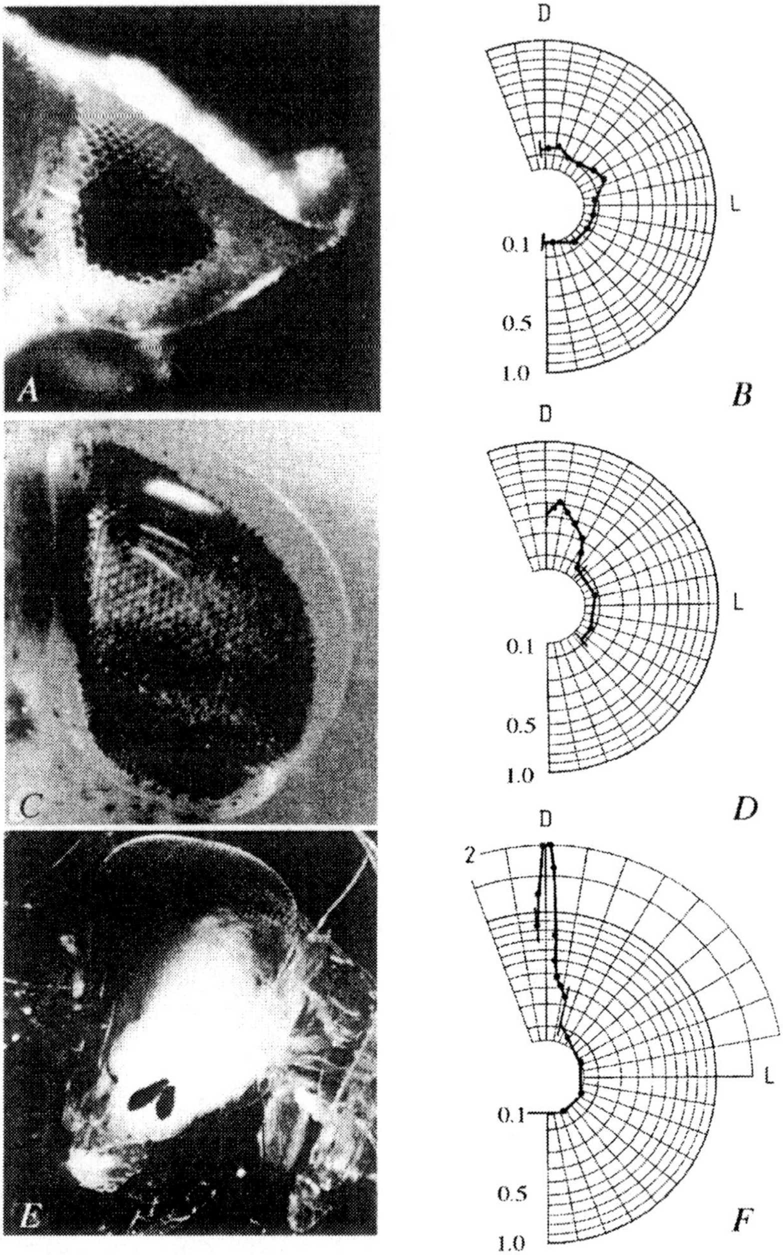

Fig. 10A-F. Eye structure and depth in hyperiid amphipods. Amphipods from different depths are shown at *left*: **A** *Thamneus* sp. (surface), **C** *Parapronoe crustulum* (200 m), and **E** *Phronima sedentaria* (< 800 m). At *right* (**B**, **D** and **F**), their respective vertical distributions of anatomical resolution (cycles deg^{-1}) are shown (here calculated as $1/2\Delta\phi$). Note how resolution in the surface species is low and rather uniform throughout the visual field (**B**). In deeper species the eye becomes more bilobed, with the dorsal lobe (viewing the downwelling light) dominating. Resolution in the dorsal lobe increases with depth (**D** and **F**) and becomes more dorsally directed. *D* dorsal; *L* lateral. (After Land 1989b)

contrast, the ventral part is quite small with facets only half as big. Resolution is not uniform either, becoming significantly better in the dorsal direction, with a clear match to the angular distribution of downwelling light (Fig. 10D). The trend becomes even more extreme in deeper-living species such as *Phronima*. In these amphipods, which live below 800 m, each eye is entirely divided, with a huge dorsal eye detached from a tiny ventral one (Fig. 10E). The dorsal facets are gigantic (146 μm wide compared to 80 μm in the ventral eyes), and resolution escalates within a narrow (15°) dorsal visual field, outside of which resolution is very poor (Fig. 10F).

What do these depth-related changes in eye design mean? Two of the trends are easy to understand: the enlargement of facets clearly improves sensitivity, and the increasingly dorsal bias of the eyes reflects the dominance of dorsal illumination. The third trend is slightly less obvious: with increasing depth dorsal resolution increases dramatically. In the three species detailed above, interommatidial angle falls from 4.8° in *Thamneus*, to 1.1° in *Parapronoe*, and to an astonishing 0.25° in *Phronima*. At the same time the total dorsal visual field narrows sharply. The explanation can be found by considering what amphipods need to see. Had the eyes been designed to view wide extended scenes, the interommatidial angle should have increased with depth. However, if the role of the dorsal eyes is to spot small silhouetted objects floating above, then the opposite would be predicted (Land 1989b). Even though the life histories of amphipods are poorly understood, the design of their eyes implies that they use them for spotting small prey animals against the dim downwelling illumination.

Of course, amphipods are not the only animals that have adapted to the aquatic light environment in this way. Many euphausiid shrimps (krill) have bilobed superposition eyes that are adapted in exactly the same manner (Fig. 11), with deeper living species having dorsal eyes with more numerous facets and sharper resolution directed into narrower dorsal visual fields (Land et al. 1979). At this point it is difficult to avoid mentioning that even in terrestrial habitats the same kinds of adaptations are found. The males of many mayfly species patrol small forest clearings at dusk in the hope that females - tiny silhouettes against a narrow and dim patch of sky - might fly over them. These males typically have divided superposition eyes, with two huge tube-like dorsal eyes of high resolving power directed into narrow overlapping dorsal fields of view. In other species the males fly over open fields and females can appear as silhouettes anywhere within the wide hemisphere of the sky. Their superposition eyes are not fully divided and have dorsal parts of medium resolution viewing much broader fields of view (Brännström and Nilsson 2000). If nothing else, these wonderful examples of convergent evolution emphasize the remarkable influence of habitat illumination on the design of compound eyes.

5 Conclusions

The animal kingdom's most widespread eye design is certainly one of its most adaptable. In this chapter I have only attempted to describe the way in which compound eyes have evolved in response to the intensity and direction of natural

Fig. 11. The divided superposition eye of the euphausiid shrimp *Stylocheiron maximum*. The huge dorsal lobe has a narrow dorsal field of view for viewing the down-welling light, while the small lateral lobe has a broader lateral field of view that could be useful for viewing bioluminescent flashes. (Photograph courtesy Dan-Eric Nilsson)

illumination. But as important as this is, there are many other qualities of light - particularly its colour and plane of polarization - that have also played a pivotal role in the evolution of compound eyes. These other qualities have likewise led to remarkable eye designs that are matched to the life styles and habitats of their owners. However, it is the intensity of light, and the direction from which it is incident, which has arguably been the strongest force in the evolution of vision, not only in compound eyes, but in all eyes. It should come as no surprise that many of the adaptations I have described here are also found in other kinds of eyes, in particular the kind that we ourselves possess. But that's another story (Warrant, in prep.).

Acknowledgments. I dedicate this review to Professor Dan-Eric Nilsson, whose friendship, support and inexhaustible inspiration have made this and many other endeavours possible. The help, encouragement and critical appraisal provided by Marie Dacke, Anna Gislén, Almut Kelber, Ronald Kröger, Karin Nordström and Peter Nordström are gratefully acknowledged. I am also particularly grateful for the generous and ongoing support of the Swedish Natural Science Research Council.

References

Aho A-C, Donner K, Hydén C, Larsen LO, Reuter T (1988). Low retinal noise in animals with low body temperature allows high visual sensitivity. Nature 334: 348-350

Barlow HB (1956) Retinal noise and absolute threshold. J Opt Soc Am 46: 634-639

Brännström PA, Nilsson D-E (2000) Gradations of eye design in the superposition eyes of male mayflies. J Comp Physiol A, in press

Dahmen H (1991) Eye specialisation in waterstriders: an adaptation to life in a flat world. J Comp Physiol A 169: 623-632

Denton EJ (1990) Light and vision at depths greater than 200 metres. In: Herring PJ, Campbell AK, Whitfield M, Maddock L (eds) Light and Life in the Sea. Cambridge

University Press, Cambridge, pp127-148

Exner S (1891) Die Physiologie der facettirten Augen von Krebsen und Insecten. Franz Deuticke, Leipzig Vienna. English translation: Hardie RC (1988) The Physiology of the Compound Eyes of Insects and Crustaceans. Springer, Berlin Heidelberg New York

Franceschini N, Hardie R, Ribi W, Kirschfeld K (1981) Sexual dimorphism in a photoreceptor. Nature 291: 241-244

Gibson JJ (1950) The Perception of the Visual World. Houghton Mifflin, Boston

Gilbert C, Strausfeld NJ (1991) The functional organization of male-specific visual neurons in flies. J Comp Physiol A 169: 395-411

Gronenberg W, Strausfeld NJ (1991) Descending pathways connecting the male-specific visual system of flies to the neck and flight motor. J Comp Physiol A 169: 413-426

Hardie R (1979) Electrophysiological analysis of fly retina. I: Comparative properties of R1-6 and R7 and 8. J Comp Physiol A 129: 19-33

Horridge GA (1978) The separation of visual axes in apposition compound eyes. Phil Trans R Soc Lond B 285: 1-59

Horridge GA, Duelli, P (1979) Anatomy of the regional differences in the eye of the mantis *Ciulfina*. J Exp Biol 80: 165-190

Hughes A (1977) The topography of vision in mammals of contrasting life style: comparative optics and retinal organisation. In: Crescitelli F (ed) Handbook of Sensory Physiology, Vol. VII/5. Springer, Berlin Heidelberg New York, pp 613-756

Jerlov NG (1976) Marine Optics. Elsevier Scientific Publishing Company, Amsterdam

Kirschfeld K (1967) Die Projektion der optischen Umwelt auf das Raster der Rhabdomere im Komplexauge von *Musca*. Exp Brain Res 3: 248-270

Kirschfeld K (1974) The absolute sensitivity of lens and compound eyes. Z Naturforsch 29C: 592-596

Krapp HG, Hengstenberg R (1996) Estimation of self-motion by optic flow processing in single visual neurons. Nature 384: 463-466

Labhart T, Nilsson D-E (1995) The dorsal eye of the dragonfly *Sympetrum*: specializations for prey detection against the blue sky. J Comp Physiol A 176: 437-453

Land MF (1976) Superposition images are formed by reflection in the eyes of some oceanic decapod crustacea. Nature 263: 764-765

Land MF (1981) Optics and vision in invertebrates. In: Autrum H (ed) Handbook of Sensory Physiology, Vol VII/6B. Springer, Berlin Heidelberg New York, pp 471-592

Land MF (1984) The resolving power of diurnal superposition eyes measured with an ophthalmoscope. J Comp Physiol 154: 515-533

Land MF (1989a) Variations in the structure and design of compound eyes. In: Stavenga DG, Hardie RC (eds) Facets of Vision. Springer, Berlin Heidelberg New York, pp 90-111

Land MF (1989b) The eyes of hyperiid amphipods: relations of optical structure to depth. J Comp Physiol A 164: 751-762

Land MF (1999) Compound eye structure: Matching eye to environment. In: Archer SN, Djamgoz MBA, Loew ER, Partridge JC, Vallerga S (eds). Adaptive Mechanisms in the Ecology of Vision. Kluwer Academic Publishers, Dordrecht, Boston, London, pp 51-71

Land MF, Burton FA, Meyer-Rochow VB (1979) The optical geometry of euphausiid eyes. J Comp Physiol 130: 49-62

Land MF, Eckert H (1985) Maps of the acute zones of flies. J Comp Physiol A 156: 525-538

Land MF, Gibson G, Horwood J (1997) Mosquito eye design: conical rhabdoms are matched to wide aperture lenses. Proc R Soc Lond B 264: 1183-1187

Land MF, Gibson G, Horwood J, Zeil J (1999) Fundamental differences in the optical structure of the eyes of nocturnal and diurnal mosquitoes. J Comp Physiol A 185: 91-103

Layne J, Land MF, Zeil J (1997) Fiddler crabs use the visual horizon to distinguish predators from conspecifics: a review of the evidence. J Mar Biol Assoc UK 77: 43-54

Lillywhite PG, Laughlin SB (1979) Transducer noise in a photoreceptor. Nature 277: 569-572

McIntyre PD, Caveney S (1998) Superposition optics and the time of flight in onitine dung beetles. J Comp Physiol A 183: 45-60

Melzer RR, Zimmermann T, Smola U (1997) Modification of branched photoreceptor axons, and the evolution of neural superposition. Cell Mol Life Sci 53: 242-247

Nilsson D-E (1988) A new type of imaging optics in compound eyes. Nature 332: 76-78

Nilsson D-E (1989) Optics and evolution of the compound eye. In: Stavenga DG, Hardie RC (eds) Facets of Vision. Springer, Berlin Heidelberg New York, pp 30-73

Nilsson D-E, Nilsson HL (1981) A crustacean compound eye adapted for low light intensities (Isopoda). J Comp Physiol 143: 503-510

Nilsson D-E, Ro A-I (1994) Did neural pooling for night vision lead to the evolution of neural superposition eyes? J Comp Physiol A 175: 289-302

Nilsson D-E, Gislén L, Brännström PA (2000) Principles and constraints in the design of superposition eyes. J Comp Physiol A, in press

Olberg R (1981) Object- and self-movement detectors in the ventral cord of the dragonfly. J Comp Physiol A 141: 327-334

Olberg R (1986) Identified target-selective visual interneurons descending from the dragonfly brain. J Comp Physiol A 159: 827-840

Schwind R (1978) Visual system of *Notonecta glauca*: a neuron sensitive to movement in the binocular visual field. J Comp Physiol 123: 315-328

Schwind R (1980) Geometrical optics of the *Notonecta* eye: adaptations to optical environment and way of life. J Comp Physiol 140: 59-68

Sherk TE (1978) Development of the compound eyes of dragonflies (Odonata). III. Adult compound eyes. J Exp Zool 203: 61-80

Snyder AW (1977) Acuity of compound eyes: physical limitations and design. J Comp Physiol 116: 161-182

Snyder AW (1979) Physics of vision in compound eyes. In: Autrum (ed) Handbook of Sensory Physiology, Vol VII/6A. Springer, Berlin Heidelberg New York. pp 225-313

Srinivasan MV, Bernard GD (1975) The effect of motion on visual acuity of the compound eye: a theoretical analysis. Vision Res 15: 515-525

Strausfeld NJ (1991) Structural organization of male-specific visual neurons in calliphorid optic lobe. J Comp Physiol A 169: 379-393

Tyler JE, Smith RC (1970) Measurement of Spectral Irradiance Underwater. Gordon and Breach, New York

Vogt K (1975) Zur Optik des Flußkrebsauges. Z Naturforsch 30: 691

Walls GL (1942) The Vertebrate Eye and its Adaptive Radiation. The Cranbrook Press, Bloomfield Hills

Warrant EJ (1999) Seeing better at night: life style, eye design and the optimum strategy of spatial and temporal summation. Vision Res 39: 1611-1630

Warrant EJ, McIntyre PD (1990a) Limitations to resolution in superposition eyes. J Comp Physiol A 167: 785-803

Warrant EJ, McIntyre PD (1990b) Maturation of optics and resolution in adult dung beetle superposition eyes. J Comp Physiol A 167: 805-815

Warrant EJ, McIntyre PD (1991) Strategies for retinal design in arthropod eyes of low F-number. J Comp Physiol A 168: 499-512

Warrant EJ, McIntyre PD (1992) The trade-off between resolution and sensitivity in compound eyes. In: Pinter RB, Nabet B (eds) Nonlinear Vision, CRC Press Inc, Boca Raton. pp 391-421

Warrant EJ, McIntyre PD (1993) Arthropod eye design and the physical limits to spatial resolving power. Prog Neurobiol 40: 413-461

Warrant EJ, Porombka T, Kirchner WH (1996) Neural image enhancement allows honeybees to see at night. Proc R Soc Lond B 263: 1521-1526

Warrant EJ, Nilsson D-E (1998) Absorption of white light in photoreceptors. Vision Res 38: 195-207

Warrant EJ, Bartsch K, Günther C (1999) Physiological optics in the hummingbird hawkmoth: a compound eye without ommatidia. J Exp Biol 202: 497-511

Wehner R (1987) "Matched filters" - neural models of the external world. J Comp Physiol A 161: 511-531

Zeil J (1983) Sexual dimorphism in the visual system of flies: the compound eyes and neural superposition in Bibionidae (Diptera). J Comp Physiol 150: 379-393

Zeil J, Nalbach G, Nalbach H-O (1986) Eyes, eye stalks and the visual world of semi-terrestrial crabs. J Comp Physiol 159: 801-811

IV.3 The Ecology of Teleost Fish Visual Pigments: a Good Example of Sensory Adaptation to the Environment?

Ron H. Douglas

Applied Vision Research Centre, Dept. Optometry and Visual Science, City University, 311-321 Goswell Road, London ECIV 7DD, UK

Abstract
The aquatic environment offers a natural laboratory for the study of visual ecology. The colour of natural bodies of water varies from the brown/reds of some freshwater lakes, through the greens of lakes and coastal waters, to the blues of the deep oceans. Teleosts have adapted the visual pigments of their rods and cones to take advantage of these different photic environments. Many shallow-living fish are probably tetrachromatic, with sensitivity extending from the near UV to the far-red and utilize the full broad daylight spectrum. Teleosts living in more green waters tend to be blue/green dichromats, having lost sensitivity to the longer and shorter wavelengths. In contrast, deep-sea teleosts generally have pure rod retinae, maximally sensitive to the dim downwelling monochromatic blue light of the ocean. In addition, their rod pigments may be spectrally tuned to be sensitive to their own bioluminescence, which in some cases may be deep red. Many fish probably modify their visual pigment complement, either during development or seasonally, as they change factors such as their feeding habits, geographical location, depth of habitat and photic environment.

Key words Visual pigment, retina, teleost, visual ecology, bioluminescence

1 Introduction

The lens and cornea of the eye produce an optical representation of the environment on the retina, where image-forming photons are absorbed by visual pigments within the photoreceptors, which convert this physical representation of the world into neurobiological activity. These electrical signals are then processed by the remaining neural cells of the retina and by the cells of the central visual pathway eventually resulting in either conscious perception of the visual stimulus or a reflex response, such as pupil closure or accommodation. The span of wavelengths an animal's visual system is able to respond to is limited partially by the wavelengths transmitted through the ocular media, which often remove short wavelengths (Douglas and Marshall 1999), but primarily by the complement of visual pigments within its photoreceptors.

The sensitivity of a visual pigment is expressed most readily by its absorption spectrum, either determined by microspectrophotometry of single photoreceptors (Bowmaker 1984), or using spectrophotometry of retinal extracts (Knowles and Dartnall 1977), or wholemounts of excised retinae (Douglas et al. 1995). All visual pigments have a bell-shaped absorption spectrum with a point of maximum absorption, the λ_{max}, at which point they are most likely to absorb a photon and hence convey maximum sensitivity (Fig. 1).

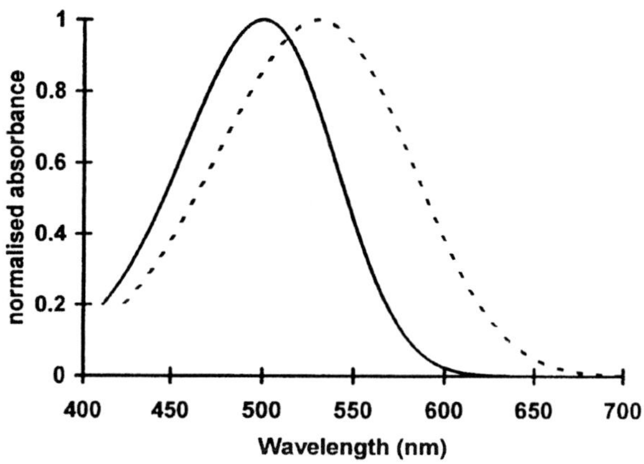

Fig. 1. Visual pigment templates for a theoretical rhodopsin with λ_{max} 500 nm and a porphyropsin based on the same opsin (λ_{max} 530 nm) forming a pigment pair. The relationship used to calculate the λ_{max} of the porphyropsin was derived by Whitmore and Bowmaker (1989). The templates fitted here and elsewhere in this chapter are those of Partridge and DeGrip (1991) for the A_1 pigment and Stavenga et al. (1993) for the A_2-based pigment

With the exception of a relatively few species, such as the majority of deep-sea fish, which are habitually exposed to low light levels, most vertebrates possess two distinct types of photoreceptor: rods and cones (Fig. 2). Rods, which due to their structure, physiology and connectivity, are the more sensitive receptors, are used at lower (scotopic) light levels. Since most animals possess only a single rod visual pigment, these receptors do not mediate colour vision, for which at least two spectrally distinct photoreceptors with overlapping absorption spectra are required. Cones, on the other hand, of which most vertebrates possess more than one spectral type, are less sensitive and therefore utilized in brighter (photopic) conditions where the number of photons is not a limiting factor. Interactions between the different cone types allow colour vision and the relative lack of synaptic convergence ensures a more acute image than that delivered by the rod system. Rods, with the exception of some deep-sea fish (see below), usually have a λ_{max} around 500-530 nm. Cone visual pigments can be much more variable. At the short wavelength end of the spectrum animals from all vertebrate groups have been shown to possess visual pigments with λ_{max} values in the

UV around 350-360 nm (Jacobs 1992). Photons of wavelengths below ca. 310 nm would be very damaging to the retina and are inevitably absorbed by structural components of the cornea and lens. What sets the limit to absorption at longer wavelengths is not so clear, but no cone visual pigment has yet been identified with a λ_{max} greater than *ca.* 630 nm (Bowmaker 1991). Given this large potential range of visual pigment λ_{max} values one might expect every animal to possess many visual pigments with λ_{max} values spanning the complete range. Although this is what invertebrate mantid shrimps appear to do, possessing up to 12 distinct spectral types of photoreceptor with λ_{max} values roughly evenly spaced between 300 and 700 nm (Marshall and Oberwinkler 1999), vertebrates are much more conservative, most possessing only 2-4 different spectral cone types and, with the exception of some amphibia and deep-sea fish, a single spectral class of rod (Bowmaker 1991).

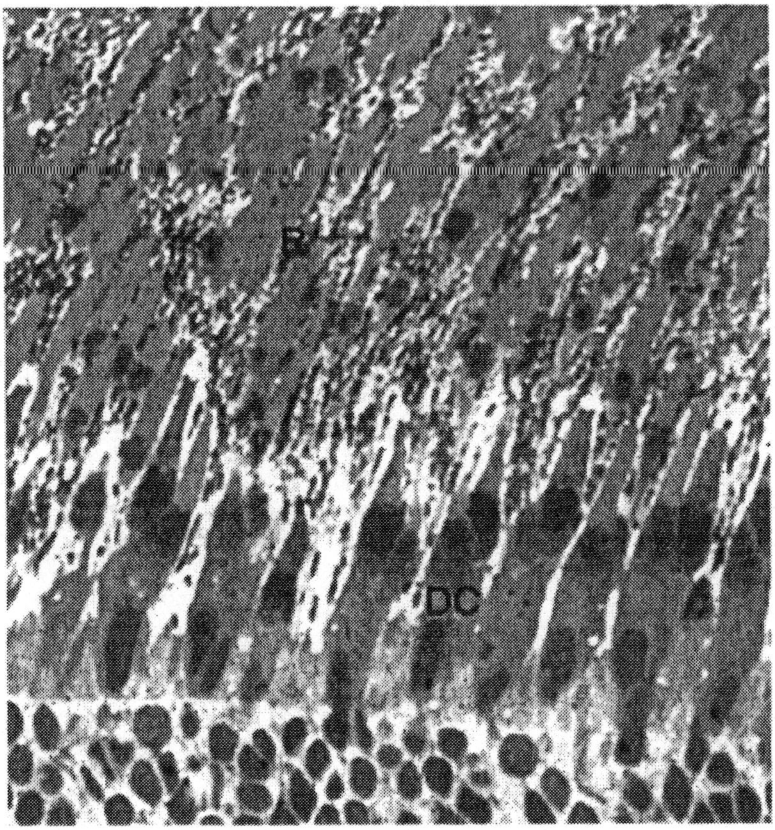

Fig. 2. Transverse section of the outer retina of the tench (*Tinca tinca*). Rods (*R*) as well as several different morphological types of cone, including double cones (*DC*) are visible

Since vertebrates appear to locate relatively few visual pigments at specific points of the spectrum, it would seem likely that these points have some specific adaptive significance to the animal. This, however, is extremely difficult to demonstrate, since it is usually not clear what precisely the animal is trying to achieve with its pigments. The pigment may, for instance, be designed to simply maximize quantum catch by matching its absorption to the surrounding spacelight, but in most instances is probably detecting a more specific visual target (Partridge and Cummings 1999).

Correlations between a pigment's λ_{max} and the visual stimulus are most apparent in fish, which of all vertebrates have the greatest visual pigment diversity. This is because the underwater environment is optically more diverse than any other. The spectral composition of bodies of water is highly variable due largely to differences in the quantity and identity of the substances dissolved or suspended within them (Douglas 1991). Thus, much of the open ocean, and certain bodies of freshwater, approach the transmission characteristics of pure water and are a clear blue, preferentially transmitting short-wave radiation. In coastal water, on the other hand, intermediate wavelengths tend to dominate, giving it a yellow/ green colour due to the presence of chlorophyll and other dissolved organic matter. Freshwater habitats are highly variable, but in contrast to oceanic and coastal waters they are often red/brown in colour, transmitting mainly longwave radiation due to suspended organic matter and run-off from the surrounding land (Lythgoe 1972; Loew and McFarland 1990). Furthermore, the spectral characteristics of the underwater environment, especially near land, may vary seasonally or over a shorter time scale, due to variations in the production of organic matter within the water and changes in the identity or quantity of material reaching it from land. Not only does the spectral content of the underwater light environment vary, so does its intensity, decreasing both with depth and turbidity. Consequently, fish live in almost every conceivable optical environment.

2 Visual Pigment Structure

All vertebrate visual pigments consist of two components: the chromophore, an aldehyde of vitamin A, which absorbs the light, and a protein, opsin, which determines the spectral absorption characteristics of the chromophore. They are located within the membranes of photoreceptor outer segment discs (Fig. 2).

The chromophore in most vertebrates is retinal, a derivative of vitamin A_1. However, some fish, reptiles and amphibians possess an additional chromophore, 3-dehydroretinal, derived from vitamin A_2. All visual pigments with retinal as their chromophore are known as rhodopsins, whereas vitamin A_2-based pigments are referred to as porphyropsins. A visual pigment consisting of a given opsin and using retinal as the chromophore, will have a narrower absorption spectrum peaking at shorter wavelengths than a pigment composed of the same opsin bound to the A_2-derived 3-dehydroretinal (Fig. 1).These two pigments, based on the same opsin but a different chromophore, are known as a pigment pair. Since porphyropsins are more longwave-sensitive, they are common in the rods and cones of freshwater fishes but are rare in the photoreceptors of marine animals. All visual pigments with λ_{max} beyond ca. 570 nm are based on 3-dehydroretinal.

All opsins have a similar structure, consisting of a chain of around 350 amino acids that crosses the outer segment disc membrane seven times in the form of α-helices. Isolated retinal and 3-dehydroretinal absorb at *ca.* 380 and 400 nm respectively. When bound to the opsin, always via a Schiff's base linkage at a lysine on the seventh transmembrane helix, the amino acids tune the chromophore to absorb at longer wavelengths. Thus, the absorption spectrum of a visual pigment depends on both the identity of the chromophore and the amino acid composition of the opsin surrounding that chromophore (Bowmaker 1991, 1995; Partridge and Cummings 1999 for reviews).

3 Shallow-Water Teleost Rod Visual Pigments

Since rods are used under scotopic conditions, they are concerned with maximising photon capture. Most simply this involves matching the absorption spectrum of the visual pigment to the ambient illumination (Partridge 1990) (sensitivity hypothesis). Consequently, fish inhabiting freshwater, which is an environment containing predominantly longwave radiation, generally have more red-sensitive visual pigments, using 3-dehydroretinal as the chromophore, than those inhabiting green-based coastal waters (Munz 1964; Schwanzara 1967; Munz and McFarland 1973, 1977; Lythgoe 1972). However, even among rods this correlation is not always good (e.g., Bowmaker et al. 1994) and animals from seemingly quite different habitats, for example, often have similar rod pigments (Lythgoe et al. 1994).

However, expecting a simple match between the background light and the visual pigment is in many cases an oversimplification because it assumes the animal is attempting to maximize sensitivity to the surrounding spacelight when in fact it is often detecting a specific target (Partridge and Cummings 1999). For example, a visual pigment with λ_{max} offset from the background will enhance the contrast of certain targets and hence increase the range at which they can be detected (Lythgoe 1968, 1972; McFarland and Munz 1975) (contrast hypothesis). An animal's behaviour and the requirements this imposes on the visual system must therefore also be considered along with the surrounding illumination. The spectral composition of the environment also changes throughout a 24-hour period and Hobson et al. (1981) have shown that scotopic visual pigments may be better matched to photopic conditions at twilight rather than the starlight and moonlight prevailing at night. However, in most instances a case can be made for explaining the λ_{max} of most rod visual pigments based on either the contrast or the sensitivity hypothesis. These arguments assume that sunlight is the only, or major source of radiation available. While this is certainly true for freshwater species and shallow oceanic animals, the situation is very different in the deep ocean.

4 Visual Pigments of Deep-Sea Fish

In fact the visual pigments of deep-sea fish are often cited as representing the classic example of visual pigment adaptation to the environment. Downwelling sunlight is rapidly attenuated with depth until, even in ideal conditions, at around 1000 m insufficient light penetrates to allow vision in even the most sensitive fish (Dentor 1990). Usually, however, sunlight becomes visually irrelevant at shallower depths. Not only is the intensity of this downwelling light reduced as it penetrates the water column, its spectral composition also becomes increasingly restricted. Due to spectral filtering by the water, at depth, light primarily consists of a narrow band of radiation between 470 and 480 nm (Kirk 1983). For this reason, it was predicted more than 60 years ago (Clark 1936; Bayliss et al. 1936) that in comparison to species living nearer the water surface, deep-sea fish should have visual pigments with their absorption maximum shifted towards shorter wavelengths. Superficially at least, such a hypsochromatic shift has been amply confirmed (Douglas et al. 1998a for review).

Figure 3 shows the λ_{max} values for all 175 species of deep-sea fish examined to date. Whereas it is certainly true that deep-sea fish on average have visual pigments absorbing around 470-490 nm, as might be expected based on the predominant wavelength in their environment, a simple matching to downwelling sunlight cannot be the whole explanation.

Firstly, the range of λ_{max} values (444 nm to in excess of 580 nm), is much greater than would be expected for a simple match to the background. Thus, animals caught in a single trawl and hence occupying approximately the same depths in the same geographical location, often have visual pigments whose λ_{max} values differ by 20 nm or more. Secondly, for simple background matching one would expect a gradual blue shift of visual pigments with depth.This is not the case (Douglas et al. 1998a). Thirdly, a simple sensitivity hypothesis cannot explain the presence of more than one visual pigment within the retinae of several deep-sea species (Douglas and Partridge 1997; Douglas et al. 1998a). Finally, theoretical modelling of the visual pigments required to optimally detect residual sunlight (Partridge, pers. comm.; Douglas et al. 1998a) indicates that visual pigments with maximum absorption at shorter wavelengths than those observed in deep-sea species are required.

Does one therefore conclude that the absorption of the visual pigment has little to do with the downwelling illumination? It is, for instance, possible that the point of maximum absorption of the pigment is determined largely by ambient pressure, phylogenetic constraints, the need to maximize the signal-to-noise ratio or thermal stability (Douglas et al. 1998a for review).

It has also been suggested that the precise location of the λ_{max} of deep-sea fish visual pigments is of little consequence due to their relatively broad absorption spectrum resulting from their unusually high concentration within the outer segments (Munz 1965; Bowmaker 1995).

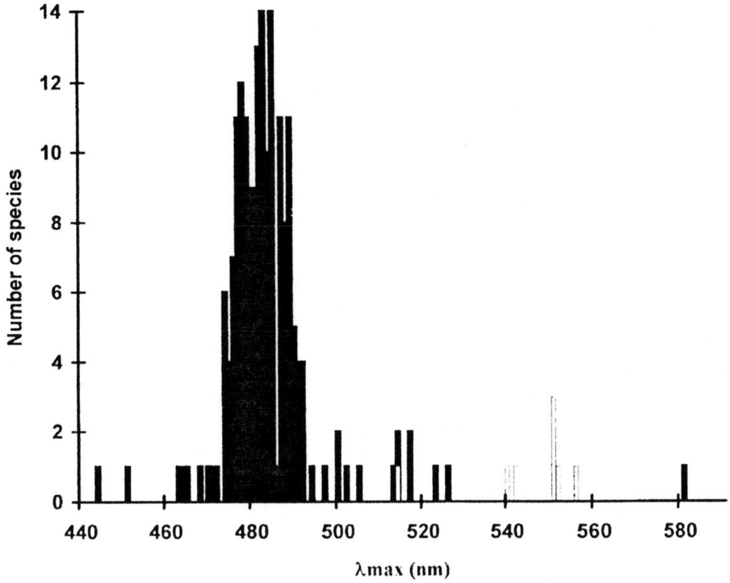

Fig. 3. Wavelength of maximum absorption for 195 visual pigments isolated from 175 species of deep-sea fish (After Douglas et al. 1998a). *Solid bars* represent rhodopsins and *open bars* porphyropsins

These factors, however, cannot explain the fact that virtually all fish living at depths greater than 1000 m, where sunlight does not penetrate in sufficient quantities to allow vision, appear to have large fully functional eyes. In fact, the explanation for this and the theoretical mismatch between deep-sea fish visual pigments and the residual downwelling spacelight noted above, the presence of multiple visual pigments and the unexpected variability of deepsea fish visual pigments, is that sunlight is not the only light available for vision in the deep-sea. More than 80% of all organisms inhabiting the deepsea produce their own light.

This bioluminescence is produced by photophores whose peak emissions usually lie between 450-500 nm, matching the wavelengths that most readily penetrate the water column, but emission maxima at longer wavelengths are not uncommon (Herring 1983; Widder et al. 1983; Latz et al. 1988; Mensinger and Case 1990, 1997). These emissions serve a variety of functions including intra- and inter-specific signalling, counter-illumination camouflage, a means of startling predators, an attractant to prey, and a way of simply illuminating their darkened world (Herring 1996: Widder 1999). Theoretical calculations indicate clearly that the absorption maxima of deepsea fish visual pigments are in most cases a good match to such biological sources of illumination (Partridge, pers. comm.; Douglas 1998a).

Unquestionably, the best illustration of how visual pigments are matched to a fish's bioluminescence is provided by three genera of deepsea stomiid dragon fish (*Malacosteus, Aristostomias and Pachystomias*). Whereas most deepsea organisms produce just one form of bioluminescence, these dragon fish have two light-producing organs around their eyes; a postorbital photophore producing blue-green bioluminescence similar to that produced by most other deepsea organisms, and a second, suborbital photophore producing far-red bioluminescence with spectral emissions peaking sharply at wavelengths beyond 700 nm (Denton et al. 1970, 1985; Widder et al. 1984; Figs. 4, 5).

Since the vast majority of deep-sea organisms have visual pigments maximally sensitive below 500 nm, they will be unable to see the far-red light produced by these dragon fish. Their visual pigments will not absorb significantly in this part of the spectrum. The stomiids, however, have evolved a number of adaptations to enable them to perceive their own far-red bioluminescence.

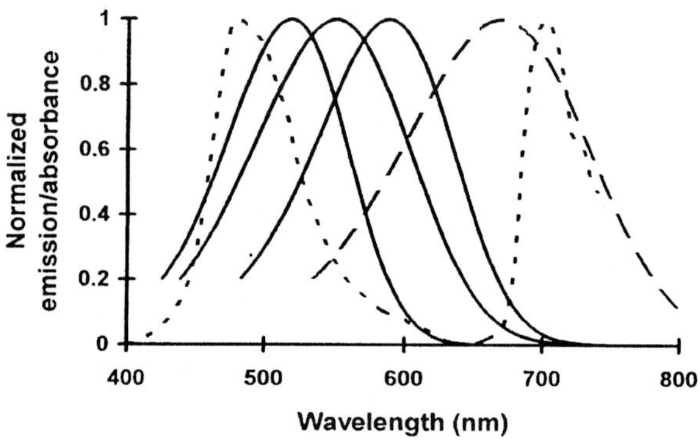

Fig. 4. Bioluminescence of *Aristostomias tittmanni* (*dotted line*; Widder et al. 1984), and the best fit templates (*solid lines*) of the three visual pigments so far identified in its retinae (a rhodopsin/porphyropsin pigment pair with λ_{max} values 520 nm and 551 nm and a rhodopsin with λ_{max} 588 nm). The *dashed line* represents a theoretical porphyropsin with λ_{max} 669 nm, which is the partner of the longwave-sensitive rhodopsin (calculated using the formula of Whitmore and Bowmaker 1989). (After Douglas et al. 1988)

The retinae of *Aristostomias* and *Pachystomias* contain at least three visual pigments that are longwave shifted compared to those of other deep-sea animals (λ_{max} values *ca.* 515-520 nm, 540-551 nm and 585-595 nm; Fig. 4) (Partridge and

Douglas 1995; Douglas et al. 1998a). The two most shortwave-sensitive pigments in each retina form a rhodopsin/ porphyropsin pigment pair in which the same shortwave opsin is bound to the two chromophores retinal and 3-dehydroretinal in different photoreceptors (Bowmaker et al. 1988; Partridge et al. 1989). The third, longer wave absorbing pigment, is a rhodopsin housed in a third class of rod and is most likely composed of a second longwave opsin bound to retinal (Partridge and Douglas 1995; Douglas et al. 1998a). This latter pigment is by far the most longwave-sensitive rod pigment ever described. However, even this pigment, although it fits this animal's bioluminescence much better than any other deep-sea visual pigment so far described, is still not a perfect match to a bioluminescent emission peaking sharply above 700 nm (Fig. 4).

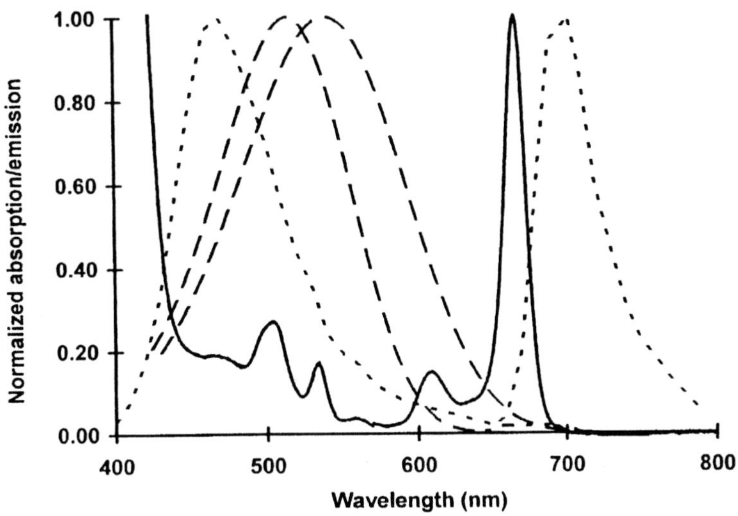

Fig. 5. Summary of absorbance spectra of *Malacosteus niger* retinal pigments and emission spectra of its bioluminescence. The *dashed curves* represent two visual pigment templates: a rhodopsin with λ_{max} 515 nm and a porphyropsin with λ_{max} 540 nm respectively. The emission spectra for the shortwave postorbital photophore (*dotted*) and the longwave emitting suborbital photophore (*dotted*) are taken from Widder et al. (1984). The absorption spectrum of the chlorophyll-derived photosensitizer is represented by a purified diethyl ether extract of a retinal suspension (*solid*). (After Douglas et al. 1999)

However, it would not be unreasonable to expect an even more red-sensitive visual pigment in the retinae of these animals made up of the longwave opsin combined with 3-dehydroretinal. Using a formula to predict the λ_{max} of a porphyropsin from a known rhodopsin utilising the same opsin (Whitmore and Bowmaker 1989), *Aristostomias* and *Pachystomias* might have a fourth, A_2-based pigment with λ_{max} values very approximately around 650 nm (this is a very rough estimate since existing formulae relating rhodopsin and porphyropsin λ_{max} values almost certainly do not apply at such long wavelengths). This would produce an excellent fit to their longwave bioluminescence (Fig. 4). If such a pigment exists, the most probable reason it has so far not been seen is that the pigments have always been isolated using dim red illumination, which is likely to bleach such a pigment.

The third genus of far-red light producing stomiid, *Malacosteus*, has achieved far-red sensitivity in a different way. Its retina contains only two pigments, which are equivalent to the shortwave pigment pair isolated in *Aristostomias* and *Pachystomias* (λ_{max} values *ca.* 515 nm and 542 nm) (Douglas et al. 1998b, 1999; Fig. 5). It cannot therefore see its own bioluminescence efficiently using just these two pigments. However, it has compensated for this apparent reduced sensitivity by utilizing a photostable, chlorophyll-derived, photosensitizer within its outer segments which absorbs light at its main absorption peak (672 nm) and in some, as yet undetermined, way, possibly involving the generation of a triplet state, isomerizing the shorter wave-sensitive visual pigments (Bowmaker et al. 1988; Douglas et al. 1998b, 1999; Fig. 5). Thus the three genera of far-red light producing stomiids have evolved two quite separate ways of ensuring visibility of their own bioluminescence. This is probably the best example of visual pigment adaptation to a specific photic stimulus.

Since stomiids are sensitive to the far-red bioluminescence produced by their suborbital photophores, which other animals in the deepsea cannot see, they have what could be regarded as a private waveband. This longwave light could therefore be used for intraspecific signalling, immune from detection by potential predators, or for the covert illumination of prey (Partridge and Douglas 1995). It would, however, not be altogether surprising if other, non-red light-producing, organisms had also evolved similar longwave sensitivity to counteract the advantage enjoyed by the stomiids.

5 Teleost Cone Visual Pigments

Most fish possess more than one spectral type of cone. While this does not prove that these animals have colour vision, which also requires the neural machinery to compare the output of these receptors, (proof of which requires extensive behavioral experiments which have only been done for a few species, e.g. Neumeyer 1992), it would be a huge surprise if most fish did not possess colour vision.

Teleost cones are morphologically distinct. There is a broad distinction between single and double cones (Fig. 2) and in general the smaller a single cone, the shorter the λ_{max} of the pigment it contains (Loew and Lythgoe 1978). While it is relatively easy to draw correlations between a fish's visual environment and its generally single-rod visual pigment, this is much harder to do with photopic pigments because each animal possesses up to four types of these receptors. Any correlations are therefore less clear-cut because different photoreceptors are likely to perform diverse visual functions, and exceptions to any rule are quite common.

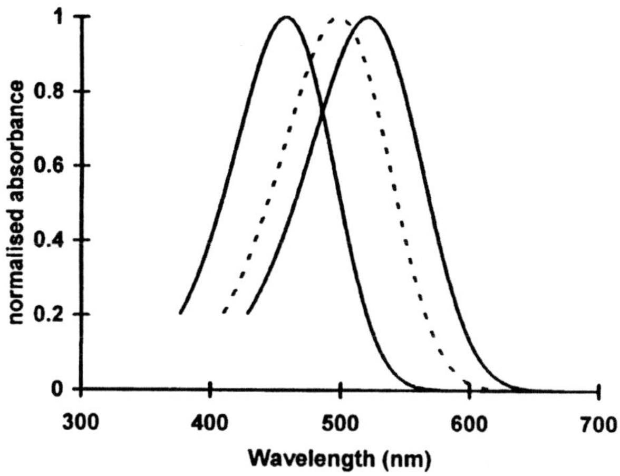

Fig. 6. Best-fitting templates to the visual pigments of the marine dwelling adult pollack, *Pollachius pollachius* (Shand et al. 1988). All pigments are rhodopsins with λ_{max} values 491 nm (rods *dotted*), 458 nm (blue cones) and 521 nm (green cones)

However, attempts have been made, with some success, to relate a fish's cone pigments to the visual environment it inhabits (e.g. Loew and Lythgoe 1978; Levine and MacNichol 1979; Lythgoe et al. 1994; Bowmaker et al. 1994; Bowmaker 1990, 1991, 1995 for reviews). In very general terms, for example, just as for rods, animals living in longwave-based freshwater generally have longer-wave sensitive visual pigments than animals inhabiting bluer oceanic environments (Lythgoe 1984). For example, many coastal species have blue/green sensitive dichromatic retinae possessing two cone types with λ_{max} values in the blue (440-460 nm) and green (520-540 nm) part of the spectrum (Fig. 6), a combination that is ideal for allowing them to make both brightness and chromaticity discriminations (Lythgoe and Partridge 1991). Even if there are more than two cone types present (e.g. Lythgoe et al. 1994) oceanic and coastal species generally lack red-sensitive receptors.

Among freshwater species, however, longwave receptors with λ_{max} values up to 630 nm are quite common (Fig. 7) and species such as the glass catfish, *Kryptopterus bicirrhus*, in fact have only longwave-sensitive cones (Douglas and Wagner 1984) (Fig. 8). Deeper living freshwater fish, such as many catfish, which inhabit turbid water devoid of short wavelengths, generally have no short-wave sensitive cones (Levine and MacNichol 1979; Sillman et al. 1993).

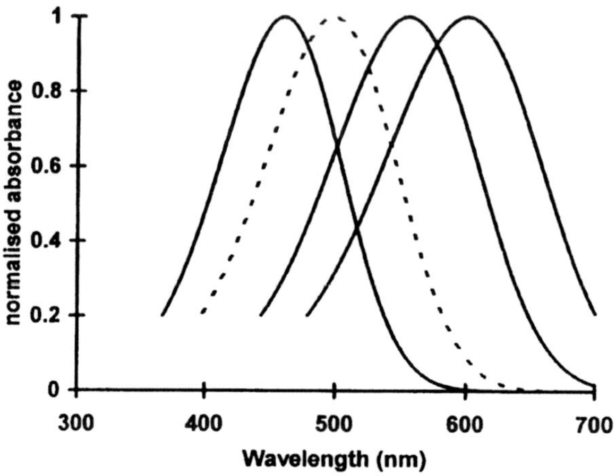

Fig. 7. Best fitting templates to the visual pigments of the freshwater cichlid *Nannacara anomala* (Ali et al. 1978).The photoreceptors contain primarily porphyropsins with λ_{max} values around 498 nm (rods *dotted*), 460 nm (blue single cones), 555 nm (green member of double cone) and 600 nm (red member of double cones)

An exception to this rule appears to be the sturgeon, one species of which exists with some catfish (Sillman et al. 1993), yet does contain a shortwave-sensitive cone (Loew and Sillman 1993; Sillman et al. 1999). This may be explained by the fact that sturgeon sometimes migrate to bluer oceanic environments (Sillman et al. 1993) while catfish remain in freshwater.

In the freshwater guppy all animals contain the same pigments within their rods (λ_{max} 503 nm), and blue (λ_{max} 410 nm) and green (λ_{max} 465 nm) cones. However, the long wavelength-sensitive cones are polymorphic with λ_{max} values ranging between 520 and 580 nm (Archer et al. 1987; Archer and Lythgoe 1990). This is because the longwave cone can contain either a 533 nm or a 572 nm rhodopsin, or a mixture of the two (Archer and Lythgoe 1990). This polymorphism will lead to individual differences in color perception which may be related to the great variety of body coloration displayed by guppies.

Although we tend to think of freshwater as a longwave environment, this is really only applicable in deeper water and species inhabiting surface layers experience the sun's full spectrum and consequently have quite a broad spectral sensitivity. Most species are at least trichromatic with receptors in the blue, green and red part of the spectrum and even ultraviolet-sensitive cones are not uncommon in species inhabiting freshwater (e.g., Avery et al. 1983; Hárosi and Hashimoto 1983; Hárosi 1985; Bowmaker and Kunz 1987; Whitmore and Bowmaker 1989; Archer and Lythgoe

1990; Hawryshyn and Hárosi 1991, 1994; Bowmaker et al. 1991; Novales-Flamarique et al. 1998) with many species having four cone visual pigments in their retinae (Fig. 9). Pigments absorbing maximally in the ultraviolet also occur in marine fish (McFarland and Loew 1994).

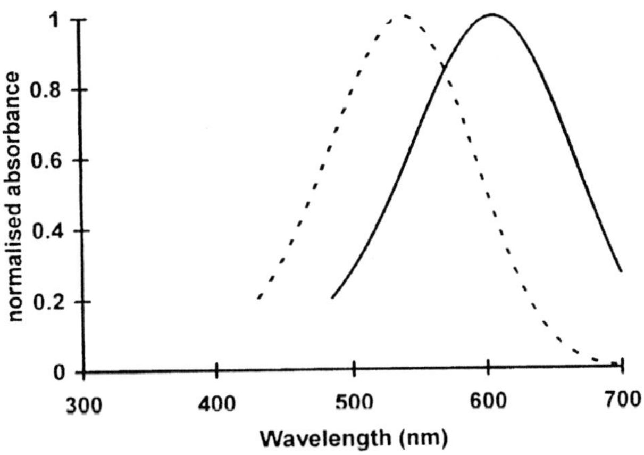

Fig. 8. Best fitting templates to the visual pigments of the freshwater glass catfish *Kryptopterus bicirrhis* (Douglas and Wagner 1984). Both pigments are porphyropsins with λ_{max} values 540 nm (rods *dotted*) and 607 nm (cones)

When comparing the visual pigments of different animals, phylogenetic factors may obscure environmental relationships. The answer is to examine closely related species inhabiting different visual environments. A recent study looked at 12 species of marine snappers belonging to the same genus (*Lutjanus*) whose habitat ranged from the relatively blue outer edge of the Great Barrier Reef to more longwave-dominated coastal habitats (Lythgoe et al. 1994). Their double cones showed a clear tendency to be more longwave sensitive in animals living closer to the shore.

Similarly, the double cones of different species of closely related cottoids of lake Baikal show a trend towards increased shortwave sensitivity with depth, possibly reflecting the increasingly bluer environment of the lower reaches of the lake (Bowmaker et al. 1994). Interestingly, in both the lake Baikal cottoids and the Australian lutjanids, the shortwave-sensitive single cones less obviously matched the environmental illumination (Lythgoe et al. 1994; Bowmaker et al. 1994) and possibly reflect more complex visual tasks than simple detection of the background (Partridge and Cummings 1999).

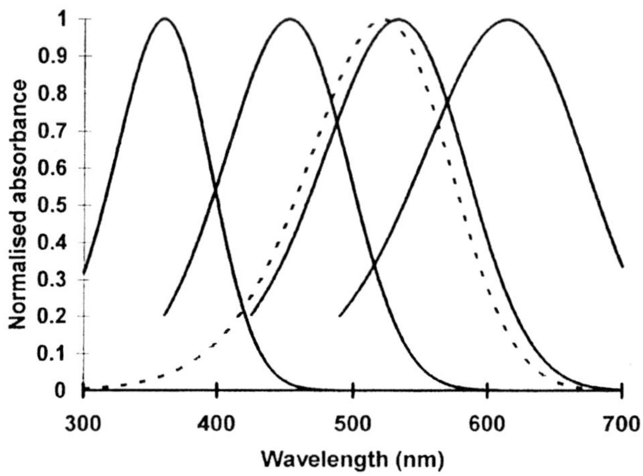

Fig. 9. Best fitting templates to the visual pigments of the goldfish *Carassius auratus* (Bowmaker et al. 1991). All pigments are best fit by porphyropsin templates with λ_{max} values 521 nm (rods *dotted*), 360 nm (UV cones), 452 nm (blue cones), 532 nm (green cones) and 614 nm (red cones)

6 Changes in Visual Pigment Content Related to Age and Environment

One of the ways of assessing visual adaptation to the environment is to examine how the visual system changes as an animal alters its photic surroundings during its lifetime (Beaudet and Hawryshyrn 1999). Developmental visual pigment changes can be brought about in three ways: development of a new class of photoreceptor, alteration of visual pigments in existing photoreceptors by the expression of a new opsin, or a change in chromophore. Fish employ all three strategies.

The visual pigment content of virtually all fish changes to some extent during their lifetime because the photoreceptors generally develop sequentially (Powers and Raymond 1990). Most fish are born with retinae consisting entirely of single cones, with rods and double cones developing later (e.g. Shand et al. 1999). This is certain to lead to changes in visual pigment content with age (Evans et al. 1993; Loew and Sillman 1993). Such changes are often related to changes in habitat and/or feeding behaviour. Often, for instance, the development of double cones, which may be more concerned with coding luminosity rather than providing information utilized in color vision (Maier and Bowmaker 1993), is associated with a shift to deeper water (Boehlert 1978; Evans et al. 1993; Beaudet and Hawryshyn 1999). Similarly, extreme shortwave sensitivity appears to be restricted in many species to younger individuals (e.g., Bowmaker and Kunz 1987; Hawryshyn et al. 1989; Whitmore and Bowmaker 1989; Loew and Wahl 1991), although in salmonids the possession of UV may be related to life in freshwater rather than simply an age effect, since older salmon, which left freshwater and migrated to a deeper existence in the ocean on smoltification, may regain their UV sensitivity on reentering shallow freshwater (Beaudet and Hawryshyn 1999 for review). Such changes in an animal's UV sensitivity may reflect a change in the animal's lifestyle and habitat away from a planktivorous surface feeding existence to life in deeper water (Browman and Hawryshyn 1994; Bowmaker 1995).

The best examples of changes in the chromophore to alter a visual pigment to suit a particular environment are provided by migratory species moving between red-based freshwater and bluer oceanic habitats. Anadromous salmon, for instance, living in the sea, have relatively shortwave-sensitive rhodopsin visual pigments, but when they migrate to freshwater to spawn, the retinal is replaced by 3-dehydroretinal to yield longer wave-sensitive porphyropsins (Beatty 1984).

Some non-migratory species, such as the freshwater rudd, also display changes in the rhodopsin/porphyropsin ratio of both their rod and cone pigments (Dartnall et al. 1961; Loew and Dartnall 1967; Beatty 1984; Whitmore and Bowmaker 1989). Although this ratio is influenced not only by age, but also by changes in season, day length, light intensity, temperature, hormonal status and diet, its significance remains unclear and there is often no obvious relationship between photic environment and the rhodopsin/ porphyropsin ratio (Muntz and Mouat 1984).

The developmental expression of new opsins to produce novel visual pigments has been less often described (although it is probably not less common) than chromophore changes. Thus, new opsins are probably expressed in both the pollack (Shand et al. 1988) and the goatfish (Shand 1993) as they mature to adapt to their new, more benthic, existence, as well as in the flounder (Evans et al. 1993) and *Chromis punctipinnis* (McFarland and Loew 1994). Why such changes occur is not always obvious (McFarland and Loew 1994). Pollack, for instance, studied in the English channel, change the spectral sensitivity of their single cones from a λ_{max} of around 420 nm, when they are surface feeding adults, to cones with λ_{max} 450-460 nm (Fig. 6), which adapts them to their deeper existence where there is less shortwave illumination. Conversely, however, a 580 nm cone pigment in planktivorous tropical goatfish is replaced by one absorbing at 530 nm for deeper-dwelling adults (Shand 1993).

Catadromous eels are particularly interesting because they change both their chromophore and express a new opsin as they migrate from relatively shallow freshwater into the deep ocean during their spawning migration to the Sargasso Sea. Individual rods contain a mixture of rhodopsin (λ_{max} 501 nm) and porphyropsin (λ_{max} 523 nm) visual pigments. In freshwater, as in other species, the A_2-based pigment predominates, but on migration towards the sea, the retina becomes primarily A_1-based (Wood et al. 1992). However, in addition to changing its chromophore during migration towards the sea, a new opsin is expressed within existing rods (Wood et al. 1992; Wood and Partridge 1993; Archer et al. 1995), resulting in a novel rhodopsin with λ_{max} 482 nm, typical of a deepsea fish.

7 Conclusions

Teleosts probably show the greatest variability of visual pigments and types of photoreceptor of any group of vertebrates. This reflects the wide range of photic environments that they occupy, from the almost total absence of daylight in the deepsea to the variable coloured waters of coastal regions and freshwater lakes and rivers. In a broad sense, there is a clear correlation between the complement of rods and cones in any given species and its photic environment. However, we have little understanding of the significant variations found between species in visual pigments and photoreceptors that occur within defined photic environments. Presumably, these reflect differences in visual behaviour related to factors such as feeding strategies, mating behaviour and predation. It will only be when all of the physiological, behavioral and ecological information has been obtained and correlated that we may begin to fully appreciate the visual ecology of teleosts.

References

Ali MA, Hárosi FI, Wagner H-J (1978) Photoreceptors and visual pigments in a cichlid fish, *Nannacara anomala.* Sensory Processes 2:130-145

Archer S, Hope A, Partridge JC (1995) The molecular basis for the green-blue sensitivity shift in the rod visual pigments of the European eel. Proc Roy Soc Lond B 262:289-295

Archer SN, Endler JA, Lythgoe JN, Partridge JC (1987) Visual pigment polymorphism in the guppy *Poecilia reticulata.* Vision Res 27:1243-1252

Archer SN, Lythgoe JN (1990) The visual pigment basis for cone polymorphism in the guppy, *Poecilia reticulata.* Vision Res 30:225-233

Avery JA, Bowmaker JK, Djamgoz MBA, Downing JEG (1983) Ultraviolet sensitive receptors in a freshwater fish. J Physiol 334:23P

Bayliss LE, Lythgoe RJ, Tansley K (1936) Some new forms of visual purple found in deep-sea fish, with a note on the visual cells of origin. Proc Roy Soc Lond B 816:95-113

Beatty DD (1984) Visual pigments and the labile scotopic visual system of fish. Vision Res 24:1563-1573

Beaudet L, Hawryshyn CW (1999) Ecological aspects of vertebrate visual ontogeny. In: Archer SN, Djamdoz MBA, Loew ER, Partridge JC, Vellerga S (eds) Adaptive Mechanisms in the Ecology of Vision. Kluwer, Dordrecht, pp 413-437

Boehlert GW (1978) Intraspecific evidence for the function of single and double cones in the teleost retina. Science 202:309-311

Bowmaker JK (1984) Microspectrophotometry of vertebrate photoreceptors. A brief review. Vision Res 24:1641-1650

Bowmaker JK (1990) Visual pigments of fishes. In: Douglas RH, Djamgoz MBA (eds) The Visual System of Fish. Chapman and Hall, London, pp 81-107

Bowmaker JK (1991) Evolution of visual pigments and photoreceptors. In: Gregory

RL, Cronly-Dillon JR (eds) Vision and Visual Dysfunction: Vol 2, Evolution of the Eye and Visual System. Macmillan, London, pp 63-81

Bowmaker JK (1995) The visual pigments of fish. Prog Ret Eye Res 15:1-31

Bowmaker JK, Dartnall HJA, Herring PJ (1988) Longwave-sensitive visual pigments in some deep-sea fishes: segregation of 'paired' rhodopsins and porphyropsins. J Comp Physiol A 163:688-698

Bowmaker JK, Govardovskii VI, Shukolyukov SA, Zueva LV, Hunt DM, Sideleva VG, Smirnova OG (1994) Visual pigments and the photic environment: the cottoid fish of lake Baikal. Vision Res 34:591-605

Bowmaker JK, Kunz YW (1987) Ultraviolet receptors, tetrachromatic colour vision and retinal mosaics in the brown trout (*Salmo trutta*): age-dependent changes. Vision Res 27:2101-2108

Bowmaker JK, Thorpe A, Douglas RH (1991) Ultraviolet-sensitive cones in the goldfish. Vision Res 31:349-352

Browman HI, Hawryshyn CW (1994) The developmental trajectory of ultraviolet photosensitivity in rainbow trout is altered by thyroxine. Vision Res 34:1397-1406

Clarke GL (1936) On the depth at which fish can see. Ecology 17:452-456

Dartnall HJA, Lander MR, Munz FW (1961) Periodic changes in the visual pigments of fish. In: Christensen B, Buchmann B (eds) Progress in Photobiology. Elsevier, Amsterdam, pp 203-213

Denton EJ (1990) Light and vision at depths greater than 200 metres. In: Herring PJ, Campbell AK, Whitfield M, Maddock L (eds) Light and Life in the Sea. Cambridge University Press, Cambridge, pp 127-148

Denton EJ, Gilpin-Brown JB, Wright PG (1970) On the 'filter' in the photophores of mesopelagic fish and on a fish emitting red light and especially sensitive to red light. J Physiol 208:72-73P

Denton EJ, Herring PJ, Widder EA, Latz MF, Case JF (1985) The roles of filters in the photophores of oceanic animals and their relation to vision in the oceanic environment. Proc Roy Soc Lond B 225:63-97

Douglas RH (1991) The aquatic environment as a natural laboratory for vision research: fish as a model system. In: Covacci R, Djamgoz MBA, Vallerga S (eds) Aspects of Marine Biology with an Emphasis on the Mediterranean. IMC Publications, Oristano, pp 13-20

Douglas RH, Marshall NJ (1999) A review of vertebrate and invertebrate optical filters. In: Archer SN, Djamgoz MBA, Loew ER, Partridge JC, Vallerga S (eds) Adaptive Mechanisms in the Ecology of Vision. Kluwer, Dordrecht, pp 95-162

Douglas RH, Partridge JC (1997) On the visual pigments of deep-sea fish. J Fish Biol 50:68-85

Douglas RH, Partridge JC, Dulai K, Hunt D, Mullineaux CW, Tauber AY, Hynninen PH (1998a) Dragon fish see using chlorophyll. Nature 393:423-424

Douglas RH, Partridge JC, Dulai KS, Hunt DM, Mullineaux CW, Hynninen PH (1999) Enhanced retinal longwave sensitivity using a chlorophyll-derived photosensitiser in *Malacosteus niger*, a deep-sea dragon fish with far red bioluminescence. Vision Res 39:2817-2832

Douglas RH, Partridge JC, Hope AJ (1995) Visual and lenticular pigments in the eyes of demersal deep-sea fishes. J Comp Physiol A 177:111-122

Douglas RH, Partridge JC, Marshall NJ (1998b) The eyes of deep-sea fish I: lens pigmentation, tapeta and visual pigments. Prog Ret Eye Res 17:597-636

Douglas RH, Wagner H-J (1984) Action spectrum of photomechanical cone contraction in the catfish retina. Invest Ophthalmol Vis Sci 25:534-538

Evans BI, Hárosi FI, Fernald RD (1993) Photoreceptor spectral absorbance in larval and adult winter flounder (*Pseudopleuronectes americanus*). Vis Neurosci 10:1065-1071

Hárosi FI (1985) Ultraviolet- and violet-absorbing vertebrate visual pigments: dichroic and bleaching properties. In: Levine JS, Fein A (eds) The Visual System. Alan Liss, New York, pp 41-55

Hárosi FI, Hashimoto Y (1983) Ultraviolet visual pigment in a vertebrate: a tetrachromatic cone system in the dace. Science 222:1021-1023

Hawryshyn CW, Arnold MG, Chaisson DJ, Martin PC (1989) The ontogeny of ultraviolet photosensitivity in rainbow trout (*Salmo gairdneri*). Vis Neurosci 2:247-254

Hawryshyn CW, Hárosi FI (1991) Ultraviolet photoreception in carp: microspectrophotometry and behaviorally determined action spectra. Vision Res 31:567-576

Hawryshyn CW, Hárosi FI (1994) Spectral characteristics of visual pigments in rainbow trout (*Oncorhynchus mykiss*). Vision Res 34:1385-1392

Herring PJ (1983) The spectral characteristics of luminous marine organisms. Proc Roy Soc Lond B 220:183-217

Herring PJ (1996) Light, colour and vision in the ocean. In: Summerhayes CP, Thorpe SA (eds) Oceanography: an Illustrated Guide. Mason Publishing, London, pp 212-227

Hobson ES, McFarland WN, Chess JR (1981) Crepuscular and nocturnal activities of Californian nearshore fishes, with consideration of their scotopic visual pigments and photic environment. Fishery Bulletin 79:1-30

Jacobs GH (1992) Ultraviolet vision in vertebrates. American Zoologist 32:544-554

Kirk JTO (1983) Light and Photosynthesis in Aquatic Ecosystems. Cambridge University Press, Cambridge

Knowles A, Dartnall HJA (1977) The Photobiology of Vision. In: Davson H (ed) The Eye: Vol 2B, Academic Press, New York, pp 1-689

Latz MI, Frank TM, Case JF (1988) Spectral composition of epipelagic organisms from the Sargasso sea. Marine Biol 98:441-446

Levine JS, MacNichol EF (1979) Visual pigments in teleost fishes: effects of habitat, microhabitat and behaviour on visual system evolution. Sensory Processes 3:95-130

Loew ER, Dartnall HJA (1976) Vitamin A1/A2-based visual pigment mixtures in cones of the rudd. Vision Res 16:891-896

Loew ER, Lythgoe JN (1978) The ecology of cone pigments in teleost fish. Vision Res 18:715-722

Loew ER, McFarland WN (1990) The underwater visual environment. In: Douglas RH, Djamgoz MBA (eds) The Visual System of Fish. Chapman and Hall, London,

pp 1-43
Loew ER, Sillman AJ (1993) Age-related changes in the visual pigments of the white sturgeon (*Acipenser transmontanus*). Can J Zool 71:1552-1557
Loew ER, Wahl CM (1991) A short-wavelength sensitive cone mechanism in juvenile yellow perch, *Perca flavescens*. Vision Res 31:353-360
Lythgoe JN (1968) Visual pigments and visual range under water. Vision Res 8:997-1012
Lythgoe JN (1972) The adaptation of visual pigments to photic environment. In: Dartnall HJA (ed) Handbook of Sensory Physiology: Vol VII/1, Photochemistry of Vision. Springer, Berlin, pp 566-603
Lythgoe JN (1984) Visual pigments and environmental light. Vision Res 24:1539-1550
Lythgoe JN, Muntz WRA, Partridge JC, Shand J, Williams DM (1994) The ecology of the visual pigments of snappers (*Lutjanidae*) on the Great Barrier Reef. J Comp Physiol A 174:461-467
Lythgoe JN, Partridge JC (1991) The modelling of optimal visual pigments of dichromatic teleosts in green coastal waters. Vision Res 31:361-371
Maier EJ, Bowmaker JK (1993) Colour vision in a passeriform bird, *Leiothrix lutea*: correlation of visual pigment absorbance and oil droplet transmission with spectral sensitivity. J Comp Physiol A 172:295-301
Marshall NJ, Oberwinkler J (1999) The colourful world of the mantis shrimp. Nature 401:873-874
McFarland WN, Loew ER (1994) Ultraviolet visual pigments in marine fishes of the family Pomacentridae. Vision Res 34:1393-1396
McFarland WN, Munz FW (1975) The evolution of photopic visual pigments in fishes. Vision Res 15:1071-1080
Mensinger AF, Case JF (1990) Luminescent properties of deep-sea fish. J Exptl Mar Biol Ecol 144:1-15
Mensinger AF, Case JF (1997) Luminescent properties of fishes from nearshore California basins. J Exptl Mar Biol Ecol 210:75-90
Muntz WR, Mouat GS (1984) Annual variations in the visual pigments of brown trout inhibiting lochs providing different light environments. Vision Res 24:1575-1580
Munz FW (1964) The visual pigments of epipelagic and rocky shore fishes. Vision Res 4:441-454
Munz FW (1965) Adaptation of visual pigments to the photic environment. In: Wolstenholme GEW, Knights J (eds) Ciba Foundation Symposium on Physiological and Environmental Psychology of Colour Vision. A. Churchill Ltd, London, pp 27-45
Munz FW, McFarland WN (1973) The significance of spectral position in the rhodopsins of tropical marine fishes. Vision Res 13:1828-1874
Munz FW, McFarland WN (1977) Evolutionary adaptations of fishes to the photic environment. In: Crescitelli F (ed) Handbooh of Sensory Physiology: Vol VII/5, The Visual System of Vertebrates. Springer Verlag, New York, pp 193-274
Neumeyer C (1992) Tetrachromatic color vision in goldfish: evidence from color

mixture experiments. J Comp Physiol A 171:639-649

Novales-Flamarique I, Hawryshyn CW (1998) The common white sucker (*Catostomus commersoni*): a fish with ultraviolet sensitivity that lacks polarization sensitivity. J Comp Physiol A 182:331-341

Partridge JC (1990) The colour sensitivity and vision in fishes. In: Herring PJ, Campbell AK, Whitfield M, Maddock L (eds) Light and Life in the Sea. Cambridge University Press, Cambridge, pp 167-184

Partridge JC, Cummings ME (1999) Adaptations of visual pigments to the aquatic environment. In: Archer SN, Djamgoz MBA, Loew ER, Partridge JC, Valerga S (eds) Adaptive Mechanisms in the Ecology of Vision. Kluwer, Dordrecht, pp 251-283

Partridge JC, De Grip WJ (1991) A new template for rhodopsin (Vitamin A_1 based) visual pigments. Vision Res 31:619-630

Partridge JC, Douglas RH (1995) Far-red sensitivity of dragon fish. Nature 375:21-22

Partridge JC, Shand J, Archer SN, Lythgoe JN, van Groningen-Luyben WAHM (1989) Interspecific variation in the visual pigments of deep-sea fishes. J Comp Physiol A 164:513-529

Powers MK, Raymond PA (1990) Development of the visual system. In: Douglas RH, Djamgoz MBA (eds) The Visual System of Fish. Chapman Hall, London, pp 419-442

Schwanzara SA (1967) The visual pigments of freshwater fishes. Vision Res 7:121-148

Shand J (1993) Changes in the spectral absorption of cone visual pigments during the settlement of the goatfish *Upeneus tragula*: the loss of red sensitivity as a benthic existence begins. J Comp Physiol A 173:115-121

Shand J, Archer MA, Collin SP (1999) Ontogenetic changes in the retinal photoreceptor mosaic in a fish, the black bream, *Acanthopagrus butcheri*. J Comp Neurol 412:203-217

Shand J, Partridge JC, Archer SN, Potts GW, Lythgoe JN (1988) Spectral absorbance changes in the violet/blue sensitive cones of the juvenile pollack, *Pollachius pollachius*. J Comp Physiol A 163:699-703

Sillman AJ, O'leary CJ, Tarantino CD, Loew ER (1999) The photoreceptors and visual pigments of two species of Acipenseriformes, the shovelnose sturgeon (*Scaphirhynchus platorynchus*) and the paddlefish (*Polyodon spathula*). J Comp Physiol A 184:37-47

Sillman AJ, Ronan SJ, Loew ER (1993) Scanning electron-microscopy and microspectrophotometry of the photoreceptors of ictalurid catfishes. J Comp Physiol A 173:801-807

Stavenga DG, Smits RP, Hoenders BJ (1993) Simple exponential functions describing the absorbency bands of visual pigment spectra. Vision Res 33:1011-1017

Whitmore AV, Bowmaker JK (1989) Seasonal variation in cone sensitivity and short-wave absorbing visual pigments in the rudd, *Scardinius erythrophthalmus*. J Comp Physiol A 166:103-115

Widder EA (1999) Bioluminescence. In: Archer SN, Djamgoz MBA, Loew ER,

Partridge JC, Valerga S (eds) Adaptive Mechanisms in the Ecology of Vision. Kluwer, Dordrecht, pp 555-581

Widder EA, Latz MI, Case JF (1983) Marine bioluminescence spectra measured with an optical multichannel detection system. Biol Bull 165:791-810

Widder EA, Latz MI, Herring PJ, Case JF (1984) Far red bioluminescence from two deep-sea fishes. Science 225:512-514

Wood P, Partridge JC (1993) Opsin substitution induced in retinal rods of the eel (*Anguilla anguilla* L.): a model for G-protein-linked receptors. Proc Roy Soc Lond B 254:227-232

Wood P, Partridge JC, de Grip WJ (1992) Rod visual pigment changes in the elver of the eel *Anguilla anguilla* (L.) measured by microspectrophotometry. J Fish Biol 41:601-611

V. 1 Selectivity in Chemical Communication Systems of Arthropods

Stefan Schulz

Institut für Organische Chemie, Technische Universität Braunschweig, Hagenring 30, 38106 Braunschweig, Germany

Abstract
The structural diversity of insect and arachnid pheromones is discussed in relation to the required uniqueness of a chemical signal to be perceived against the background of chemical noise. Various levels of structural elaboration can be found, depending on the requirements of a species its environment. The physical properties needed for proper function and the biosynthetic routes used for the production of these compounds are also discussed. The wide variability of communication systems is presented, showing the difficulties of drawing general rules useful for understanding the structural diversity in arthropod pheromone communication.

Key words Pheromones, structural diversity, biosynthesis, physical properties of pheromones, spiders

1 Introduction and Terminology

Compounds employed in chemical communication systems between living organisms are termed semiochemicals. According to their function, they can be divided into pheromones and allelochemicals (see Fig. 1), based on the system of Nordlund (Nordlund and Jones 1981).

Semiochemicals

Pheromones		**Allelochemicals**	
intraspecific	change in	interspecific	advantage for
Primer	physiology	**Kairomone**	receiver
Releaser	behavior	**Allomone**	emitter
		Synomone	both

Fig. 1. Terms used in chemical communication systems.

Pheromones are compounds used for communication between members of the same species. They can be further divided into releasers, which elicit a modification of behavior in the receiver, and primers, which induce changes in the physiology of the receiver. The vast majority of pheromones identified so far are releasers. Typical members of this class are sex pheromones of female night-flying moths, which attract conspecific males over long distances. Further examples are alarm pheromones of ants or aggregation pheromones of bark beetles. A typical primer is (*E*)-9-oxodec-2-enoic acid, the so-called queen substance of the honey bee, which inhibits queen rearing and diminishes the release of juvenile hormone in worker bees, but also acts as a releaser, being involved in the retinue response (vander Meer et al. 1998).

Unlike pheromones, allelochemicals are used for communication between different species. They are divided into kairomones, which are advantageous for the receiver, allomones, which are advantageous for the emitter, and synomones, which are useful for both sender and receiver. Allomones can be defensive compounds given off by a species to deter predators, whereas kairomones can be compounds that are given off by a prey species and that attract a predator. For example, the alarm pheromone 6-methyl-5-hepten-2-one (sulcatone) of the meat ant *Iridomyrmex purpureus* is used by the ant-feeding spider *Habronestes bradleyi* to find its prey (Allan et al. 1996). This shows that a compound given off by one species can, at the same time, be a pheromone (releaser) and a kairomone, depending on the situation in which it is acting. The volatile compounds given off by many flowers are typical synomones. They attract pollinating insects, which obtain a reward in the form of pollen or nectar.

A term also used frequently in recent times is infochemical, which is less clearly defined. It can be used to denote compounds that are used by an organism to gather information and that happen to be present in the environment, but that are not produced by the emitter for a specific reason. All these definitions are not very precise and sometimes cause confusion when used with different meanings.

The number of arthropod species for which the structures and composition of pheromones have been elucidated reaches about 1500 (see Table 1). Most of the work has been carried out on specific insect orders, often driven by the desire of applying pheromones in pest control (e.g., Lepidoptera). Nevertheless, compared with the nearly 850000 insect species described so far (Daly et al. 1998), this number is only a limited sample (0.17%) and may even be biased against certain orders. General rules derived therefrom have to be treated with caution, and this is even more true of pheromones from arachnids.

All semiochemicals used in various biological contexts carry by definition information. To use this chemical channel successfully, an organism must employ compounds that are sufficiently specific to overcome the surrounding chemical noise. This noise is always present, as in other sensory systems, and consists of compounds given off into the environment by living organisms, decaying material, anthropogenic sources, geochemical processes, and so on. Different strategies might be used to achieve selective signaling. Generally, specificity can be achieved on the emitter side by producing a specific signal, or on the receiver side by a selective receiving mechanism. Mixtures of compounds are most often used to obtain a selective signal. They allow greater flexibility in the composition of the signal, often by means of less specific compounds that are also employed by other species. In

Table 1. Estimated numbers of species for which pheromones have been identified

Order	Number
Insecta	
Lepidoptera	1000
Coleoptera	200
Hymenoptera	100
Diptera	70
Arachnida	
Acari	25
Araneae	3
Scorpiones	0

contrast, a specific single compound needs more biosynthetic effort. The ecological niche in which a species lives often determines the level of specificity required. Related species, which are separated through space or by different patterns of activity, can use identical compounds without any difficulty. Close-range signals, in which the partners have already recognized each other, require less specificity. Defensive compounds have to induce strong effects on a receptor or other target areas in the attacker. Semiochemicals can be produced either de novo by biosynthesis starting from common precursors present in the primary metabolism of a species, or by ingestion and derivatization of unique compounds (secondary metabolites) present in the environment of the emitter. Monophagous or oligophagous species can successfully use specific secondary metabolites of their hosts as precursors. Highly specific signaling compounds can be formed by this process. An example are the male pheromone components, derived from pyrrolizidine alkaloids (compounds **5**, **44** and **45** in Figs. 3 and 13), of several lepidopteran genera.

In the present article, the emitting side of the chemical channel, and particularly the carrier of the information, viz. the involved molecules, will be discussed, with special emphasis on the uniqueness of semiochemicals in conjunction with physical properties and biosynthetic routes. For most of the semiochemicals identified so far, the biosynthetic pathway leading to them is not known. Nevertheless, well-established pathways to similar compounds in other organisms allow the discussion of the transformations needed to synthesize these semiochemicals.

2 Volatility and Structure

Compounds used for chemical communication through space have to be volatile or must be absorbed on small particles that will easily travel in agitated air and transport the actual signaling components. The term volatile is not clearly defined, and is often used in a quite different context. The volatility of a compound depends on the size (molecular weight), polarity, and structure of the molecule and can be expressed as the vapor pressure. Molecules with high molecular weight and high polarity have a low vapor pressure. Most salts are, for example, not volatile, but apolar compounds with a relatively high molecular weight may still be volatile enough to be used in chemical communication systems.

Arguably, the most volatile pheromone component identified so far is ethyl acetate, which is a constituent of the male sex pheromone of the mediterranean fruit fly, *Ceratitis capitata* (Jang et al. 1994). Its boiling point of 77°C and the vapor pressure of 76 Torr at room temperature indicate that the substance easily evaporates. If a reservoir (e.g., a gland or surface) containing **1** is exposed, most of the molecules will evaporate immediately, giving a short signal of high intensity. A high mean emission rate of the pheromone, up to 11.4 ng min^{-1} per fly, has been measured (Flath et al. 1993). The flies assemble in small groups or leks for calling, and within this group, every male calls only for short periods. This may reflect the fast evaporation of the compound. To maintain a steady evaporation of a substance, either a biosynthetic process that matches the evaporation rate of the pheromone (which must be very fast in the case of **1**), or a reservoir that contains the pheromone is required. Such a reservoir, e.g., a gland, can be filled either by a continuous biosynthetic process or by uptake with food, when food components are used as pheromones. The range of a very volatile signal is relatively small, because the high diffusion coefficient of small molecules leads to rapid dilution in air below the threshold level of a given receptor. In addition, the concentration of low molecular weight compounds in the atmosphere is relatively high (Kesselmeier and Staudt 1999), because of their high volatility. The possible structural variability rises exponentially with an increasing number of atoms in the molecule. Both effects favor larger molecules for signal transmission.

If the vapor pressure decreases with increasing size or polarity of the molecule, the evaporation rate is reduced, and the signal can be given off for a longer period of time, because the time needed for biosynthesis or translocation now is a less limiting factor. Larger molecules also diffuse more slowly, and thus the signal stays longer in the environment. At the other end of the vapor pressure scale of pheromones is nonacosane (**2**), which is used as a pheromone component by the solitary digger bee, *Andrena nigroaenea* (Schiestl et al. 1999). It has a vapor pressure of 1.33×10^{-4} Torr at 105°C (calculated to be 5.0×10^{-10} Torr at room temperature!) and a boiling point of 714°C (Piacente et al. 1991). Such compounds will be given off in small amounts only. They might be suitable to be carried openly all the time, e.g., on the cuticle. The loss of material by evaporation is so low that it can be substituted without the need for accelerated biosynthesis. On the other hand, the low concentration in the air requires highly sensitive receptors and prohibits action over long distances, because of diffusional dilution. The active range of **2** has been estimated to be about 0.3 m (M. Ayasse, pers. comm.). Many flies use cuticular pheromones, which evaporate very slowly (Howard 1993; Nelson and Blomquist 1995), depending on their size. As an example, the housefly *Musca domestica* carries its sex pheromone, (*Z*)-9-tricosene (**3**), openly within its normal cuticular lipids (Carlson et al. 1971).

The well-known moth sex pheromones given off by females to attract males are of intermediate volatility. About 80-90% of the species investigated so far use long

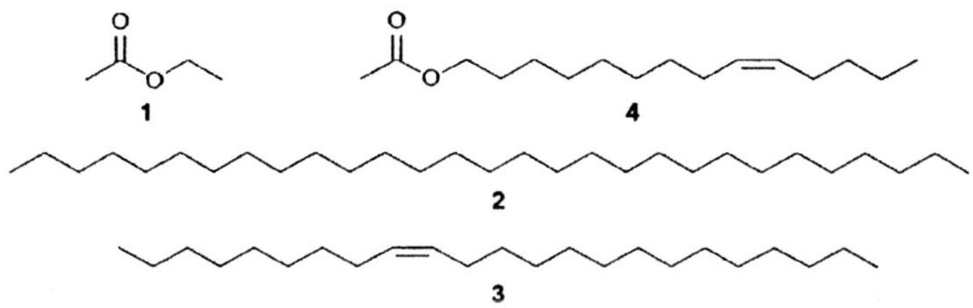

Fig.2. Chemical structures of pheromone components with high (**1**), medium (**4**), and low (**2**, **3**) volatility

chain acetates, alcohols, or aldehydes with 12 to 18 carbon atoms in an unbranched chain and up to 3 double bonds at specific positions. One of the most frequently employed compounds is (*Z*)-9-tetradecenyl acetate (**4**). It has been identified in pheromone glands of over 220 species and established as a pheromone component in over 50 species (Arn et al. 1992). Its vapor pressure is 6.6×10^{-4} Torr and that of the respective alcohol 1.7×10^{-3} Torr (Heath and Tumlinson 1986). These relatively small values are sufficient to lure conspecific males over long distances. The diffusional dilution is relatively slow, ensuring a long lasting signal. These properties of the molecules — a medium vapor pressure and a molecular weight between 150 and 300 — are ideally suited to attract individuals over long distances.

The amount of material needed for calling matches the rate of synthesis, so that the pheromone is produced in larger amounts only when needed. This can be seen in the tobacco hornworm, *Manduca sexta*, which uses a complex mixture of eight aldehydes as its sex pheromone (see Fig. 6). Only during the scotophase, when the female is calling, is any pheromone produced (Tumlinson et al. 1996). When a brain regulatory peptide (PBAN, pheromone biosynthesis activating neuropeptide) is injected into the pheromone gland, the production of the pheromone is stimulated, independently of the scotophase (Tumlinson et al. 1996). Such hormonally controlled biosynthesis is present in most moth species investigated so far, but other factors, such as juvenile hormones, can also play a role (Raina and Menn 1993; Tillman et al. 1999).

Apart from volatilization, chemicals can also be transferred through space by adsorption on small particles released by the emitter. Some male danaine butterflies employ so-called pheromone-transfer particles (PTP). Their pheromone glands contain evertible brushes of hairs (hairpencils), which can be ruptured upon contact to release the PTP (Boppré and Vane-Wright 1989). Males of the Queen butterfly, *Danaus gilippus*, brush the antennae of the female with the hairpencils. Their PTP, loaded with the pheromone danaidone (**5**; see Fig. 3), are transferred and adhere to the antennae by a sticky substance, the bishomoterpene **6**. Thus, the signal compound stays longer on the antennae, ensuring a longer-lasting stimulus, which is required for successful mating (Pliske and Eisner 1969; Schneider and Seibt 1969; Meinwald et al. 1969). The PTP of another danaine butterfly, *Amauris niavius*, contain methyl salicylate (**7**), 3,4-dimethoxyacetophenone (**8**), and **5**, embedded in a matrix of alkanes (Schulz and Boppré, unpubl. results). The occurrence of PTP has also been described in other butterflies (Boppré 1984).

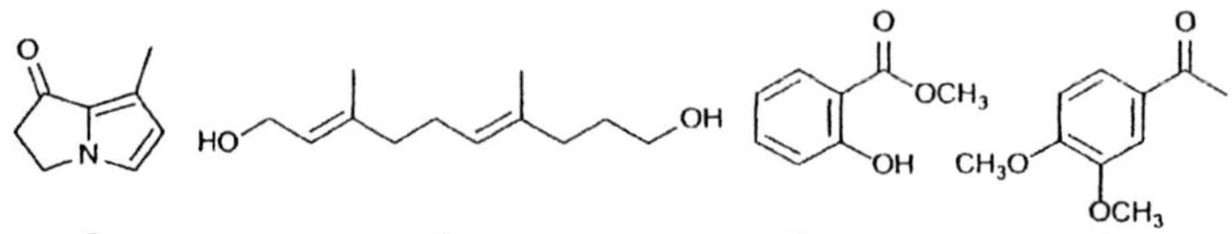

5 6 7 8

Fig.3. Chemical structures of compounds present on the pheromone-transfer particles of *Danaus gilippus* (**5**, **6**) and *Amauris niavius* (**5**, **7**, **8**)

Nonvolatile pheromones, which require direct contact of the receiver with the emitter or with a substratum on which they are deposited in order to be perceived, are also frequently used for communication. Signals can remain active for a long period of time on a surface, being better protected against degradation than in air. They are often embedded in a matrix that also has protective properties. The nymph recognition pheromone of the cockroach *Nauphoeta cinerea* (Mori and Argade 1994) consists of a mixture of very long chain aliphatic epoxides, such as (9*Z*,25*S*, 26*R*,43*Z*)-25,26-epoxy-9,43-henpentacontadiene (**9**; see Fig. 4), which inhibits wing raising in adult males. This high molecular weight compound is embedded in the cuticular wax layer of the nymphs. Glycosides can also be used as long-lasting signals. The carbohydrate parts of the molecules are highly polar, ensuring no volatility of the molecule. The oviposition-deterring pheromone of the cherry fly, *Rhagoletis cerasi*, is another signaling glucoside (**10**); it is deposited with the egg on the surface of a cherry and inhibits oviposition of other flies at this cherry (Ernst and Wagner 1989). The pheromone is a mixture of the (8*R*,15*R*)- and (8*S*,15*R*)-diastereomers.

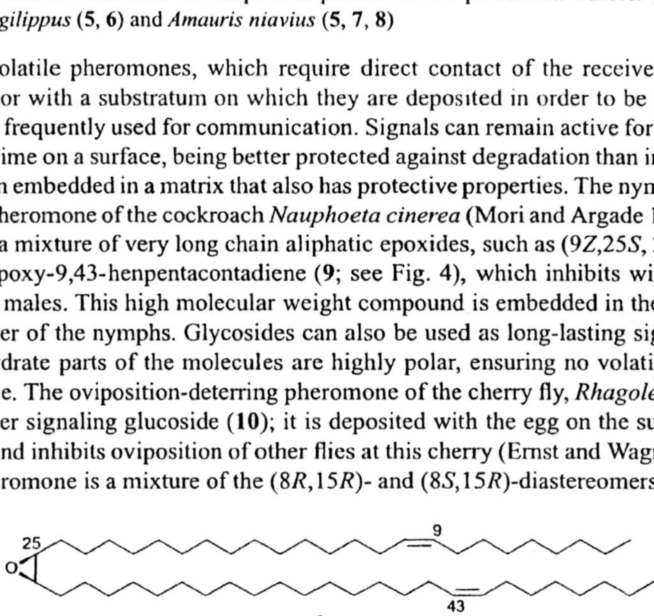

9

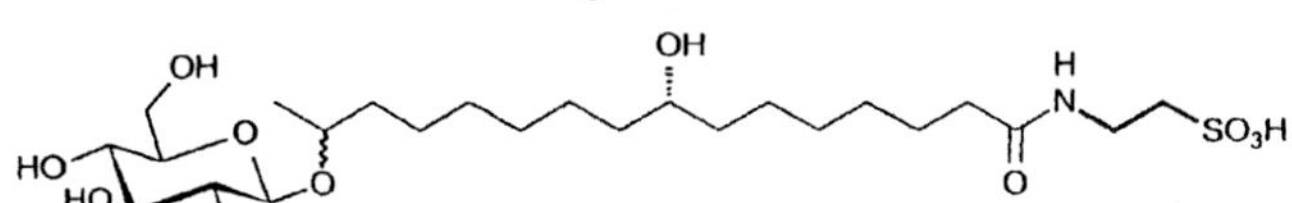

10

Fig.4. Chemical structures of nonvolatile insect pheromone components

3 Selectivity and Structure

In this section, we shall explore the structural uniqueness needed by a semiochemical to be recognizable by a receiver against the background of chemical noise. Because of the large number of pheromone components identified so far (Francke and Schulz 1999), we shall focus on selected examples.

Males of the spider *Linyphia triangularis* (Linyphiidae) are able to distinguish between a web constructed by a conspecific mated or unmated female. In the latter case, it starts to roll up the web prior to courtship in order to reduce evapo-

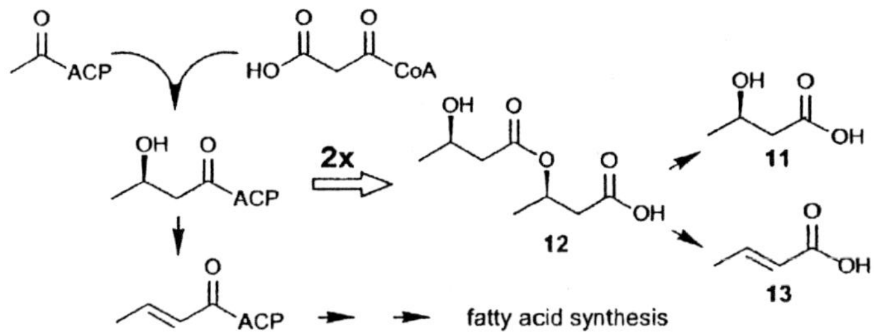

Fig.5. Proposed biosynthesis of the web reduction-inducing pheromone **12** of the spider *Linyphia triangularis* and its transformation into the more volatile pheromone components **11** and **13**

ration of a male-attracting pheromone emitted from the silk. The web reduction-inducing pheromone has been elucidated to be the condensation product of two (*R*)-3-hydroxybutyric acid (**11**) units, (*R,R*)-3-(3-hydroxybutyryloxy)-butyric acid (**12**) (Schulz and Toft 1993). This compound is closely related to primary metabolism products, because monomer **11** is an intermediate in the biosynthesis of fatty acids (see Fig. 5). It can be assumed that **12** will be formed by simple condensation of two molecules of **11**.

Pheromone **12** is not species-specific. The compound occurs on silk from unmated adult females of seven species of the genera *Linyphia, Neriene,* and *Microlinyphia* (Schulz, Toft, and Watson, unpubl. results). The inherent properties of **12** are used elegantly by the spider. This compound is not very stable and is degraded slowly and autocatalytically when exposed to light and humidity. Cleavage of the ester bond leads to **11**, whereas elimination of **11** furnishes crotonic acid (**13**). Both compounds are more volatile than the parent **12**. Biotests with *Neriene montana* indicate that **13** is a potent attractant for this species (Toft and Schulz, unpubl. results). The spider places the pheromone **12** on the silk during web construction. Here, it slowly degrades and attracts males, which, after contact with the silk, start web reduction and courtship. By means of such a system, the female spider need not call actively like, for example, female Lepidoptera, and is able to perform other tasks while maintaining information transmission. Despite the use of the same pheromonal compound, *L. triangularis* and *L. tenuipalpis* spiders, which live in the same habitat under similar conditions, are able to distinguish between silk from their own species. The discrimination mechanism probably involves complex mixtures of methyl ether lipids, which are present on the silk of linyphiids (see Schulz 1997, for a discussion). This pheromone is an example of a nonspecific compound; the ecological separation of the different species allows them to use the same component.

A more complex system is used by many night-flying Lepidoptera with respect to their female sex pheromones. These pheromones usually consist of blends of several components belonging to the chemical classes mentioned in Section 2 and are also often specific mixtures of geometrical and positional isomers. The best trap catches are usually obtained with the complete blend. The two aldehydes (10*E*,12*Z*)-10,12-hexadecadienal (**14**) and (10*E*,12*E*,14*Z*)-10,12,14-hexadecatrienal (**15**, ratio

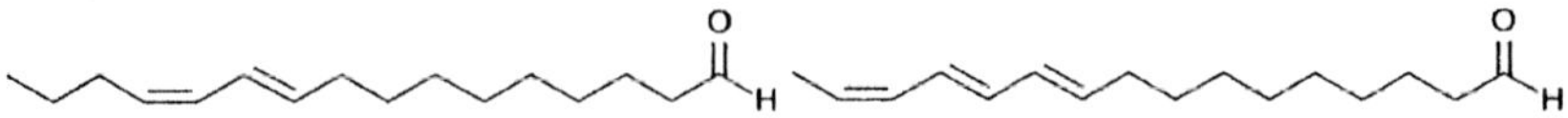

14 15

Fig.6. Chemical structures of the major sex pheromone components of *Manduca sexta*

2 : 1) are essential for trapping *M. sexta*, but the complete pheromone containing 8 components gives optimized trap catches (Tumlinson et al. 1994).

The biosynthesis of these compounds is closely related in many species and starts from the common fatty acids, palmitic or stearic acid (see Fig. 7). The two key steps are chain shortening and desaturation, which are species-specific. Frequently, a Δ11-desaturase is involved, but desaturation at other positions also occurs (Roelofs 1995; Tillman et al. 1999). The sequence of the different steps differs and determines the composition of the blend. The precursor palmitoyl-CoA (**16**) may be chain-shortened by β-oxidation to form tetradecanoyl-CoA (**17**), which is then desaturated by a Δ11-desaturase to form (*Z*)-11-tetradecenoyl-CoA (**18**). Reduction would directly produce the pheromonal alcohol (*Z*)-11-tetradecenol (**19**), which can be either acetylated to the corresponding acetate **20** or oxidized to the respective aldehyde **21**. In certain cases, direct reduction of **18** to the aldehyde **21** seems to take place (Prestwich and Blomquist 1987). In the turnip moth, *Agrotis segetum*, the sequence of desaturation and chain shortening is reversed. Desaturation of **16** results in formation of the precursor (*Z*)-11-hexadecenoyl-CoA (**22**), which is subsequently chain-shortened to the respective C_{14}, C_{12}, and C_{10}-acyl-CoA. These precursors are converted by reduction and acetylation to the three pheromonal acetates

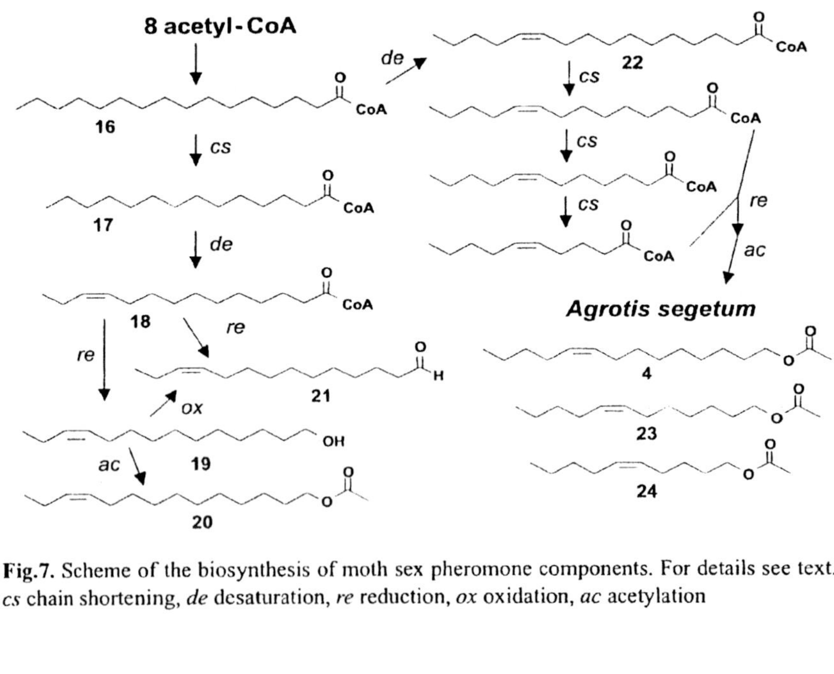

Fig.7. Scheme of the biosynthesis of moth sex pheromone components. For details see text. *cs* chain shortening, *de* desaturation, *re* reduction, *ox* oxidation, *ac* acetylation

(*Z*)-9-tetradecenyl acetate (**4**), (*Z*)-7-dodecenyl acetate (**23**), and (*Z*)-5-decenyl acetate (**24**) (Löfstedt et al. 1986).

In general, every species has its own characteristic blend. Thus, the need for the development of unique chemical structures is reduced. Selectivity can be achieved by using extant biosynthetic routes to known compounds or by slight modifications thereof. For example, the specific Δ11-desaturase may have evolved from the ubiquitous Δ9-desaturase. Within a species, variations in the blends between populations (see e.g., Löfstedt 1993) and even individuals have been observed (see e.g., Barrer et al. 1987). The specific blends will become less defined with distance from the source, because of dilution, whereas minor components will fall below the threshold level of the receiver. Nevertheless, some males will still respond to nonoptimized blends. This allows bolas spiders to use incomplete moth sex pheromone blends for prey attraction (Yeargan 1994). In some cases, blends are not enough for ensuring a unique signal; this suggests the existence of antagonists. The oblique-banded leafroller, *Choristoneura rosaceana*, uses a sex pheromone consisting of a 100 : 2 : 1 : 1 mixture of **20**, **21**, **19**, and (*E*)-11-tetradecenyl acetate. The sex pheromone of the sympatric leafroller *Pandemis limitata* is made up of a 91 : 9 mixture of the two positional isomers **20** and **4**. If as little as 1% of the minor *Pandemis* component **4** is added to the complete blend of *C. rosaceana*, trap catches of this species are completely inhibited. Specialized receptor cells probing for this antagonist are present in *C. rosaceana* (Evenden et al. 1999).

Sometimes, the elaboration of a biosynthetic pathway leads to unique compounds, which obliterate the need to use mixtures. This is exemplified in the processionary moth, *Thaumetopoea pityocampa*, which is unlike other lepidopteran genera in being able to synthesize triple bonds. Its pheromone is (*Z*)-13-hexadecen-11-ynyl acetate (**25**; see Fig. 8), which is unaccompanied by other components. Nevertheless, this species has receptors for antagonists that are used by related *Thaumetopoea* spp. as pheromone components (Quero et al. 1997).

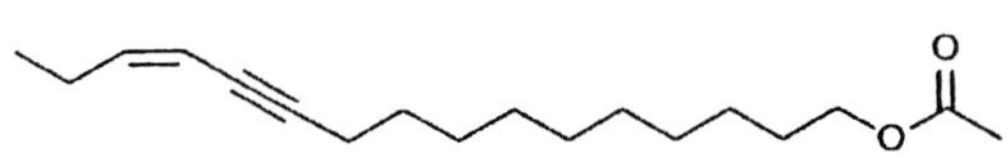

25

Fig.8. Chemical structure of the single sex pheromone component **25** of *Thaumetopoea pityocampa*

The compounds discussed so far in this section are all acetogenins, being made up of acetate (C_2) units, and are directly linked to the primary fatty acid metabolism and katabolism. A common deviation from the pure acetogenins is to employ starter units other than acetyl-CoA. The desert funnel web spider, *Agenelopsis aperta*, attracts males by using the ketone 8-methylnonan-2-one (**26**, Papke, Riechert, and Schulz, subm.). The biosynthesis of this compound can be rationalized as starting from 3-methylbutyrate (derived from the amino acid leucine by transamination and decarboxylation), which in turn is elongated by acetate units to form 9-methyl-3-oxodecanoate (**27**). The loss of the terminal carboxyl group would then lead to **26** (see Fig. 9). Amino acid derivatives are used as starters in the biosynthesis of branched lepidopteran pheromones (Charlton and Roelofs 1991).

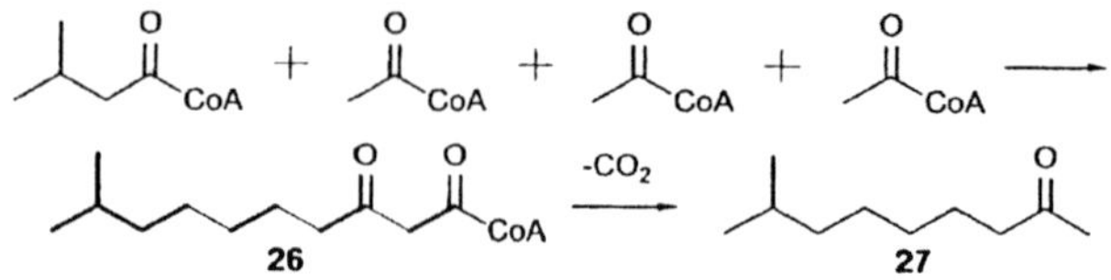

Fig.9. Proposed biosynthesis of the sex pheromone (**26**) of the spider *Agenelopsis aperta*

A mixture of unique structures is used by the European primitive moth, *Hepialus hecta*. The calling system in this species is reversed; the males actively call by giving off a sex pheromone consisting of the dihydropyranone **28** and the two bicyclic acetals **29** and **30** (Schulz et al. 1990). These are polyketides that differ from those of normal fatty acid biosynthesis by retention of oxygen atoms during the chain elongation steps and intraspecific reactions. Additionally, acetate units can be substituted by propionate (C_3). Polyketide semiochemicals are relatively rare compared with acetogenins in arthropods. The formation of **29** can be explained by the fusion of three acetate and two propionate units and several internal reactions including a typical decarboxylation step. An oxidation of the pyran ring finally leads to the ketone **30**. The biosynthetic scheme depicted in Fig. 10 is supported by the occurrence of minor derivatives with different patterns of ethyl and methyl side chains at C-1 and C-3. The pyranone **28** is composed of three C_2 units and one C_3 unit. One possible intermediate, **31**, would then be enantioselectively reduced at the C-3 keto group and finally cyclized by loss of water and CO_2. Interestingly, the American relative, *Hepialus californicus*, contains the isomeric pyranone **32** as the only compound present in its pheromone glands (Kubo et al. 1985), pointing to a acetate-propionate interchange at the beginning and end of the biosynthesis, or a reduction at C-7 instead at C-3.

It is tempting to speculate that other hepialid pheromones have similar structures; surprisingly, the pheromone glands of a third hepialid, *Hepialus humili*, contain only the sesquiterpene (*E,E*)-α-farnesene (**33**), originating from an unrelated terpene

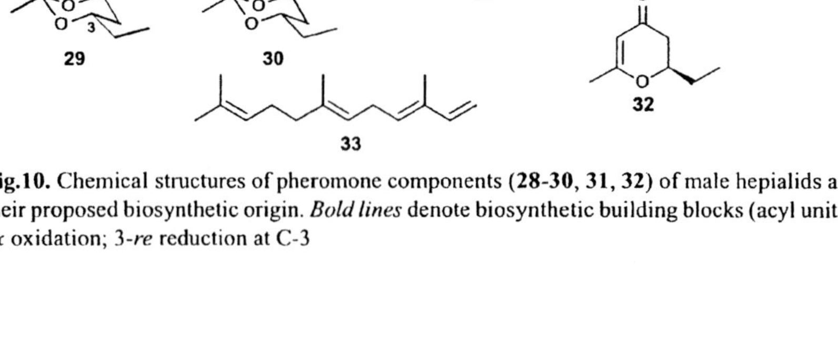

Fig.10. Chemical structures of pheromone components (**28-30, 31, 32**) of male hepialids and their proposed biosynthetic origin. *Bold lines* denote biosynthetic building blocks (acyl units); *ox* oxidation; 3-*re* reduction at C-3

biosynthesis (Schulz 1987). This may reflect different ecological constraints of the species.

Biosynthetically, the pheromones of *Carpophilus* (Nitidulidae) beetles are even more different from acetogenins. The acetate units are replaced by propionate and butyrate (C_4) units (see Fig. 11). Either four or five acyl units are coupled together, and all oxygen atoms are lost by the removal of water furnishing double bonds. A general scheme can be observed: after starting with either C_2, C_3, or C_4 (1), a C_3 unit is attached (2), chain extension is performed by C_3 or C_4 units (3), the resulting fatty acyl compound is then decarboxylated to become the pheromonal hydrocarbon (4) (Bartelt and Weisleder 1996). The preference of certain units in the biosynthesis leads to species specificity. The two major hydrocarbons in *C. davidsoni*, out of 15 identified, are the polyenes **34** and **35**. The male beetles give off the pheromone in species-specific mixtures to attract both sexes. Food odors act synergistically and can increase a trap catch 100-fold compared with pure pheromone or host odor (Bartelt and Weisleder 1996).

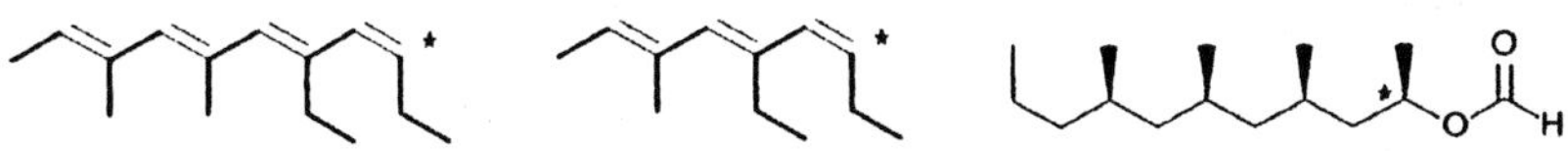

34 **35** **36**

Fig. 11. Chemical structures of pheromone components of male nitidulid beetles (**34** and **35**) and the mite *Lardoglyphus konoi* (**36**). *Bold lines* denote biosynthetic building blocks (acyl units). *The star* shows the position of the carboxyl group, which is lost during biosynthesis, in the precursor acid

Polyketide biosynthesis also occurs in arachnids. The storage mite *Lardoglyphus konoi* uses the formate **36** as an aggregation pheromone (Kuwahara et al. 1982). Its structure can be explained by the condensation of five propionate units followed by the loss of CO_2 and oxidation. The ester is unique among mite pheromones; most of these are terpenes. The aldehyde neral (**37**) and the related neryl formate (**38**) (see Fig. 12) are alarm pheromones for several related and unrelated mite species (Kuwahara 1991; Franz et al. 1999). These common compounds are derived from geranyl pyrophosphate (**39**), the general precursor of monoterpenes (composed of 2 C_5 units, see Fig. 12). They also have antimicrobial activity and are probably used by the mites to avoid the attack by microorganisms (Kuwahara et al. 1989). In this case, a special signal is not needed, and a compound, probably invented for another reason (defense), is sufficient to fulfil the desired communicative role.

Bark beetle pheromones are often derived from terpenes that are taken up by the beetles from the host. To obtain a more specific signal, small modifications of the original compounds are introduced. Bark beetles of *Ips* spp. use ipsdienol (**40**) in their aggregation pheromone. They can convert the precursor myrcene (**41**), which occurs in the resin of their host plant, into this component by oxidation (Hendry et al. 1980; see Fig. 12). However, they are also able to produce **40** de novo (Ivarsson and Birgersson 1995). The original biosynthesis and modification of host-derived precursors can take place simultaneously. Selectivity can also be obtained by the use of various enantiomers (two compounds that are mirror images of each other) as

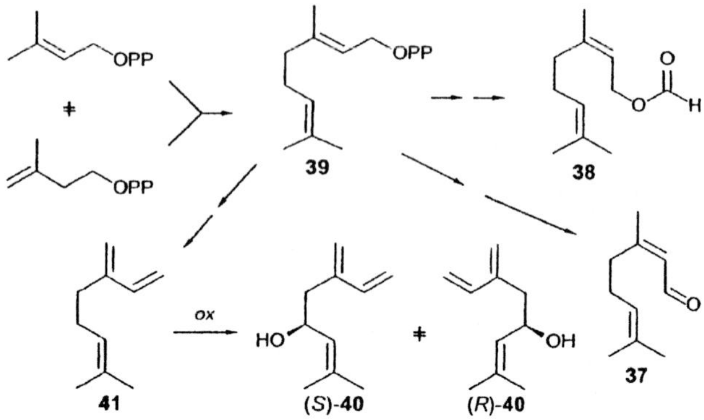

Fig. 12. Chemical structures and biosynthetic origin of monoterpenes (**37, 38, 42, 40**) used as pheromone components by arthropods

pheromones. Several modes of enantiomer usage in chemical communication systems can be found in nature and have been discussed in depth by Mori (1996). Different compositions of the two enantiomers (*R*)-**40** and (*S*)-**40** are used by different races of *I. pini*. The California race is attracted by (*R*)-**40**; as little as 5% of the (*S*)-enantiomer, which is the aggregation pheromone of *I. paraconfusus*, inhibits attraction, because both species use the same host tree. The New York race, facing no competition from another species, is attracted by (*R*)/(*S*)-mixtures ranging from 60 : 40 to 30 : 70 (Miller et al. 1996). This case exemplifies not only the ecological impact on the composition of a pheromone, but also the flexibility of mixtures.

The males of the giant danaine butterfly, *Idea leuconoe*, use a complex blend of chemicals for pheromonal communication (see Fig. 13). One of the three principal components of its sex pheromone is geranyl methylsulfide (**42**) (Nishida et al. 1996). This terpenoid compound carries the same carbon skeleton as **37**, but also a thiomethyl functional group. It is the only sulfur compound so far known as an arthropod pheromone. The other two components are the alkaloid **5** and the β-lactone **43**. They are derived from host pyrrolizidine alkaloids that are transformed into the pheromone by the butterflies (Nishida et al. 1991). In the moth *Utetheisa ornatrix*, the amount of the related pheromone hydroxydanaidal (**44**) advertises the amount of protective alkaloid that the females can obtain, as a nuptial gift during copulation, with the sperm to protect their eggs (Dussourd et al. 1991). A large number of male Lepidoptera of different families obtain pyrrolizidine alkaloids and derive a closely related family of pheromones therefrom, including **5** and **44** (Schulz 1998). The exact function of the dihydropyrrolizine in most species is unknown, but their use during courtship points to similar functions as that in *U. ornatrix*. Nevertheless, long-range attraction of females to **44** seems to operate in *I. leuconoe* (Nishida et al. 1996) and in the arctiid moth *Creatonotos transiens* (Boppré and Schneider 1989). In the latter case, the uptake of the pheromone precursor even has a morphological effect, regulating the size of the coremata and the amount of **44** they contain (Boppré and Schneider 1985).

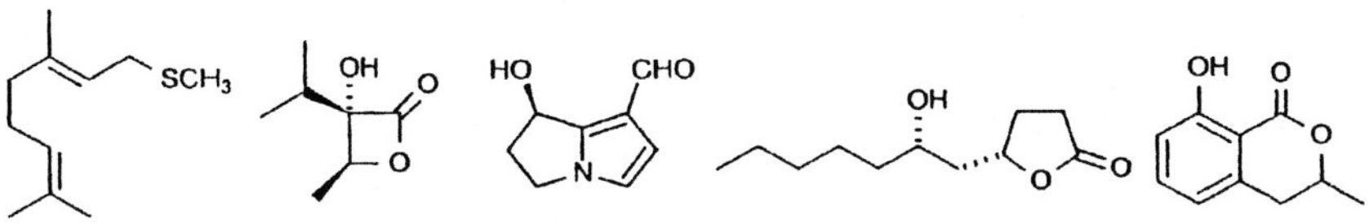

Fig.13. Chemical structures of pheromone components of male *Idea leuconoe* butterflies

The pheromone glands of male butterflies are often characterized by complex blend of chemicals, belonging to different unrelated chemical classes (Francke and Schulz 1999). The two types of hairpencils of *I. leuconoe* contain more than 150 components, serving various functions (Schulz and Nishida 1996). A number of characteristic hydroxylactones such as **45** seem to have a synergistic function, enhancing the attractive power of the former compounds. They are biosynthetically derived from the acetate pool. Phenols are taken up pharmacophagously, for example the microbial metabolite mellein **46**, and most probably serve as a warning odor. Additional compounds are derived from cuticular alkanes, such as long-chain saturated and unsaturated hydrocarbons, ketones, alcohols, and 2,5-dialkyltetrahydrofurans. They may provide protection against degradation, adjust evaporation rates, serve as solvents, and as fixatives during contact with female antennae. Nevertheless, they may carry information in direct contact interactions, because similar compounds have been identified as copulation stimulants in the fly *Cochliomyia hominivorax* (Pomonis et al. 1993). The known antioxidant α-tocopherol, which is very often found in hairpencil secretions, may play this role in these mixtures. In summary, a mixture of compounds originating from different biosynthetic pools and obtained by either de novo synthesis or uptake from the outside, modified or unmodified, serve the various purposes needed by the butterfly for chemical communication.

4 Final Remarks

The given examples, excluding the even more complex systems found in social insects an excellent book by vander Meer et al. 1998 has recently been devoted to this subject, show the structural diversity of compounds used in chemical communication systems and provide some hints regarding the reasons for which certain components are used. Several constraints lead to the selection of a compound as a signal molecule. The requirements of their physical properties and the availability of precursors are significant. Such precursors can be synthesized from common building blocks of primary metabolism, e.g., amino acids or acetyl-CoA, or taken up externally. In the latter case, further transformation is normally needed to obtain a signal distinct from the food plant or another source. Ecological factors also affect the selection of the compounds but remain in many cases unknown. The chemical stability of a signal compound has not been covered here, but it should be noted that compounds given off into the atmosphere are diluted relatively rapidly and are degraded surprisingly quickly. Terpenes such as myrcene **41** have a life span of only about 40–80 minutes in air before they decompose (Kesselmeier and Staudt 1999). On the receptor side, enzymes exist that degrade signal molecules and thus enable the receptor to respond once more (Prestwich and Blomquist 1987). Despite considerable progress achieved during the past 30 years, more effort is needed to unravel basic structural principles underlying chemical communication systems.

References

Allan RA, Elgar MA, Capon RJ (1996) Exploitation of an ant chemical alarm signal by the zodariid spider *Habronestes bradleyi* Walckenaer. Proc R Soc London Ser B 263: 69–73

Arn H, Tóth M, Priessner E (1992) List of sex pheromones of Lepidoptera and related attractants. International Organization for Biological Control, West Palearctic Regional Section, Wädenswil, 2 ed., see also the pherolist database: http://www-pherolist.slu.se

Barrer PM, Lacey MJ, Shani A (1987) Variation in the relative quantities of airborne sex pheromone components from individual female *Ephestia cautella* (Lepidoptera: Pyralidae). J Chem Ecol 13: 639–653

Bartelt RJ, Weisleder D (1996) Polyketide origin of pheromones of *Carpophilus davidsoni* and *C. mutilatus* (Coleoptera: Nitidulidae). Bioorg Med Chem 4: 429–438

Boppré M (1984) Chemically mediated interactions between butterflies. In: Vane-Wright RI, Ackery PR (eds.) The Biology of Butterflies. Academic Press London, London, pp. 259–275

Boppré M, Schneider D (1985) Pyrrolizidine alkaloids quantitatively regulate both scent organ morphogenesis and pheromone biosynthesis in male *Creatonotos* moths (Lepidoptera: Arctiidae). J Comp Physiol A 157: 569–577

Boppré M, Schneider D (1989) The biology of *Creatonotos* (Lepidoptera: Arctiidae) with special reference to the androconial system. Zool J Linn Soc 96: 339–356

Boppré M, Vane-Wright RI (1989) Androconial systems in Danainae (Lepidoptera): Functional morphology of *Amauris*, *Danaus*, *Tirumala* and *Euploea*. Zool J Linn Soc 97: 101–133

Carlson DA, Mayer MS, Silhacek DL, James JD, Beroza M, Bierl BA (1971) Sex attractant pheromone of the housefly: Isolation, identification and synthesis. Science 174: 76–78

Charlton RE, Roelofs WL (1991) Biosynthesis of a volatile, methyl-branched hydrocarbon sex pheromone from leucine by arctiid moths (*Holomelina* spp.). Arch Insect Biochem Physiol 18: 81–97

Daly HV, Doyen JT, Purcell AH (1998) Introduction to Insect Biology and Diversity. Oxford University Press, Oxford

Dussourd DE, Harvis CA, Meinwald J, Eisner T (1991) Defense mechanisms of arthropods. 107. Pheromonal advertisement of a nuptial gift by a male moth *Utetheisa ornatrix*. Proc Natl Acad Sci USA 88: 9224–9227

Ernst B, Wagner B (1989) Synthesis of the oviposition-deterring pheromone (ODP) in *Rhagoletis cerasi* L. Helv Chim Acta 72: 165–171

Evenden ML, Judd GJR, Borden JH (1999) Simultaneous disruption of pheromone communication in *Choristoneura rosaceana* and *Pandemis limitata* with pheromone and antagonist blends. J Chem Ecol 25: 501–517

Flath RA, Jang EB, Light DM, Mon TR, Carvalho LA, Binder RG, John JO (1993) Volatile pheromonal emissions from the male mediterranean fruit fly: Effects of fly age and time of day. J Agric Food Chem 41: 830–837

Francke W, Schulz S (1999) Pheromones. In: Barton D, Nakanishi K, Meth-Cohn O, Mori K (eds.) Comprehensive Natural Products Chemistry. Elsevier, Amsterdam, vol. 8, pp. 197–261

Franz JT, Schulz S, Wegener R, Masuch G, Bergmann KC, Müsken H (1999) Pheromone — Die Sprache der Milben. Allergo J 8: 70–77

Heath RR, Tumlinson JH (1986) Correlation of retention times on liquid crystal capillary column with reported vapor pressures and half-lives of compounds used in pheromone formulations. J Chem Ecol 12: 2081–2088

Hendry LB, Piatek B, Browne LE, Wood DL, Byers JA, Fish RA, Hicks RA (1980) In vivo conversion of a labelled host plant chemical to pheromones of the bark beetle, *Ips paraconfusus*. Nature 284: 485

Howard RW (1993) Cuticular hydrocarbons and chemical communication. In: Stanley-Samuelson DW, Nelson DR (eds.) Insect Lipids: Chemistry, Biochemistry and Biology. University of Nebraska Press, Lincoln, pp. 179–226

Ivarsson P, Birgersson G (1995) Regulation and biosynthesis of pheromone components in the double spined bark beetle *Ips duplicatus* (Coleoptera: Scolytidae). J Insect Physiol 41: 843–849

Jang EB, Light DM, Binder RG, Flath RA, Carvalho LA (1994) Attraction of female Mediterranean fruit flies to the five major components of male-produced pheromone in a laboratory wind tunnel. J Chem Ecol 20: 9–20

Kesselmeier J, Staudt M (1999) Biogenic volatile organic compounds (VOC): An overview on emission, physiology and ecology. J Atmos Chem 33: 23–88

Kubo I, Matsumoto T, Wagner DL, Shoolery JN (1985) Isolation and structure of hepialone; principal component from male sex scales of *Hepialus californicus*. Tetrahedron Lett 26: 563–566

Kuwahara Y (1991) Pheromonal communication of mites and ticks. In: Dusbabek F, Bukva V (eds.), Modern Acarology Academia. Prag, vol. 1, pp. 43–52

Kuwahara Y, Leal WS, Suzuki T, Maeda M, Masutani T (1989) Antifungal activity of *Caloglyphus polyphyllae* sex pheromone and other mite exudates. Naturwissenschaften 76: 578–579

Kuwahara Y, Yen LTM, Tominaga Y, Matsumoto K, Wada Y (1982) 1,3,5,7-tetramethyldecylformate, Lardolure: Aggregation pheromone of the acarid mite, *Lardoglyphus konoi* (Sasa et Asanuma) (Acarina: Acaridae). Agric Biol Chem 46: 2283–2291

Löfstedt C (1993) Moth pheromone genetics and evolution. Phil Trans R Soc Lond B 340: 167–177

Löfstedt C, Elmfors A, Sjögren MWE (1986) Confirmation of sex pheromone biosynthesis from (16-D_3)palmitic acid in the turnip moth using capillary gas chromatography. Experientia 42: 1059–1061

Meinwald J, Meinwald YC, Mazzocchi PH (1969) Sex pheromone of the Queen butterfly: Chemistry. Science 164: 1174–1175

Miller DR, Borden JH, Slessor KN (1996) Enanantiospecific pheromone production and response for populations of pine engraver, *Ips pini* (Say) (Coleoptera: Scolytidae), in British Columbia. J Chem Ecol 22: 2157–2172

Mori K (1996) Molecular asymmetry and pheromone science. Biosci Biotech Biochem 60: 1925–1932

Mori K, Argade NP (1994) Synthesis of (9*Z*,25*S*,26*R*,43*Z*)-25,26-epoxy-9,43-henpentacontadiene and its antipode, components of the nymph recognition pheromone produced by the nymphs of the cockroach *Nauphoeta cinerea*. Liebigs Ann Chem : 695–700

Nelson DR, Blomquist RJ (1995) Insect waxes. In: Hamilton RJ (ed.), Waxes: Chemistry Molecular Biology and Functions. The Oily Press, Dundee, pp. 1–90

Nishida R, Kim CS, Fukami H, Irie R (1991) Ideamine N-Oxides: Pyrrolizidine alkaloids sequestered by the danaine butterfly, *Idea leuconoe*. Agric Biol Chem 55: 1787–1792

Nishida R, Schulz S, Kim CH, Fukami H, Kuwahara Y, Honda K, Hayashi N (1996) Male sex pheromone of a giant danaine butterfly, *Idea leuconoe*. J Chem Ecol 22: 949–972

Nordlund DA, Jones RL (1981) In: Nordlund DA (ed.) Semiochemicals: Their Role in Pest Control. Wiley, New York, p. 13

Piacente V, Pompili T, Scardala P, Ferro D (1991) Temperature dependence of the vaporization enthalpies of *n*-alkanes from vapor pressure measurements. J Chem Thermodynamics 23: 379–396

Pliske TE, Eisner T (1969) Sex pheromone of the Queen butterfly: Biology. Science 164: 1170–1172

Pomonis JG, Hammack L, Hakk H (1993) Identification of compounds in an HPLC fraction from female extracts that elicit mating response in male screwworm flies *Cochliomyia hominivorax*. J Chem Ecol 19: 985–1008

Prestwich GD, Blomquist GJ (1987) Pheromone Biochemistry. Academic Press, Orlando

Quero C, Malo EA, Fabriás G, Camps F, Lucas P, Renou M, Guerrero A (1997) Reinvestigation of female sex pheromone of processionary moth (*Thaumetopoea pitycampa*): No evidence for minor components. J Chem Ecol 23: 713–726

Raina AK, Menn JJ (1993) Pheromone biosynthesis activating neuropeptide: From discovery to current status. Arch Insect Biochem Physiol 22: 141–151

Roelofs WL (1995) Chemistry of sex attraction. Proc Natl Acad Sci USA 92: 44–49

Schiestl FP, Ayasse M, Paulus HF, Lofstedt C, Hansson BS, Ibarra F, Francke W (1999) Orchid pollination by sexual swindle. Nature 399: 421–422

Schneider D, Seibt U (1969) Sex pheromone of the Queen butterfly: Electroantennogram response. Science 164: 1173–1174

Schulz S (1987) Die Chemie der Duftorgane männlicher Lepidopteren. Ph.D. thesis, University of Hamburg

Schulz S (1997) The chemistry of spider toxins and spider silk. Angew Chem Int Ed Engl 36: 314–326

Schulz S (1998) Insect-plant interactions — metabolism of plant compounds to pheromones and allomones by Lepidoptera and leaf beetles. Eur J Org Chem : 13–20

Schulz S, Francke W, König WA, Schurig V, Mori K, Kittmann R, Schneider D (1990) Male pheromone of swift moth, *Hepialus hecta* L. (Lepidoptera: Hepialidae). J Chem Ecol 16: 3511–3521

Schulz S, Nishida R (1996) The pheromone system of the male danaine butterfly, *Idea leuconoe*. Bioorg Med Chem 4: 341–349

Schulz S, Toft S (1993) Identification of a sex pheromone from a spider. Science 260: 1635–1637

Tillman JA, Seybold SJ, Jurenka RA, Blomquist GJ (1999) Insect pheromones — an overview of biosynthesis and endocrine regulation. Insect Biochem Molec Biol 29: 481–514

Tumlinson JH, Mitchell ER, Doolittle RE, Jackson DM (1994) Field tests of synthetic *Manduca sexta* sex pheromone. J Chem Ecol 20: 579–591

Tumlinson JH, Teal PEA, Fang N (1996) The integral role of triacyl glycerols in the biosynthesis of the aldehydic sex pheromones of *Manduca sexta*. Bioorg Med Chem 4: 451–460

vander Meer RK, Breed MD, Winston ML, Espelie KE (1998) Pheromone Communication in Social Insects. Westview Press, Boulder

Yeargan KV (1994) Biology of bolas spiders. Annu Rev Entomol 39: 81–99

V.2 Sensory Ecology of Arthropods Utilizing Plant Infochemicals

Joop J.A. van Loon, Marcel Dicke

Laboratory of Entomology, Wageningen University, PO Box 8031, 6700 EH Wageningen, The Netherlands

Abstract
Herbivorous and carnivorous arthropods are confronted with the highly diverse chemistry that the plant world offers. The majority of herbivores displays very specialized host-plant selection behavior which has its sensory basis in specialized olfactory and gustatory cells that specifically respond to plant taxon-specific substances. The production of volatile compounds by plants is prone to variation caused by a range of abiotic and biotic factors. Predators and parasitoids, acting as the major natural enemies of herbivores, are likewise confronted with variability in food-related volatile stimuli. Rather than utilizing cues directly emanating from their food, they predominantly rely on plant-produced cues to locate prey or host from a distance. Plants respond to damage by herbivores by changes in composition of the volatile blend released. These herbivore-induced plant volatiles are used by carnivores to increase the efficiency of their foraging behavior. Although their chemosensory ecology is largely unexplored, recent studies suggest that also carnivorous arthropods possess chemoreceptors that are sensitive to herbivore-induced plant volatiles. Analyzing the degree of chemosensory specificity and plasticity in carnivorous arthropods in the context of their foraging behavior poses an appealing challenge for research in sensory ecology.

Key words Insect-plant interactions, olfaction, searching mechanisms, herbivore-induced synomones

1 Introduction

1.1 Dietary Specialization in Arthropods

Interactions between plants and arthropods are fruitfully studied under the perspective of sensory ecology as it has been revealed that arthropods make extensive use of their sensory systems to find plants and plant-feeding arthropod hosts and prey. Arthropods fulfill vital roles in food webs of most terrestrial ecosystems. Together with higher plants they constitute the majority of species described (more than a million known species; Wilson 1992). For arthropod communities inhabiting trees, about 25% of the species are herbivores, 25% are predators, 25% are parasitoids, 10% are scavengers and 15% are tourists (non-predatory arthropods having no intimate or lasting association with the plant; Moran and Southwood 1982). In terms of biomass, herbivores represent 54%. A prominent feature of plant utilization by herbivorous arthropods is the dominance of dietary specificity. Over 80% of the herbivorous species for which the host plant range is well documented are specialist feeders, confining their food intake to a few closely related plant species (monophagous) or to plant species belonging to one or a few phytochemically similar plant families (oligophagous) (Schoonhoven et al. 1998). The remainder has generalist or polyphagous feeding habits, accepting dozens of plant species belonging to unrelated families, though certainly not all plant species offered. Similarly, carnivorous species differ largely in their dietary range.

Most parasitic carnivorous species are specialized on a limited range of host species, while predatory species are generally less specialized (Slansky and Rodriguez 1987) yet prey preferences are found (Dicke et al. 1988) and several species of specialized predators have been documented (Sabelis 1992). The interactions between herbivorous insects and plants have long been considered from a bilateral perspective. From a trophic point of view this seems adequate but from an informational point of view a proper understanding of herbivore food selection behavior necessitates the study of effects of signals from competing herbivores and natural enemies as well (Dicke 2000). Likewise, over the past decade it has become clear that the food searching and acceptance behavior of carnivorous arthropods can only be understood from a multitrophic perspective (Vet and Dicke 1992; Turlings et al. 1993; Dicke and Vet 1999; Sabelis et al. 1999).

1.2 Sensory Information for Food Selection Behavior

1.2.1 Behavioral Chains

A description of the behavior leading to food intake or oviposition takes the form of a fixed sequence of distinct behavioral elements during which the insect responds to a succession of stimuli of different modalities. When the outcome of sensory evaluation is to reject the particular plant, prey or host as food or as oviposition site, the animal may "jump back" to one of the earlier steps in the sequence. Modification of selection behavior as a result of previous experience can be observed as faster decision making or as changes in preference, but the sequence remains the same. Such sequences of behavioral phases and of elements within each phase can be quite elaborate. Examples of sequences displayed by a parasitoid and a predator are given in Fig. 1.

The sequence of food selection behavior can be subdivided into two main consecutive phases: (1) searching and (2) contact-testing. The searching phase ends by establishing contact with the host plant, host insect, or prey, the second phase ends by actual ingestion of plant or insect material or with the deposition of eggs. Going through the sequence, the number of sensory modalities and intensity of the stimuli are increasing. The final decisions to accept or reject plant tissue, host, or prey are based not only on sensory information, but also on the insect's physiological state (satiety, sexual maturity, egg maturation, etc.; Barton Browne 1993). Sensory information from the food, information on internal state, and information on previous experiences stored in the insect's memory are integrated in the central nervous system and in concert determine food acceptance and preference (Dethier 1982; Miller and Strickler 1984; Bell 1991; Papaj and Lewis 1993).

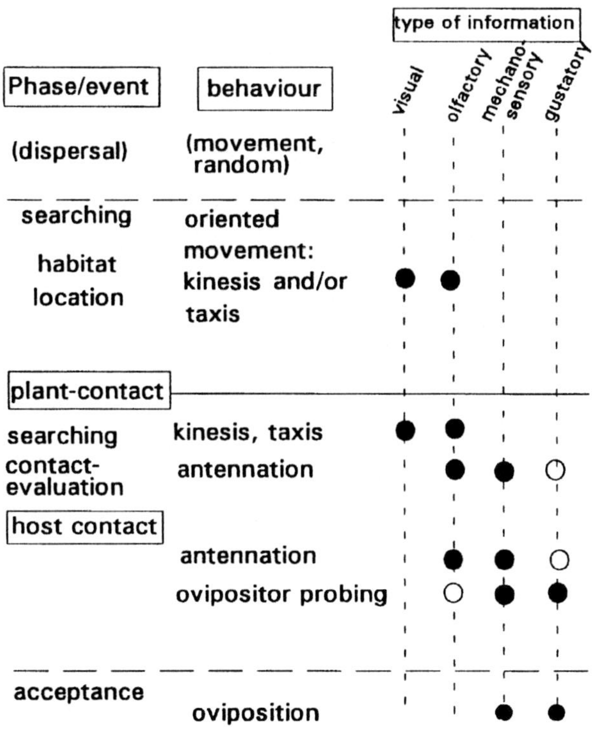

Fig. 1. Behavioral sequences of carnivorous arthropods during host or prey selection. *Left column* Behavioral phase or event, *middle column* common behavioral elements occurring within a behavourial phase; *right column* main plant-derived stimuli affecting the behavior; *solid circles* well documented cues for several species; *open circles* suggested or probable (based on Lewis et al. 1975). Not all elements were found in each species studied. For similar outlines of host-plant selection behavior of herbivorous arthropods, the reader is referred to Schoonhoven et al. (1998)

1.2.2 Searching Phase

In the searching phase, arthropods rely on optical and/or odoros information (Fig. 1). It is generally assumed that odors released from plant, host, or prey convey a more specific signal to the arthropod than optical cues (color, hue, contrast), which are more uniformly distributed, especially in the plant kingdom. For several species it has been demonstrated that both sensory modalities are employed in an integrated way (Prokopy 1986; Aluja and Prokopy 1993; Harris and Foster 1995,). In this chapter we will confine the discussion to chemical information.

1.2.3 Searching Mechanisms

In arthropods, the whole range of possible searching mechanisms has been found, varying from random search to highly directed search patterns. Searching movements are considered random when frequency, rate, and direction of movement appear unrelated to the suitability of plants, hosts, or prey that are within the perceptual range, i.e., the range within which food-derived cues are detectable by the sensory system. During such searching, kinetic responses are commonly observed. This searching mechanism may be the best possible when either environmental cues provide no directionality or when the insect cannot perceive the required stimuli.

Directed movement, on the other hand, becomes possible when the food emits chemical cues that allow, either alone or in combination with a second cue like air flow, a direction to be derived by the sensory system of the searching arthropod, either by spatial or temporal comparisons of stimulus intensity or intermittency. A commonly observed type of oriented searching mechanism is odor-conditioned anemotaxis. Because under natural circumstances gradients in odor concentrations are absent (Murlis et al. 1992), chemotaxis cannot be used except maybe over distances of a few centimetres from the odor source. Anemotaxis, however, has been demonstrated for both herbivorous and carnivorous arthropods (Fig. 1; van Loon et al. 2000). From an evolutionary point of view, oriented search is assumed to increase fitness as it allows a more efficient use of time and energy compared with random search.

1.2.4 Contact Phase

Once the food is contacted, different forms of explorative behavior are exhibited which involve repeated contacting by antennae, mouthparts, legs, and/or ovipositor that bear mechanosensory and contact-chemosensory sensilla. Only at the end of the contact phase, does acceptance of the food source occur. At this point mechanosensory and gustatory information have become available in addition to olfactory information (Fig. 1). The accumulation of volatile and contact chemical information has reached its maximum. The final decision is in the majority of cases based on taste information. This is apparent from the acceptance of artificial substrates containing nonvolatile cues devoid of naturally occurring optical, odoros or mechanosensory cues. The crucial role of contact chemoreception in host-plant selection of herbivorous insects has recently been reviewed by Bernays and Chapman (1994), van Loon (1996), and Schoonhoven et al. (1998). In this chapter we will confine the discussion to the use of olfactory information during the searching phase.

2 Chemoreception of Volatile Plant Infochemicals by Arthropods

2.1 Diversity and Taxonomic Patterns of Chemical Signals Produced by Plants

Plants offer a huge biochemical diversity to herbivorous arthropods. Over 100,000 different low molecular weight (MW < 1000 Da) organic chemicals have been identified thus far from plants (Buckingham 1993) and this number is rapidly increasing. Parasitic and predatory arthropods are potentially confronted with an even greater diversity of compounds as they encounter plant compounds as well as those produced by their hosts or prey. Although much less is known about the chemistry of arthropod-produced chemicals than about plant-produced chemicals, this seems not so much due to the scarcity of new chemicals that could be detected in arthropods but rather to the technical difficulty to obtain sufficient quantities for chemical identification.

The chemical fingerprints, either as bouquets of volatiles or as mixtures of non-volatile compounds present on the surface or in the interior of plants, hosts, or prey, can be extremely complex in a qualitative sense, in terms of total number of compounds and chemical classes represented, and also variable in the quantitative sense, due to differences in the ratios in which they can occur in the mixtures. For plants, taxonomic patterns in nonvolatile chemistry can be discerned.

Most plan families are characterized by nonvolatile secondary plant substances that do not occur in other families. For example, Brassicaceae and other families belonging to the order Capparales contain glucosinolates, a family of over 100 different sulfur-containing glycosides that are toxic to many animals (Chew 1988).

The family of Solanaceae contains a group of steroid alkaloids which seem to occur uniquely in this family (Ripperger and Schreiber 1981). These two families can unambiguously be told apart by phytochemical analysis, which is feasible for many other families as well (Hegnauer 1962-1994). For this particular example, the chemical distinction would also turn up from analyses of the volatile compounds as Brassicaceae produce hydrolysis products of glucosinolates called isothiocyanates that are entirely absent from emissions collected from solanaceous plants. By now it is well established that specialist herbivorous arthropods make use of these taxonomic patterns during host plant selection.

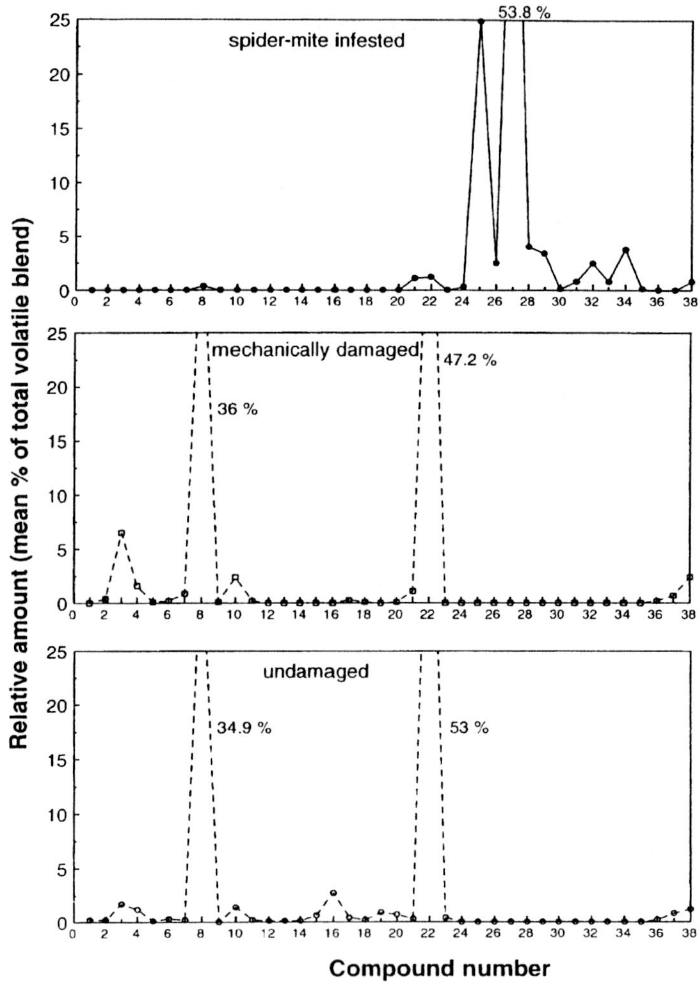

Fig. 2 A. Profiles of volatile production by cucumber plants (after Takabayashi et al. 1994). For details on the *numbered compounds* the reader is referred to the original references. (Takabayashi et al. 1994; Bolter et al. 1997; Dicke and Vet 1999)

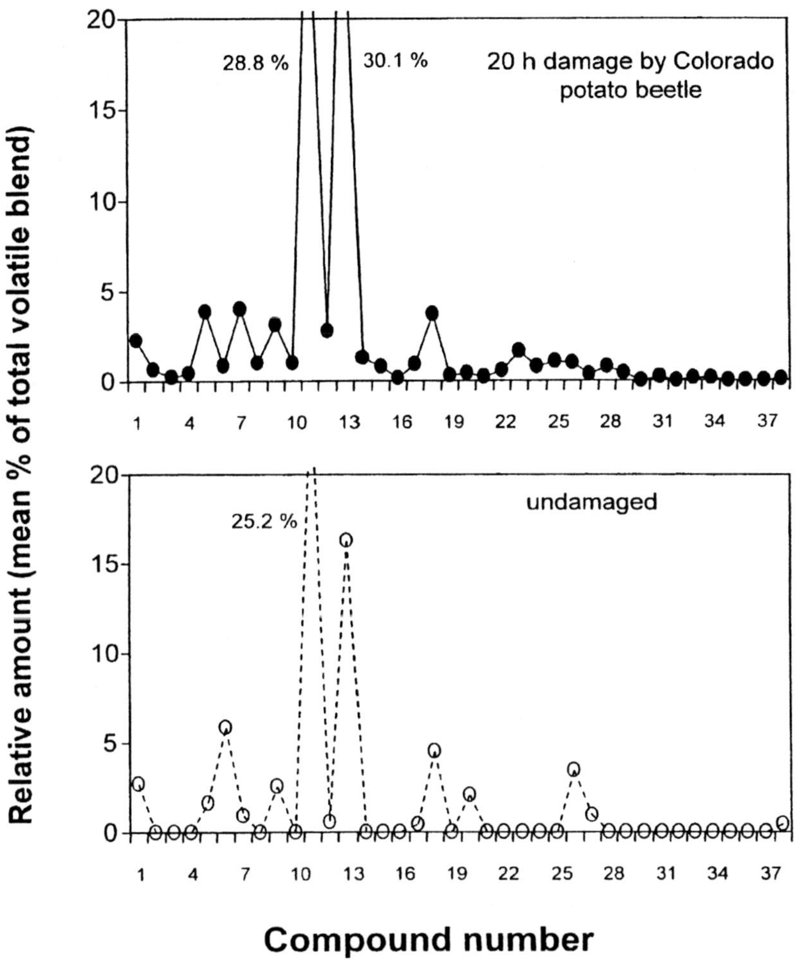

Fig. 2 B. Profiles of volatile production by potato plants (after Bolter et al. 1997). For details see Fig. 2 A

2.2 Volatile Infochemicals from Plants

2.1.1 Composition of Volatile Bouquets from Intact and Damaged Plants

In the headspace of intact plants generally between 10 to over 100 volatile compounds occur (Schoonhoven et al. 1998). The major groups of compounds released are six-carbon alcohols, aldehydes and esters of acetate and other short-chain aliphatic acids, produced through the lipoxygenase pathway (together often designated as green leaf volatiles; GLV) and a variable range of mono- and sesquiterpenoids. When damaged by arthropods, not only may the absolute quantities of the volatiles as emitted from undamaged plants increase, but also two different clear-cut changes in the composition of the volatile bouquet released may occur: (1) a quantitative change due to shifts in the relative quantities and (2) a qualitative change seen as the damage-induced release of new compounds that are not emitted from undamaged or from mechanically damaged plants (Fig. 2).

For some herbivorous species it has been shown that they secrete elicitors into the damaged plant tissue that induce a volatile emission that differs from that induced by mechanical damage (Mattiacci et al. 1994; Turlings and Fritzsche 1999). In a few cases the chemical nature of these elicitor molecules has been identified (Mattiacci et al. 1995; Alborn et al. 1997).

2.2.2 Factors Causing Variability in Plant Volatile Emissions

For any plant species, the composition of the bouquet of volatiles emitted as well as the amount released per gram tissue is prone to a considerable degree of variation (Dicke and Vet 1999). The most important sources of variation are genotypic differences, developmental stage, and biotic factors such as competing plants, infection by pathogens, and damage by arthropods. Abiotic factors that affect volatile production are, e.g., shading and soil nutrient level.

2.3 Responses to Plant Volatiles: Herbivores

2.3.1 Responses of Herbivores to Volatiles from Undamaged Plants

For a range of herbivorous species, plant volatiles have been shown to elicit odor-conditioned anemotaxis (reviewed by Visser 1986; Metcalf and Metcalf 1992; Bernays and Chapman 1994; Schoonhoven et al. 1998). Regarding the chemical identity of the kairomones utilized during host-plant searching, knowledge has increased rapidly relative to the limited information on nonvolatile chemicals that trigger acceptance. The reasons are of a practical rather than a scientific nature: olfactory activity of volatiles can be tested on-line, using so-called coupled gas chromatography–electroantennography or single-cell recording techniques (Arn et al. 1975; Wadhams 1984) and analytical chemistry of volatiles is more feasible compared to chemistry of nonvolatiles in view of the fact that the number of potentially active compounds in the latter category is much greater. Cases have been documented in which the detector of the gas chromatograph was much less sensitive than the biological detector, being the olfactory receptors producing electrophysiological responses (Pickett 1990; Weissbecker et al. 1997).

Several specialized species respond strongly to purified volatile compounds characteristic for the host-plant family with which they are associated (Schoonhoven et al. 1998). Although only few studies have explicitly addressed this issue, behavioral specialization has its equivalent at the sensory level as specialized olfactory receptors for such taxon-specific compounds have been found (Honda et al. 1987; Blight et al. 1989; Nottingham et al. 1991). In addition, all species investigated have olfactory receptor neurons (ORNs) that respond to the generally occurring GLVs (see above). All higher green plants emit the generally occurring GLVs (see above and Schoonhoven et al. 1998). Both the precise range of compounds and the ratios of the different components in the blend differ between families, however. This leaves the possibility that, despite the fact that the compounds as such are produced by many plant species, still the blends can convey specific information to the arthropod because of quantitative differences which allow a distinction between blends from different plant species or families (Visser 1986).

It has been repeatedly concluded that chemoreception of GLVs is achieved by so-called generalist ORNs (Ma and Visser 1978; Boeckh 1984; de Jong and Visser 1988). Such sensory neurons are assumed to respond to a wide range of compounds, which may even belong to different chemical categories. Several recent reports on herbivorous arthropods, however, identified receptors that are orders of magnitude more sensitive to only some GLVs (Dickens 1990; Hansson et al. 1999).

For the other major chemical category of generally occurring plant volatiles, i.e., terpenoids, specialized ORNs responsive to individual plant-derived terpenoids have been found (Dickens 1990; Anderson et al. 1993; Wibe and Mustaparta 1996; Jönsson and Anderson 1999). Several arguments have been put forward to question the existence or role of generalist receptors. One is that in older studies on olfaction unrealistically high concentrations of odorants were applied that cause a non-specific response in many cells. Also, only few studies determined dose-response relationships which are necessary to evaluate differences in sensitivity. A third argument is that the range of compounds emitted by plants is broad. No study tested all candidate compounds exhaustively. Thus, key compounds exciting specific ORNs may have been missed. It is important to stress that especially the latter argument reflects the main difference with pheromone chemoreception: the chemical complexity of the pheromone message is less complex compared to that contained in the blend of GLVs and terpenoids emitted by plants.

To understand the role of ORNs with low specificity to GLVs, more studies into the specificity of olfactory cells in response to plant volatiles are required. However, attention paid to this and other physiological aspects of arthropod chemoreception has declined over the past decade. The focus of olfactory research has shifted to the molecular characterization of olfactory receptors and their transduction pathways (Breer and Shepherd 1993; Clyne et al. 1999). The behavioral effects of plant taxon-specific volatiles differs between generalist and specialist herbivores. Specialist herbivores may show a strong attraction to plant-taxon specific volatiles. In generalists a weak or no attraction is seen, whereas in several instances generalist herbivores are repelled by these infochemicals.

The generally occurring GLVs often attract both specialists and generalists. For several specialist herbivores, despite explicit attempts no taxon-specific plant volatiles were found that could act as kairomones. As emission patterns of GLVs differ quantitatively between plant taxa, it is conceivable that information on the ratio of these generally occurring components is conveying information with sufficient specificity. The Colorado potato beetle seems to make use of such information during its odor-conditioned anemotactic response to potato plants; individual GLVs added in excess amounts to the total blend abolished its anemotactic response (Visser and Avé 1978).

2.3.2 Responses of Herbivores to Volatiles from Damaged Plants

When the role of plant volatiles for host-plant location by herbivores is investigated, compounds are commonly collected as the headspace of intact plants. However, under field conditions most, if not all, plants experience damage or infection. These are known to result in volatile emissions which in several cases differ qualitatively from those of intact plants (Dicke 2000).

Different species of herbivorous arthropods indeed respond differently to the volatile bouquets emitted by damaged plants. Some species avoid plants damaged by conspecific or heterospecific herbivores, while others are attracted to such plants (reviewed by Dicke and Vet 1999). These differences in behavioral responses between species might be correlated with adaptations to the dominant ecological constraints imposed upon them. Avoidance of damaged plants may be functional to avoid competition, while attraction, leading to aggregated feeding, may be adaptive as a strategy of overwhelming induced plant defence.

2.4 Responses to Volatile Plant Infochemicals: Carnivores

2.4.1 Detectability and Reliability of Plant-Produced Olfactory Information

In their search for herbivorous prey, carnivorous arthropods are confronted with the problem of locating small organisms hidden in a large green biomass. As argued by Vet and Dicke (1992), carnivores face a reliability-detectability problem. Although volatile chemical information emitted by plants is abundantly available in the environment and thus readily detectable, the reliability in signaling the presence of herbivorous hosts or prey feeding somewhere in the vegetation is limited. Conversely, volatiles emitted by the prey or host are reliable signals but are embedded in the much larger amounts of plant volatiles. In contrast, it is well established for a range of tritrophic systems that the plant is the source of volatile information that guides the carnivore to its host or prey (reviewed in Dicke and Vet 1999).

Apart from the detectability problem, the inherent variability in plant volatile production as described in Section 2.2.2 may complicate oriented searching of prey or hosts even further. The olfactory system of carnivores therefore has to accomplish the task of discriminating irrelevant volatile information (noise) from relevant information that reliably indicates the presence of host or prey (signal). The arthropod olfactory system can assess both quality and quantity of odor bouquets (de Jong and Visser 1988, Smith and Getz 1994). Results at the behavioral level suggest that parasitoids discriminate qualitative differences better than quantitative differences (Vet et al. 1998).

We will subsequently focus on answering two crucial questions that follow from the foregoing discussion: does herbivory result in qualitative differences in plant volatile production and do carnivores possess olfactory receptors (specifically) sensitive to such compounds?

2.4.2 Plant Volatiles Produced in Response to Herbivory: Qualitative vs. Quantitative Differences

The occurrence of qualitative differences in plant volatile production in response to herbivory depends on the plant-herbivore interaction. Based on the information presently available three categories can be distinguished that differ in the specificity and consequently reliability of the volatile blend available to the carnivore: (1) plant species which release the same compounds when damaged by different herbivore species as when damaged artificially, with variation only in the relative contribution of each compound to the blend; (2) plants that produce novel compounds when damaged by herbivores compared to artificial damage while different herbivore species lead to quantitative differences in the blend; (3) plants that emit novel compounds when damaged by herbivores compared to artificial damage and that emit qualitatively different blends in response to damage by different herbivore species. Evidently, the informational specificity of the interactions belonging to (3) and thus the reliability of the signal to the carnivore is highest. Of ten tritrophic systems for which the composition of the volatile blends in response to damage by different herbivore species have been determined, four yielded qualitative differences. In some cases the novel compounds make up the largest part of the bouquet (Fig. 2A).

The distinction of these three categories stresses differences in the specificity of the olfactory information available to carnivores. It should not be seen as absolute, however. An inherent problem in deciding on the chemical uniqueness of a certain blend component is that any chemical analysis has its limits of detection and, as noted above, the arthropod olfactory system may defy the sensitivity of our analytical equipment. Carnivores foraging for herbivores on plants in category (1) and (2) are expected to base their searching behavior on quantitative chemical information. In order to overcome the limited specificity contained in such a message a neural link has to be established by associative learning between actual host or prey encounters (eliciting contact-chemosensory and tactile activity) with the quantitative olfactory profile perceived just prior to and during this encounter. This type of learning is widespread in hymenopteran parasitoids (Vet et al. 1995). It effectively leads to "temporary specialization".

From behavioral studies it appears that several carnivorous species are well able to discriminate between odor mixtures that differ quantitatively even when lacking previous experience (for review see Dicke 1999). In the majority of cases, carnivores were shown to discriminate between plants infested by different herbivores, either innately or after previous feeding or oviposition experience (reviewed by Dicke and Vet 1999 and Dicke 1999). Specialist carnivores may rely largely on innate responses to olfactory cues while generalist feeders may employ associative learning (Vet et al. 1995), a difference that holds even for closely related species (Geervliet et al. 1998).

2.4.3 Olfactory Characteristics of Arthropod Carnivores

The olfactory system of arthropod carnivores has hardly been explored in any detail for the aspects of specificity and sensitivity. We found six studies on parasitoids (Lecomte and Pouzat 1985; Baehrecke et al. 1989; Ramachandran and Norris 1991; Salom et al. 1991; Li et al. 1992; Du et al. 1998) and five studies on predators (Tømmerås 1985; Sant'ana and Dickens 1998; Dickens 1999; Weissbecker et al. 1999, 2000). All these studies but one (Tømmerås 1985) have employed electro-antennogram (EAG) recording, a valuable tool for a first screening of olfactory responses to volatiles. In these studies, the choice of the compounds has partly been based on the commercial availability or from the viewpoint of a structure-activity approach rather than on their actual occurrence in the headspace of herbivore-damaged plants or on their ecological significance. Several damage-related compounds have been identified in a growing number of studies over the past five years. These include butyrate- and valerate-esters of (Z)-3-hexen-1-ol (Whitman and Eller 1990, 1992); homoterpenes (Dicke and Vet 1999; Turlings and Fritzsche 1999) and methylsalicylate (Dicke et al. 1990; Du et al. 1998, Dickens 1999). These have not been extensively tested yet. In several cases parasitoids or predators display a significantly higher olfactory sensitivity to some damage-related compounds than the herbivore they attack (Ramachandran and Norris 1991; Dickens 1999; Weissbecker et al. 1999, 2000).

The EAG technique does not allow conclusions on the specificity of single olfactory neurons. Until now only one study addressed predator ORN specificity. Also, this study focussed on prey (bark beetle) pheromones rather than on host-plant volatiles (Tømmerås 1985). Although it has been demonstrated for quite a few tritrophic systems that parasitoids or predators are attracted to herbivore-damaged plants and that the plant is the producer of the synomones (Dicke and Vet

1999), the nature of the active chemicals is known in only few cases. A "minimal blend" that elicits attraction (out of the often chemically complex odor blend emitted) could be defined for the specialized predaceous mite *Phytoseiulus persimilis* (Dicke et al. 1990), the braconid wasp *Cotesia marginiventris* (Turlings and Fritzsche 1999) and the aphid parasitoid *Aphidius ervi* (Du et al. 1998). For the braconid *Microplitis croceipes* individual GLVs were found to elicit an upwind response in windtunnel assays (Whitman and Eller 1990, 1992). In these cases, however, the complex blend was found to be more attractive than the reduced "synthetic blend", indicating that the complete signal is not yet known. Indeed, in a detailed analysis it was found that some compounds in the blend may mask the attractive components (Turlings and Fritzsche 1999).

3 Conclusions

Plants produce an overwhelming amount of so-called secondary compounds, both volatiles and nonvolatiles. These compounds can be exploited by arthropods when searching for food or oviposition sites. Effects of plant chemicals on arthropods have long been studied for herbivorous arthropods exclusively. Accordingly, there is extensive knowledge on both the sensory perception and behavioral responses. Herbivorous arthropods are predominantly specialists and plant chemistry is one of the important factors - though not the only factor - mediating host plant selection (e.g., Bernays and Chapman 1994; van Loon 1996; Schoonhoven et al. 1998). Most of the information available is on non-volatile plant chemicals. Carnivorous arthropods also use plant chemicals when searching for their herbivorous arthropod victims. Herbivore-damaged plants emit volatiles utilized by carnivorous arthropods to locate the herbivores that inflicted the damage. The blend emitted by herbivore-damaged plants may differ quantitatively (blends of same compounds but at different ratios) or qualitatively (blends composed of different sets of chemicals).

The specificity of herbivore-induced plant volatiles can convey information that enables carnivores to discriminate among herbivore species. The ability to discriminate often is acquired through learning. Learning seems to be more important in the case of quantitative differences than in the case of qualitative differences in volatile blends (Vet et al. 1998). Although some information is accumulating on the sensory physiology of carnivorous arthropods, very little information is available yet on the perception of and on the behavioral response to blends that differ in composition. However, this information is essential if we want to better understand how carnivorous arthropods evaluate complex odor blends so as to efficiently locate their herbivorous victims.

References

Alborn, HT Turlings TCJ, Jones TH, Steinhagen G, Loughrin JH, Tumlinson JH (1997) An elicitor of plant volatiles from beet armyworm oral secretion. Science 276: 945-949

Anderson P, Hilker M, Hansson BS, Bombosch S, Klein B, Schildknecht H (1993) Oviposition deterring components in larval frass of *Spodoptera littoralis* (Boisd.) (Lepidoptera: Noctuidae): a behavioral and electrophysiological evaluation. J Insect Physiol 39: 129-137

Aluja M, Prokopy R (1993) Host odor and visual stimulus interactions during intratree host finding behavior of *Rhagoletis pomonella* flies. J Chem Ecol 11: 2671-2696

Arn H, Städler E, Rauscher S (1975) The electroantennographic detector - a selective and sensitive tool in gas chromatographic analysis of insect pheromones. Z Naturforsch C 30: 722-725

Baehrecke EH, Williams HJ, Vinson SB (1989) Electroantennogram responses of *Campoletis sonorensis* (Hymenoptera: Ichneumonidae) to chemicals in cotton (*Gossypium hirsutum*, L.). J Chem Ecol 15: 37-45

Barton Browne L (1993) Physiologically induced changes in resource oriented behavior. Annu Rev Entomol 38: 1-25

Bell WJ (1991) Searching Behavior: The Behavioral Ecology of Finding Resources. Chapman & Hall, London, 358 pp

Bernays EA, Chapman, RF (1994) Host-plant Selection by Phytophagous Insects. Chapman & Hall, New York

Blight MM, Pickett JA, Wadhams LJ, Woodcock CM (1989) Antennal responses of *Ceutorhynchus assimilis* and *Psylloides chrysocephala* to volatiles from oilseed rape. Aspects Appl Biol 23: 329-334

Breer H, Shepherd GM (1993) Implications of the NO-cAMP system for olfaction. Trends Neurosci 16: 5-9

Boeckh J (1984) Neurophysiological aspects of insect olfaction. In: Lewis T (ed) Insect Communication. Proc 12th Symp. Royal Entomol Soc London, Academic Press, London, pp 83-104

Bolter CJ, Dicke M, Van Loon JJA, Visser JH, Posthumus MA (1997) Attraction of the Colorado potato beetle to herbivore-damaged plants during herbivory and after its termination. J Chem Ecol 23: 1003-1023

Buckingham J (1993) Dictionary of Natural Products. Chapman & Hall, London

Chew FS (1988) Searching for defensive chemistry in the Cruciferae, or do glucosinolates always control interaction of Cruciferae with their potential herbivores and symbionts? No! In: Spencer KC (ed) Chemical Mediation of Coevolution. Academic Press, San Diego, pp 81-112

Clyne PJ, Warr CG, Freeman MR, Lessing D, Kim JH, Carlson JR (1999) A novel family of divergent seven-transmembrane proteins: candidate odorant receptors in *Drosophila*. Neuron 22: 327-338

De Jong R, Visser JH (1988) Integration of olfactory information in the Colorado potato beetle brain. Brain Research 447: 10-17

Dethier VG (1982) Mechanism of host-plant recognition. Entomol Exp Appl 31: 49-56

Dicke M (1999) Are herbivore-induced plant volatiles reliable indicators of herbivore identity to foraging carnivorous arthropods? Entomol Exp Appl 91: 131-142

Dicke M (2000) Chemical ecology of host-plant selection by herbivorous arthropods: a multitrophic perspective. Bioch Syst Ecol (in press)

Dicke M, Sabelis MW, De Jong M (1988) Analysis of prey preference of phytoseiid mites as determined with an olfactometer, predation models and electrophoresis. Exp Appl Acarol 5: 225-241

Dicke M, Beek TA van, Posthumus MA, Ben Dom N, Bokhoven H van, Groot Æ de (1990) Isolation and identification of volatile kairomone that affects acarine predator-prey interactions. Involvement of host plant in its production. J Chem Ecol 16: 381-396

Dicke M, Vet LEM (1999) Plant-carnivore interactions: evolutionary and ecological consequences for plant, herbivore and carnivore. In: Olff H, Brown VK, Drent RH (eds) Herbivores: Between Plants and Predators. The 38th Symposium of the British Ecological Society. Blackwell Science, pp 483-520

Dickens JC (1990) Specialized receptor neurons for pheromone and host plant odors in the boll weevil, *Anthonomus grandis* Boh. (Coleoptera: Curculionidae). Chemical Senses 15: 311-331

Dickens JC (1999) Predator-prey interactions: Olfactory adaptations of generalist and specialist predators. Agric Forest Entomol 1: 47-54

Du Y, Poppy GM, Powell W, Pickett JA, Wadhams LJ, Woodcock CM (1998) Identification of semiochemicals released during aphid feeding that attract parasitoid *Aphidius ervi*. J Chem Ecol 24: 1355-1368

Geervliet JBF, Ariens S, Dicke M, Vet LEM (1998) Long-distance assessment of patch profitability through volatile infochemicals by the parasitoids *Cotesia glomerata* and *C. rubecula* (Hymenoptera: Braconidae). Biol Contr 11: 113-121

Hansson BS, Larsson MC, Leal, WS (1999) Green leaf volatile detecting olfactory receptor neurones display very high sensitivity and specificity in a scarab beetle. Physiol Entomol 24: 121-126

Harris MO, Foster SP (1995) Behavior and integration. In: Cardé RT, Bell WJ (eds) Chemical Ecology of Insects 2. Chapman & Hall New York, pp 3-46

Hegnauer R (1962-1994) Chemotaxonomie der Pflanzen. Birkhäuser, Basel

Honda I, Ishikawa Y, Matsumoto Y (1987) Electrophysiological studies on the antennal olfactory cells of the onion fly, *Hylemya antiqua* Meigen (Diptera, Anthomyidae). Appl Entomol Zool 22: 417-423

Jönsson M, Anderson P (1999) Electrophysiological response to herbivore-induced host plant volatiles in the moth *Spodoptera littoralis*. Physiological Entomology 24: 377-385

Lecomte C, Pouzat J (1985) Réponses électroantennographiques de deux parasitoïdes Ichneumonidés, *Diadromus pulchellus* et *D. collaris*, aux odeurs de végétaux, du phytophage-hôte *Acrolepiopsis assectella* et du partenaire sexuel. Entomol Exp Appl. 39: 295-306

Lewis WJ, Jones RL, Nordlund DA, Gross HR (1975) Kairomones and their use for management of entomophagous insects. II. Mechanisms causing increase in rate of parasitization by *Trichogramma* spp. J Chem Ecol 1: 349-360

Li Y, Dickens JC, Steiner WWM (1992) Antennal olfactory responsiveness of *Microplitis croceipes* (Hymenoptera: Braconidae) to cotton plant volatiles. J Chem Ecol 18: 1761-1773

Ma W-C, Visser JH (1978) Single unit analysis of odour quality coding by the olfactory antennal receptor system of the Colorado beetle. Entomol Exp Appl 24:520-533

Mattiacci L, Dicke M, Posthumus MA (1994) Induction of parasitoid attracting synomone in brussels sprouts plants by feeding of *Pieris brassicae* larvae: role of mechanical damage and herbivore elicitor. J Chem Ecol 20: 2229-2247

Mattiacci L, Dicke M, Posthumus, MA (1995) ß-Glucosidase: an elicitor of herbivore-induced plant odor that attracts host-searching parasitic wasps. Proc Natl Acad Sci USA 92: 2036-2040

Metcalf RL, Metcalf ER (1992) Plant Kairomones in Insect Ecology and Control. Chapman & Hall, New York

Miller JR, Strickler KL (1984) Finding and accepting host plants In: Bell WJ, Cardé RT (eds) Chemical Ecology of Insects. Chapman & Hall, New York, pp 127-157

Moran VC, Southwood TRE (1982) The guild composition of arthropod communities in trees. J Animal Ecol 51: 289-306

Murlis J, Elkinton JS, Cardé RT (1992) Odor plumes and how insects use them. Annu Rev Entomol 37: 505-532

Nottingham SF, Hardie J, Dawson GW, Hick AJ, Pickett JA, Wadhams LJ, Woodcock CM (1991) Behavioral and electrophysiological responses of aphids to host and nonhost plant volatiles. J Chem Ecol 17: 1231-1242

Papaj DR, Lewis AC (eds) (1993) Insect Learning: Ecological and Evolutionary Perspectives. Chapman & Hall, New York, 398 pp

Pickett, JA (1990) Gas chromatography-mass spectrometry in insect pheromone identification: three extreme case histories. In: McCaffery AR, Wilson, ID, Chromatography and Isolation of Insect Hormones and Pheromones. Plenum Press, New York, pp 299-309

Prokopy RJ (1986) Visual and olfactory stimulus interaction in resource finding by insects. In: Payne TL, Birch MC, Kennedy CEJ (eds) Mechanisms in Insect Olfaction. Oxford University Press, Oxford, pp 81-89

Ramachadran R, Norris DM (1991) Volatiles mediating plant-herbivore-natural enemy interactions: electroantennogram responses of soybean looper, *Pseudoplusia includens*, and a parasitoid, *Microplitis demolitor*, to green leaf volatiles. J Chem Ecol 17: 1665-1690

Ripperger H, Schreiber K (1981) Solanum steroid alkaloids. In: Manske RHF, Holmes HL (eds) The Alkaloids, VII. Academic Press, New York, pp 247-312

Sabelis MW (1992) Predatory arthropods. In: Crawley MJ (ed) Natural Enemies. The Population Biology of Predators, Parasites and Diseases. Blackwell, London, pp 225-264

Sabelis MW, Baalen M van, Bakker FM, Bruin J, Drukker B, Egas M, Janssen ARM, Lesna IK, Pels B, Van Rijn PCJ, Scutureanu P (1999) The evolution of direct and indirect plant defense against herbivorous arthropods. In: Olff H, Brown VK, Drent RH (eds) Herbivores: Between Plants and Predators. The 38th Symposium of the British Ecological Society. Blackwell Science pp 109-166

Sant'ana J, Dickens JC (1998) Comparative electrophysiological studies of olfaction in predaceous bugs, *Podisus maculiventris* and *P. nigrispinus*. J Chem Ecol 24: 965 - 984

Salom SM, Birgersson G, Payne TL, Berisford CW (1991) Electroantennogram responses of the southern pine beetle parasitoid *Dinotiscus dendroctoni* (Ashmead) (Hymenoptera, Pteromalidae) to potential semiochemicals. J Chem Ecol 17: 2527-2538

Schoonhoven LM, Jermy T, Van Loon JJA (1998) Insect-plant Biology. From Physiology to Evolution. Chapman & Hall, London, 409 pp

Slansky F jr, Rodriguez JG (eds) (1987) Nutritional Ecology of Insects, Spiders and Related Invertebrates. John Wiley and Sons, New York

Smith BH, Getz WM (1994) Nonpheromonal olfactory processing in insects. Annu Rev Entomol 39: 351-375

Takabayashi J, Dicke M, Posthumus MA (1994) Volatile herbivore-induced terpenoids in plant-mite interactions: variation caused by biotic and abiotic factors. J Chem Ecol 20: 1329-1354

Tømmerås BA (1985) Specialization of the olfactory receptor cells in the bark beetle Ips typographus and its predator *Thanasimus formicarius* to bark beetle pheromones and host tree volatiles. J Comp Physiol A 157: 335-341

Turlings TCJ, Fritzsche MA (1999) Attraction of parasitic wasps by caterpillar-damaged plants. In: Chadwick DJ, Goode JA (eds) Insect-plant Interactions and Induced Plant Defence. Novartis Foundation Symposium 223, pp 21-38

Turlings, TCJ, Wäckers FL, Vet LEM, Lewis WJ, Tumlinson JH (1993) Learning of host-finding cues by hymenopterous parasitoids. In: Papaj DR, Lewis AC (eds.) Insect Learning. Chapman & Hall, New York, pp 51-78

Van Loon JJA (1996) Chemosensory basis of oviposition and feeding behavior in herbivorous insects. Entomol Exp Appl 80: 7-13

Van Loon JJA, De Vos EW, Dicke M (2000) Orientation behavior of the predatory hemipteran *Perillus bioculatus* to plant and prey odours. Entomol Exp Appl (in press)

Vet LEM, Dicke M (1992) Ecology of infochemical use by natural enemies in a tritrophic context. Annu Rev Entomol 37: 141-172

Vet LEM, Lewis WJ, Cardé RT (1995) Parasitoid foraging and learning. In: Cardé RT and Bell WJ (eds) Chemical Ecology of Insects 2. Chapman & Hall, New York, pp 65-101

Vet LEM, De Jong AG, Franchi E, Papaj DR (1998) The effect of complete versus incomplete information on odour discrimination in a parasitic wasp. Anim Behav 55: 1271-1279

Visser JH (1986) Host odor perception in phytophagous insects. Ann Rev Entomol 31: 121-144

Visser JH, Avé DA (1978) General green leaf volatiles in the olfactory orientation of the Colorado beetle, *Leptinotarsa decemlineata*. Entomol Exp Appl 24: 538-549

Wadhams LJ (1984) The coupled gas chromatography-single cell recording technique. In: Hummel HE, Miller TA (eds) Techniques in Pheromone Research. Springer Verlag, New York pp. 179-189

Weissbecker B, Schütz S, Klein A, Hummel HE (1997) Analysis of volatiles emitted by potato plants by means of a Colorado potato beetle electroantennographic detector. Talanta 44: 2217-2242

Weissbecker B, Dicke M, Van Loon JJA (1999) Electroantennogram responses of a predator, *Perillus bioculatus*, and its prey, *Leptinotarsa decemlineata*, to plant volatiles. J Chem Ecol 25: 2313-2325

Weissbecker B, Van Loon JJA, Posthumus MA, Bouwmeester HJ, Dicke M (2000) Identification of volatile sesquiterpenoids and their olfactory detection by the twospotted stinkbug *Perillus bioculatus*. J Chem Ecol (in press)

Whitman DW, Eller FJ (1990) Parasitic wasps orient to green leaf volatiles. Chemoecology 1: 69-75

Whitman DW, Eller FJ (1992) Orientation of *Microplitis croceipes* (Hymenoptera: Braconidae) to green leaf volatiles: dose-response curves. J Chem Ecol 18: 1743-1753

Wibe A, Mustaparta H (1996) Encoding of plant odours by receptor neurons in the pine weevil (*Hylobius abietis*) studied by linking gas chromatography-electrophysiology. J Comp Physiol A 179: 331-344

Wilson EO (1992) The Diversity of Life. Harvard University Press, Cambridge

VI.1 Problems in Hygro- and Thermoreception

Harald Tichy and Ewald Gingl

Institut for Zoology, University of Vienna, 1090 Vienna, Althanstrasse 14, Austria

Abstract
Hygroreceptors have been found only in insects and a spider. They associate in antagonistic pairs of a moist and a dry cell in the same sensilla with a thermoreceptor. In insect hygro-thermoreceptive sensilla all of the three cells respond to changes in humidity and two of the three cells respond in addition to changes in temperature. Assigning the individual cells to a particular modality thus becomes a problem. Structural features of thermoreceptors have a bearing on their sensitivity. The size of the dendritic membrane area is related to the sensitivity to slowly changing and steady temperatures. It may determine the number of molecular receptors which in turn sets signal-to-noise ratio. Sensitivity to rapid changes in temperature seems to be increased by positioning the dendritic tips above the surface of the body wall. This is believed to reduce damping of heat transfer. Structural diversity among insect and spider hygro-thermoreceptive sensilla suggests different transduction mechanisms. Three models for hygroreception are discussed. In mechanical hygrometers activity is initiated by swelling or shrinking in hygroscopic sensillum structures, whereas in psychrometers the degree of cooling due to evaporation is used to measure humidity. In electrochemical hygrometers humidity affects electrolyte concentration outside the dendrites.

Key words Hygro- and thermoreceptor identification, structure-function relationships, humidity transduction models

1 Introduction

The environment is to a large extent determined by the average climate and its seasonal variations. Superimposed is the capricious weather with frequent shifts in air temperature, precipitation in the form of rain or snow, and solar radiation. The ranges of temperature and humidity which are found even within restricted habitats are considerable. Animals are continually challenged to regulate and maintain functional capacities as ambient temperature and humidity change. It is therefore not surprising that, in addition to metabolic adaptations, most animals show active temperature and humidity preference or avoidance responses that enhance or optimize their chances of survival and reproduction. In these animals, the existence of hygro- and thermoreceptors is very likely. When an animal exhibits a locomotor response to temperature or humidity stimulation, temperature and humidity serve as proximate factors in orientation. In the case of behavioral regulation of body temperature and water content, temperature and humidity serve as ultimate ecological factors as well. However, the proximate and ultimate factors governing preference or avoidance responses to temperature and humidity must not always be the same. Nonthermal and nonhumidity cues may provide the ultimate ecological value of a temperature or humidity response; some examples are habitat selection, interspecific niche differentiation, predator avoidance, prey occurrence, escape reactions, and migrations (diel, seasonal, spawning), abundance of food, absence of predators and competitors, light intensity, or depth in soil. The associations between proximate orientation and ultimate ecological factors conferring adaptive value on preference or avoidance behavior immediately lead to an awkward paradox. It is obviously desirable to separate the physical state of an animal's environment from its physiological responses to that environment; but unless the nature of the response is recognized and understood, the components of the thermal and humid environment cannot be specified in a physiologically meaningful way. A way to resolve this circular problem is by assessing the sensitivity of thermo- and hygroreceptive sensory cells to temperature and humidity stimuli occurring in the animal's biotope. A comparison of the differentiation of hygro- and thermoreceptive sense organs in animals with different ecology may give information on their adaptation to serve that kind of life and the knowledge of how certain structures of the sense organs vary because of different functional demands.

The number of animals which have been shown to respond behaviorally to temperature and humidity stimulation is much greater than the number in which hygro- and thermoreceptors have been identified electrophysiologically and investigated by electronmicroscopic techniques. Most of our knowledge on hygro- and thermoreception is based on studies of a few insects and a spider (reviewed by Loftus 1978; Altner and Loftus 1985; Tichy and Loftus 1996; Steinbrecht 1998; Yokohari 1999). The reasons are that in arthropods hygro- and thermoreceptors can be localized by removing body appendages, and, in addition, they occur together in the same cuticular sensillum. Due to their superficial position they are acessible to the electrophysiological experiment.

This chapter brings together three lines of study. The first of these is on the fine structure of hygro- and thermoreceptive sensilla and the second on their electrophysiological properties. The third line refers to ideas about mechanisms of improving sensitivity to temperature stimuli and to the transduction of humidity stimuli into action potentials.

2 Identification of Hygro- and Thermoreceptive Sensory Cells

The experimental procedure used in most electrophysiological studies on hygro- and thermoreceptive sensilla is summarized in Fig. 1 (Waldow 1970; Loftus 1976; Yokohari and Tateda 1976; Yokohari 1978; Tichy 1987; Tichy and Loftus 1990). The insect is immobilized and the antenna is held securely (Fig. 1A). One electrode is inserted lengthwise into the tip of the antenna (Fig. 1A), and the other into the base of the sensillum (Fig. 1A,C). With this electrode placement biphasic action potentials are recorded from the three sensory cells contained within the sensillum (Fig. 1B). The sensillum is a small cuticular peg which is enclosed by a cuticular wall (Fig. 1D). The dendrites of two sensory cells extend into the lumen of the sensillar peg. The difference in their diameters is small, as are the amplitudes of the recorded action potentials of two cells (Fig. 1B). Two of the three cells are bimodal, responding to changes in both temperature and vapor pressure. Stimulation is provided by three air streams emerging at 2.5 ms^{-1} from jets 7 mm in diameter (Fig. 1E). Each air stream can be directed separately onto the preparation. Two of the air streams (A,B) are at the same temperature (T) but at different partial pressure of water vapor (Pw), another pair (A,C) is at the same Pw but at different T. A change from stream A to C produces a change in T at constant Pw, a change from A to B produces a change in Pw at constant T.

Simultaneous recordings from the three receptor cells in the hygrothermoreceptive sensillum revealed that each cell responds to changes in Pw and two of them to changes in T. Nevertheless the differences in the responses, especially in the direction of change in impulse frequency (F), provide criteria for their individual identification. In Fig. 1F 1, when T rapidly decreases at constant Pw, two cells respond with an increase in F. These two cells could be regarded as cold cells. However, they could equally well be considered receptors for relative humidity (Hr). Because Hr increases when T decreases and Pw remains constant. Rapid warming causes both cells to interrupt their discharge (Fig. 1F 2). These responses likewise are consistent with both interpretations because the increase in T implies a decrease in Hr. If the increase in F of these two cells is indeed a response to an increase in Hr, one would expect that F increases when Hr increases as a result of an increase in Pw with T held constant. However, when Pw increases at constant T, only one of the two cells responds as expected; impulse frequency of the other cell decreases with this increase in Hr (Fig. 1F 3).

These opposite responses reveal fundamental differences between the two cells. If it should turn out that the responses to the two modalities are governed by a single parameter in each case, they could be regarded as unimodal. In fact, only one of the two cells meets the criterion for a unimodal hygroreceptor. Unlike the first cell, the second cell responds with an increase in impulse frequency to an increase in Hr regardless of whether the increase in Hr is brought about by a decrease in T at constant Pw or an increase in Pw at constant T. This cell is designated moist cell.

Its function as a hygroreceptor, however, is conceivable only if it is capable of taking up water from the air or giving it off to the air. The sensillum surface must be hygroscopic. A hygroscopic surface of the sensillum is important for the interpretation of the responses of the first cell as well. This cell behaves like a cold cell in that impulse frequency is increased upon cooling the sensillum and decreased by warming it.

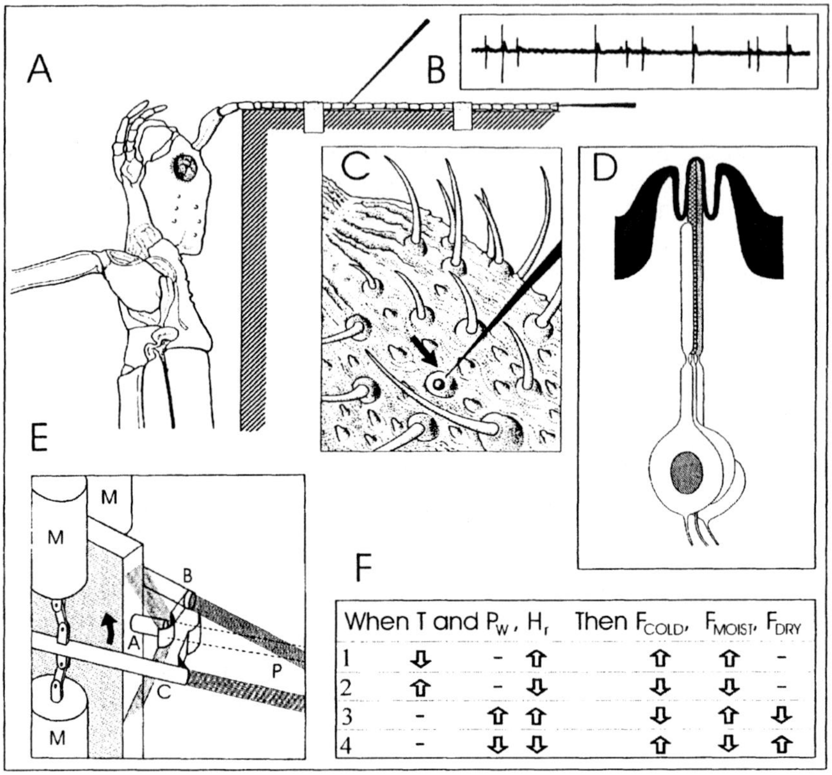

Fig. 1A-F. Diagram of the experimental arrangement (A), the electrophysiologically recorded activity of 3 sensory cells (B), the location (C) and anatomy of the hygro-thermoreceptive sensillum (D), the 3 air streams used for stimulation (E), and the direction of changes in the frequency of the 3 sensory cells in the sensillum (F), following rapid changes of either temperature T or water vapor pressure Pw with the other of the two parameters held constant. **A** The experimental animal (a stick insect) is fixed on a Plexiglas holder by strips of adhesive tape. The indifferent electrode is inserted lengthwise about 2 mm into the tip of the antenna, and the recording electrode at an angle of about 45° into the margin of the sensillum which is innervated by a moist, a dry, and a cold cell. **B** Recordings of the action potentials of the three receptor cells. **C** Surface view of the antennal segment showing recording electrode inserted into the margin of the sensillum. The hygro-thermoreceptive sensillum is surrounded by olfactory hair sensilla. **D** Longitudinal section through the sensillum shown in C illustrating the cuticular peg inserted at the bottom of a pit but not rising above the cuticular walls which surround it. The unbranched dendrites of two cells completely fill the lumen of the peg. The lamellated dendrite of the third cell does not extend to the tip of the peg but terminates at its base. **E** Each of the 3 stimulating air streams (A-C) can be directed separately onto a common point of intersection (*P*), where the preparation is located. When another air stream is brought into position by electromagnets (*M*), a gate attached to the nozzle simultaneously deflects the first air stream without interrupting it. **F** Changes in temperature (*T*) or water vapor pressure (Pw) lead to changes in relative humidity (*Hr*) until Pw reaches saturation for the particular T. *Arrow pointing*

upward discharge frequency (*F*) increases. *Arrow pointing downward* discharge frequency (*F*) decreases. *Slash,* value remains constant. Read as follows, e.g., line 2: when T increases and Pw is constant, Hr decreases. Then the discharge frequency of the cold cell decreases, the discharge frequency of the moist cell decreases, and the discharge frequency of the dry cell does not change

The cell could also be acting as a cold cell in its responses to changes in Pw. Changes in Pw of a hygroscopic surface would be expected to produce changes in T. When Pw is reduced some of the water bound to the surface will evaporate; evaporation of water will cool the sensillum. When Pw increases, the binding of water will cause warming. A decrease in Pw will increase impulse frequency of a cold cell (Fig. 1F 4) and an increase in Pw will decrease it in a cold cell (Fig. 1F 3). The responses of the cell to changes in Pw can be interpreted as responses to changes in enthalpic temperature. This cell is a cold cell.

The third cell exhibits an increase in F when Pw is decreased (Fig. 1F 4) and a decrease in F when Pw is increased (Fig. 1F 3). These changes in impulse frequency are in the same direction as those of the cold cell. However, the response of the cold cell is transitory, lasting less than a second. The response of the third cell to the same stimulus rises rapidly to a maximum and then slowly levels off during time periods of many seconds. This response is opposite to that of the moist cells. The third cell is therefore designated as a dry cell.

The cold, the moist and the dry cells respond clearly to changes in Pw, the cold and the moist cells to changes in T as well. These responses are important because insects are generally exposed to with changes in both T and Pw as they range about. Hr is changed by both T and Pw, and the cold and the moist cell respond to both. However, they respond differently to both parameters (Fig. 1F). Thus the ambiguity in the response could be eliminated by taking into consideration the responses of all three cells. However, next to nothing is known about central processing of sensory information on ambient humidity and temperature.

3 Thermoreceptor Sensitivity: Significance of Dendritic Membrane Area and Dendritic Tip Position

Although the number of arthropod hygro- and thermoreceptive sensilla examined for both their fine structure and physiology is quite small, considerable structural differences are emerging which have a bearing on the response to temperature. Before discussing structure-function relationships, the internal structures of insect and spider sensilla must be elucidated (for review see Tichy and Loftus 1996).

Hygro-thermoreceptive sensilla occur as two morphological types. Those with an unperforated cuticular wall form type 1 (Figs. 2A-C, 3A-E); they occur in insects. Type 2 sensilla have a pore opening at the tip and are known from a spider (Figs. 2D-F, 3F). Both types of sensilla contain an antagonistic pair of hygroreceptors together with a thermoreceptor. The thermoreceptor is a cold cell in insects and a warm cell in the spider.

The position of hygro-thermoreceptive sensilla in the exosceleton appears to reflect a tradeoff between exposure and protection. Some sensilla rise from the floor of a crater (Figs. 2AB, 3D, *Carausius morosus*, Altner et al. 1978), or are entrenched in a groove (Fig. 3E, *Bombyx mori*, Steinbrecht 1989; Steinbrecht et al. 1989; Steinbrecht and Müller 1991; Zimmermann 1991) or located in a heavily walled pit (Fig. 3B, *Apis mellifera*, Yokohari et al 1982; Yokohari 1983). Some are hidden in a hole (Fig. 3C, *Locusta migratoria*, Altner et al. 1981). The sensillum of the cockroach displays the tightest fitting protective wall, so tight that it can be taken for a thin outer wall of the sensillum with a ring-shaped opening just under the cap (Fig. 3A, *Periplaneta americana*, Yokohari 1981; Tominaga and Yokohari 1982). In the spider, the tarsal organ encloses seven nipple-shaped sensilla as a group (Figs. 2D-F, 3F, *Cupiennius salei*, Anton and Tichy 1994) in its tiny air-filled cavity.

In insects, the dendrites of two of the three sensory cells are unbranched. In some cases they extend out to the apex of the peg and fill its lumen completely (Figs. 3A-D, 4B); in others, they terminate well short of it (bombycid moth, Fig. 3E). The dendrite of the third sensory cell takes different forms and extensions. It is unbranched in the honeybee, ending at some distance from the apex of the peg (Fig. 3B) and branched in the migratory locust, ending below the base of the peg (Figs. 3C, 4B). In the cockroach, the third dendrite branches and sends numerous slender extensions into the cuticular peg (Figs. 3A, 4B). It is folded or even transformed into lamellae that terminate below the cuticle in the stick insect (Fig. 3D) and the bombycid moth (Fig. 3E).

In the moth, a few of the slender lamellae may enter the cuticular peg. A special location is found in the saturniid moth *Antheraea*, where the lamellae lie within an elongated socket which rises up to 60 μm above the antennal surface (Fig. 4B, Haug 1986; Zimmerman 1991). For the spider sensilla the situation is different. The apical pore is filled with dense material (Figs. 3F, 4B). All sensory cells have unbranched dendrites that terminate at the tip pore. Which of the dendrites is the source of the thermoreceptive activity observed in the recordings? In insects, two of the three cells are quite long, lie very close together, and extend far out into the peg. It is reasonable to assume that they have similar function and belong to the two hygroreceptors. Consequently, the third dendrite is considered to be thermoreceptive. Transduction in thermoreceptors has been attributed to the temperature dependence of both a sodium/potassium pump and the conductances of sodium and potassium channels (Braun et al. 1990). The number of these molecular receptors is expected to rise with dendrite size. A larger number will increase the signal-to-noise ratio and thereby improve absolute temperature sensitivity. Apart from the size the functionally most significant feature of the thermoreceptive dendrite is its position. Whereas the dendrite of the spider warm cell extends to the tip of the sensillum and terminates just inside a single pore opening (Figs. 2F, 3F, 4B), the dendrites of insect cold cells end beneath the base of poreless sensilla (Figs. 2B, 3A-E, 4B). That is, their tips are not positioned above the surface of the antenna. In the saturniid moth *Antheraea*, however, the cold-cell dendrite rises above the antennal surface by being contained in an elongated socket that bears the sensillum (Fig. 4B).

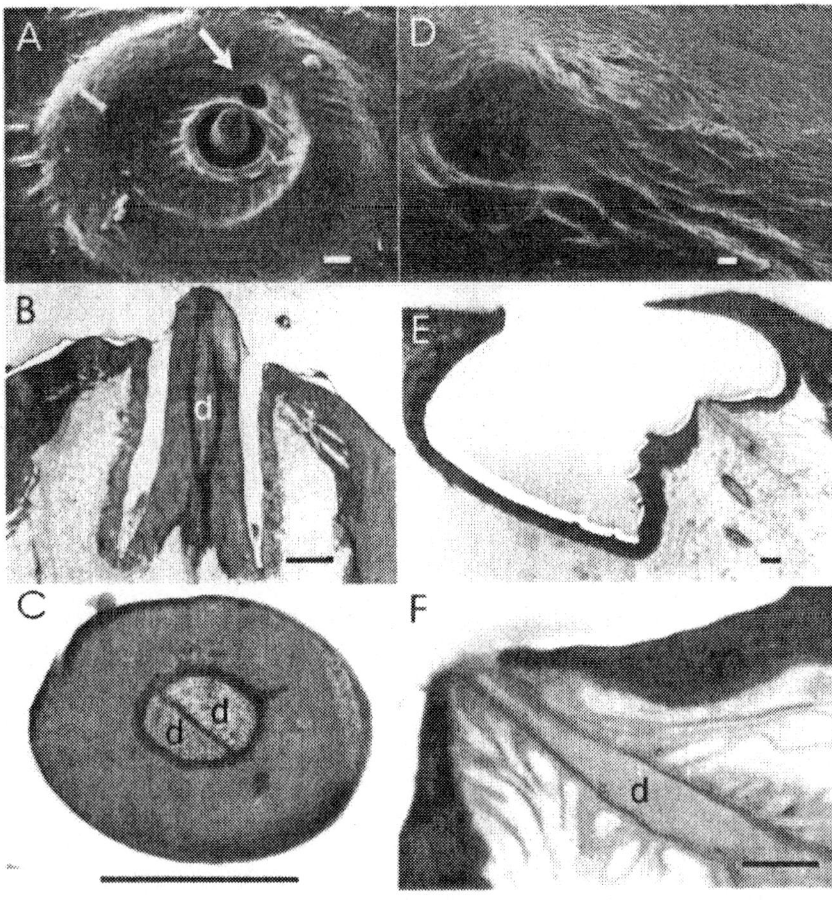

Fig. 2 A-F. Hygro-thermoreceptive sensilla of the stick insect *Carausius morosus* (from Altner et al 1978) and the wandering spider *Cupiennius salei* (from Anton and Tichy 1994). **A** The insect sensillum consists of a small cuticular peg inserted at the bottom of a pit and surrounded by an inflated collar. The collar was pierced by the recording electrode (arrow) during electrophysiological investigation. **B,C** Longitudinal and cross-sections through the peg showing the lumen completely filled by the dendrites of two sensory cells. **D** The spider sensilla are found in an air-filled capsule with an elliptical aperture (Blumenthal's tarsal organ). **E** Longitudinal section of the capsule showing nipple-shaped spider tip-pore sensillum. **F** Tip pore in longitudinal section. The dendrite extends almost to the very boundary of the opening. It is surrounded and covered by amorphous material. The dendritic sheath is enclosed by cell lamellae. Bars, 1 μm.

Therefore, one would expect that at rapid temperature changes the exposed dendrites of the spider's warm cell and the cold cell of the saturniid moth would reach a new temperature faster than the dendrites which are covered by cuticle and positioned beneath the base of the sensillum. This would imply that in the spider and the moth the thermoreceptor sensitivity to rapid temperature changes is increased by a reduced damping of the heat transfer. The more superficial dendrites should take on the temperature of the ambient air faster than the deeper ones, if only because of the additional amount of material in which the latter are embedded. During slowly changing temperature, on the other hand, the temperature of the thermoreceptor should not lag behind that of the ambient air. Thus, during slow temperature change the size of the dendritic membrane area rather than the position of the dendritic tip should characterize sensitivity of the thermoreceptors.

To elucidate the relationship between thermoreceptor structure and sensitivity, the rank order for sensitivity to rapid step-like temperature change, slowly oscillating temperature change, and steady temperature was compared with the size of the dendritic membrane area and the position of the dendritic tip (Fig. 4 A,C, Ehn and Tichy 1996). The improvement in the signal-to-noise ratio caused by the increase in the size of the dendritic membrane area indeed increases with the square root of the outline of the dendrites (Fig. 4A 4). The resulting square root ratios are closely reflected in the ratios of both the values of sensitivity to slowly changing and steady temperature (Fig. 4A 2,3).

The position of the dendritic tip seems to characterize the thermoreceptors in terms of sensitivity to rapid temperature changes (Fig. 4A,B). The cold cell of the saturniid moth, which has a considerably smaller dendritic membrane area than the cold cell of cave beetle, displays the highest dynamic sensitivity; but in contrast to the cold cell of the cave beetle, the dendrite of the saturniid moth's cold cell projects into a slender elongate socket which rises 60 µm above the antennal surface. Therefore, at rapid temperature changes the dendrite of the saturniid moth is supposed to reach a new temperature level faster than that of the cave beetle which terminates beneath the sensillum base. Further support for such a view is provided by the warm cell of the spider which shows the second highest sensitivity to rapid temperature changes and has a dendrite that terminates at a pore opening on the tip of a small (height 5 µm) sensillum. Also, the low values of sensitivity described for the cold cells of the cockroach and the migratory locust are in good agreement with the position of their dendrites beneath the surface of the body wall.

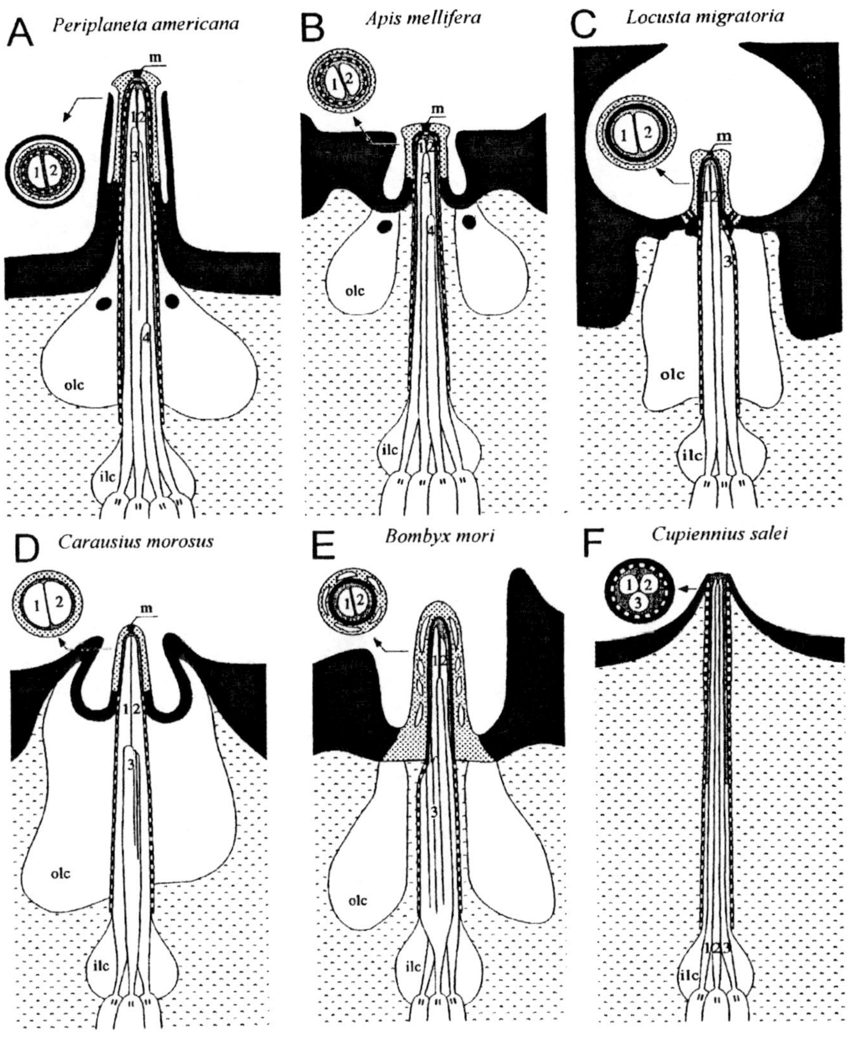

Fig. 3A-F. Diagrams of the cuticular processes, cellular associations and lymph cavities in hygro-thermoreceptive sensilla (after Tichy and Loftus 1996): **A** *Periplaneta americana* (after Yokohari 1981; Tominaga and Yokohari 1982); **B** *Apis mellifera* (after Yokohari 1983; Yokohari et al. 1982); **C** *Locusta migratoria*, (after Altner et al. 1981); **D** *Carausius morosus* (after Altner et al. 1978); **E** *Bombyx mori* (after Steinbrecht and Müller 1991; Steinbrecht et al. 1989; Zimmermann 1991); **F** *Cupiennius salei* (after Anton and Tichy 1994). The main feature the spider sensillum has in common with insects is a thin-walled sensillum cuticle that lacks a pore system to connect the lumen inside the sheath with ambient air (**A-E**). The molting pore (*m*) at the tip appears plugged by dense material (**A-D**). Of the three or four sensory cells innervating the sensillum, two possess unbranched outer dendritic segments (*1,2*) extending up to the apex of the cuticular peg. The outer segment of the third takes on various forms. It is shorter; sometimes unbranched (**B,C**), sometimes forked with slender extensions (**A**) or even with lamellae (**D,E**) that project into the peg (**E**) or end below its base (**D**). The fourth cell, when found (**A,B**), maintains its ciliary structure for the length of its outer segment and terminates well short of the peg (4). A small inner receptor lymph cavity (*ilc*) encloses the central portion of the dendrites, between their inner and outer segments (*1-4*), and is surrounded by sheath cells. Beside these cells, or just outside the dendritic sheath, the more voluminous outer receptor lymph cavity (*olc*) is present. In *Cupiennius*, a pore at the tip of the nipple-shaped sensilla leads directly into the lumen of the dendritic sheath (**F**). Each sensillum is capped by a thin layer of amorphous material, found also in the pore and then encasing the terminals of the three unbranched dendrites (*1-3*). This layer merges with the dense lymph bathing the outer segments further inside

4 Humidity Transduction: Functional Interpretation of Sensillum Structure

Of paramount interest in hygroreception is the manner in which humidity acts on the hygroreceptive sensory cells. Hygroreceptors were proposed to operate as mechanical hygrometers, psychrometers, and electrochemical hygrometers (for reviews see Tichy and Loftus 1996; Steinbrecht 1999). The underlying models are quite different and were developed for different sensilla. Might it just be that the structural diversity even among insect hygroreceptive sensilla is too great for a single model to apply in all cases? Structural characteristics of the interior of the sensillum wall, the dendritic sheath, the receptor lymph cavities, and the dendritic segments are essential to the understanding of the models.

Pores are generally absent in the thermo-hygroreceptive sensilla of insects. In the spider, the sensillum displays an apical pore filled with dense material. The material is piled up on the outside, apparently discharging through the pore (Fig. 3F).

In insect sensilla, a sheath encloses the dendrites in a lumen and, together with the surrounding sheath cells, separates them from the outer lymph cavity (Fig. 3A-E). The sheath may have an important function. It may seal off the hygroreceptors from the water inside the body. In the spider's tip-pore sensilla, the dendritic sheath extends up to the pore, where it fuses laterally with the cuticle wall (Fig. 3F).

In insect sensilla two lymph cavities, filled with extracellular fluid, are present (Fig. 3A-E). The inner cavity surrounds the sensory cells in the region of their cilia. From there the cavity proceeds distally and becomes the lumen of the dendritic sheath. In this narrow canal a lymph surrounds the dendritic outer segments from their ciliary region up to the tips. The lymph appears to form layers, differing in electron density and wrapping the dendrites.

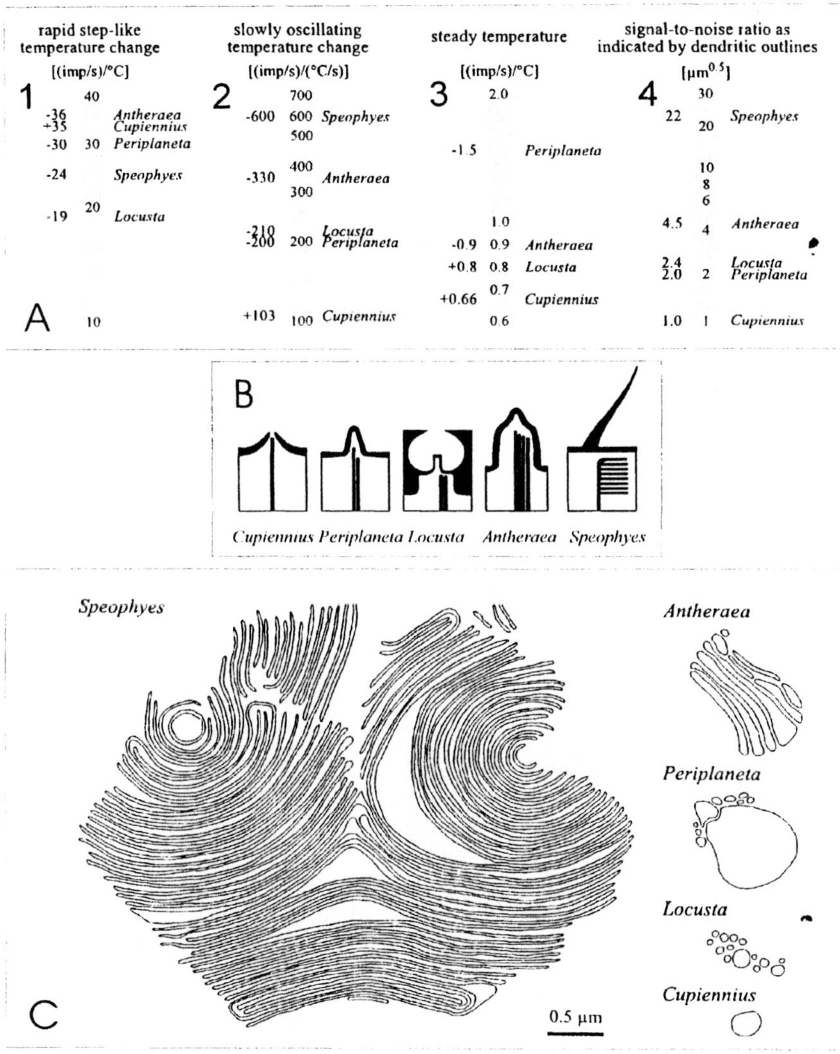

Fig. 4A-C. Structure-function relationships. **A** Values of sensitivity of various arthropod thermoreceptors to 1 rapid step-like temperature change, 2 slowly oscillating temperature change, and 3 steady temperature. 4 Square root of the outlines of the dendritic cross-sections indicating the improvement in the signal-to-noise ratio caused by the increase in the size of the dendritic membrane areas. Sensitivity values: *Periplaneta americana* (Loftus 1968, 1969), *Speophyes lucidulus* (Loftus and Corbière-Tichané 1981, 1987; Corbière-Tichané and Loftus 1983), *Antheraea pernyi* (Haug 1986; Ameismeier and Loftus 1988),

Locusta migratoria (Ameismeier and Loftus 1988), *Cupiennius salei* (Anton and Tichy 1994). Outlines of dendrites taken from **C**. Note log-scale within bars; on the left side of the bars, values are given numerically. **B** Sketch of form and position of thermoreceptive dendrite in insects and spider hygro-thermoreceptive sensilla. **C** Digitized tracings from published transmission electronmicrographs of transverse sections of the dendrites of different arthropod thermoreceptors. The outlines are drawn to scale, so that differences in size of the membrane area can be appreciated. *Speophyes lucidulus (*Corbière-Tichané 1971, Fig. 3a), *Antheraea pernyi* (Zimmermann 1991, Fig. 4d), *Periplaneta americana* (Yokohari 1981, Fig. 13c), *Locusta migratoria* (Ameismeier and Loftus 1988, Fig. 1), *Cupiennius salei* (Anton and Tichy 1994, Fig. 2b)

The outer receptor lymph cavity is closer to the cuticle and more spacious. Nowhere is an observable avenue of easy communication between the outer and the inner cavities. The sheath and the cells surrounding it appear to be a barrier between the two. In the tip-pore sensilla of the spider there is only one receptor lymph cavity, the inner cavity (Fig. 3F). The cavity narrows as it enters the dendritic sheath and proceeds outwardly towards the pore. The electron density of the lymph increases along the way but thins again at the pore opening. The fluid would seem to discharge slowly there and pile up amorphous material at the entrance, thus placing a limit on the access of the lymph to the atmosphere. The pore may be involved in stimulus conduction and/or transduction.

4.1 Mechanical Hygrometer Model

For insects the most favored model views the cuticular wall as a hygro-mechanical transducer. Humidity-dependent shrinking due to water loss and swelling due to water uptake are believed to cause changes in the geometry of the wall that in turn leads to deformation of the dendritic membranes and voltage changes across them (Fig. 5). Here, the intimate association of the dendritic membranes with the cuticular wall would be important.

If changes in humidity indeed lead to structural changes of the wall, surface structures should change as well. We used the scanning force microscope to analyze the effect of changes in humidity on the surface of the peg sensilla in the honeybee *Apis mellifera* (Resch et al. 1998). Female hair, which was used as reference material, was found to increase in length by 1.0% when Hr was increased from 0 to 100%. These data are in agreement with the original specification of the supplier (Lambrecht Company 1996). In the honeybee this same technique revealed no humidity induced structural changes of the surface on top of the peg. Recent experiments on the cockroach (Yokohari 1999) displayed a 65-nm increase in both the diameter and length of the peg when relative humidity was raised from 0 to 100%. This value was obtained by placing a piezoelectric element on the apex of the peg. Unfortunately, no full account of this work was published, so that procedure and analysis cannot be evaluated.

4.2 Psychrometer Model

Psychrometers measure the humidity (or the dryness) of the air by means of evaporative temperature depression. To determine the depression, two separate temperature measurements in ambient air are needed, one by a thermometer with a dry surface and one by a thermometer with a wet surface. The dry-surface thermometer shows the air temperature. The wet-surface thermometer indicates a lower value by reason of the cooling effect of water evaporating from its surface. The temperature depression is the difference between the two readings. As absolute humidity decreases and temperature rises, the greater the temperature depression and the greater the power of the air to desiccate. Together, evaporation cooling and air temperature are a quite precise measure for humidity, both absolute and relative. While the hygroreceptors in the mechanical hygrometer model are mechanoreceptors and the movement of water can be in either direction, out of the air into the sensillum (inward) or the reverse (outward), in the psychrometer model the hygroreceptors are thermoreceptors and the water moves only towards the outside. Only evaporation occurs. Water from inside will reach the surface where it evaporates and causes cooling of the sensillum. The question is, where is the reservoir of water that is evaporated? Interestingly, in the bombycid moth it was found that after prolonged exposure to dry air the dendrites shorten considerably (Steinbrecht and Müller 1991). This is suggested to be due to evaporation of water to the air. The water may come from the receptor lymph cavity, the dendrites themselves, and the cells surrounding them.

4.3 Electrochemical Hygrometer Model

In the case of the tarsal organ of the spider the electrochemical hygrometer model has much to offer. It views the concentration of electrolytes in the thin layer of lymph just inside the sheath and immediately surrounding the dendrites as varying with humidity. The drier the air, the greater the evaporation rate of water out of the lymph and the greater the change in electrolyte concentration. The concentration in turn would affect the potential across the dendritic membranes and thereby govern the responses of the hygroreceptors. As with the psychrometer model, a hygroscopic material would appear unnecessary. Rather lymph is viewed here as moving in one direction only, slowly towards the outside, where its water content is exposed to controlled evaporation in ambient air. The electrochemical hygrometer shares evaporation as a constitutive feature with the psychrometer model. Concentration of the electrolytes is affected by the humidity of the air. The drier the air, the greater the evaporation rate of water out of the lymph and the greater the change in electrolyte concentration.

Is the electrolyte concentration model feasible for insects? It might be, provided their structures permit a slight and controlled flow of lymph along the dendrites towards the outside where humidity could affect the concentration of its electrolytes. None of the sensilla in Fig. 3A-E displays a large, "open" apical pore like the sensilla of the spider tarsal organ. Many insects probably encounter drier air than the examined spider (Barth et al. 1988) does, however. Here a smaller pore with tighter "plugging" might be advantageous. Further, if a drier environment were to demand a reservoir of lymph larger than that of the spider in order to maintain a slight flow with a constant electrolyte concentration in the background, the large outer lymph cavity of insects might serve as a source. In this case both the cuticle just outside the outer lymph cavity and the dendritic sheath would be needed as water barriers.

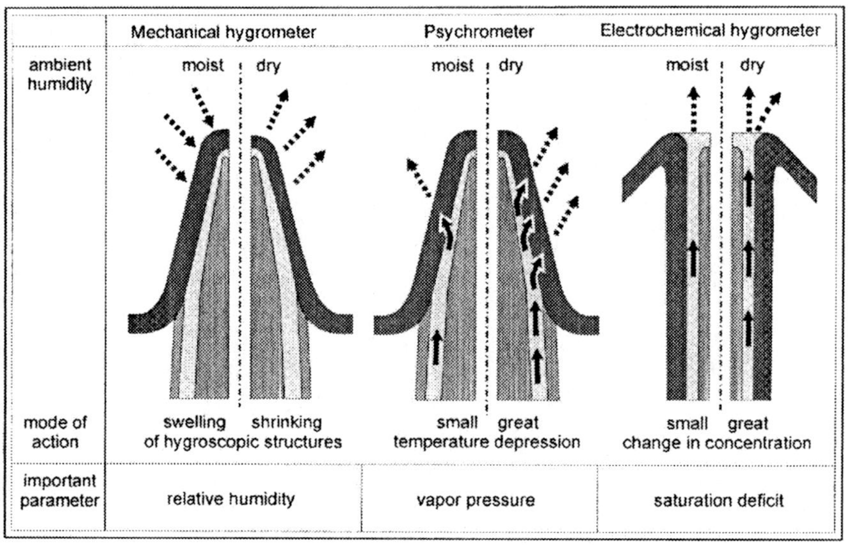

Fig. 5. Models for hygroreception. In mechanical hygrometers activity is initiated by swelling or shrinking of hygroscopic sensillum structures, in psychrometers the degree of cooling due to evaporation is used to measure humidity, and in electrochemical hygrometers humidity affects electrolyte concentration just outside the dendrites. In the mechanical hygrometer model, movement of water is from the air into the sensillum and from the sensillum into the air, and in both the psychrometer and the electrochemical model it is from the sensillum towards the outside. Mechanical hygrometers would response to the relative humidity, psychrometers to the partial pressure of water vapor and electrochemical hygrometers to the saturation deficit of the air.

These models are interesting not only in relation to structural features. They also pose some intriguing questions as to the parameters governing the responses. If it is assumed that the recorded responses of hygroreceptors are not affected by temperature per se, it follows that the different types of hygroreceptors would respond to different parameters when tested with humidity stimuli at different temperatures. Hygroreceptors acting as mechanical hygrometers would produce the same response to the same relative humidity, regardless of the ambient temperature. Those acting as psychrometers would produce the same response to the same vapor pressure. Hygroreceptors acting as electrochemical hygrometers would produce the same response to the same saturation deficit. Consequently, animals with mechanical hygrometers would select environments of the same relative humidity. Those with psychrometers would show a preference to environments with the same vapor pressure. The humidity responses of animals with chemical hygrometers would be related to the saturation deficit of the air.

Acknowledgments. Supported by a grant of the Austrian Science Foundation (FWF) to H. Tichy.

References

Altner H, Loftus R (1985) Ultrastructure and function of insect thermo- and hygroreceptors. Annu Rev Entomol 30:273-295

Altner H, Tichy H, Altner I (1978) Lamellated outer dendritic segments of a sensory cell within a poreless thermo- and hygroreceptive sensillum of the insect *Carausius morosus*. Cell Tissue Res 191:287-304

Altner H, Loftus R, Schaller-Selzer, L, Tichy H (1983) Modality-specificity in insect sensilla and multimodal input from body appendages. In: Horn E (ed) Multimodal convergence in sensory systems. Fortschritte der Zoologie, vol 28. Fischer, Stuttgart, New York, pp. 17-31

Ameismeier F, Loftus R (1988) Response characteristics of cold cell on the antenna of *Locusta migratoria* L. J Comp Physiol A 163:507-516

Anton S, Tichy H (1994) Hygro- and thermoreceptors in tip pore sensilla of the tarsal organ of the spider *Cupiennius salei*: innervation and central projection. Cell Tissue Res 278:399-407

Barth FG, Seyfarth EA, Bleckmann H, Schüch W (1988) Spiders of the genus *Cupiennius* Simon 1891 (Aranea, Ctenidae). I. Range distribution, dwelling plants, and climatic characteristics of the habitats. Oecologia 77:187-193

Braun HA, Schäfer K, Wissing H (1990) Theories and models of temperature transduction. In: Bligh J, Voigt K (eds) Thermoreception and Temperature Regulation. Springer, Berlin Heidelberg New York, pp 19-29

Corbière-Tichané G (1971) Structure nerveuse énigmatique dans l'antenne de la larve du *Speophyes lucidulus* Delar. (Coléoptère cavernicole de la sous-famille des Bathysciinae). Étude au microscope électronique. J Microscopie 10:191-202

Corbière-Tichané G, Loftus R (1983) Antennal thermal receptors of the cave beetle, *Speophyes lucidulus* Delar. II. Cold receptor response to slowly changing temperature. J Comp Physiol 153:343-351

Ehn R, Tichy H (1996) Response characteristics of a spider warm cell: temperature sensitvities and structural properties. J Comp Physiol A 178:537-542

Haug T (1986) Struktur, Funktion und Projektion der antennalen Thermo- und Hygrorezeptoren von *Antheraea pernyi* (Lepidoptera: Saturniidae). Doct Diss, Universität Regensburg

Loftus R (1968) Response of the antennal cold receptor of *Periplaneta americana* to rapid temperature changes and to steady temperature. Z vergl Physiol 59:413-455

Loftus R (1969) Differential thermal components in the response of the antennal cold receptor of *Periplaneta americana* to slowly changing temperature. Z vergl Physiol 63:415-433

Loftus R (1976) Temperature-dependent dry receptor on antenna of *Periplaneta americana*. Tonic Response. J Comp Physiol 111:153-170

Loftus R (1978) Peripheral thermal receptors. In: Ali MA (ed) NATO advanced study institutes series, vol 18. Sensory Ecology: Reviews and Perspectives. Plenum Press, New York, pp 439-466

Loftus R, Corbière-Tichané G (1981) Antennal warm and cold receptors of the cave beetle, *Speophyes lucidulus* Delar., in sensilla with a lamellated dendrite. I. Response to sudden temperature change. J Comp Physiol 143:443-452

Loftus R, Corbière-Tichané G (1987) Response of antennal cold receptors of the catopid beetles, *Speophyes lucidulus* Delar. and *Choleva angustata* Fab. to very slowly changing temperature. J Comp Physiol A 161:399-405

Resch R, Ehn R, Tichy H, Friedbacher G (1998) In-situ investigation of humidity-induced changes on human hair and antennae of the honeybee, *Apis mellifera* L., by scanning force micorscopy. Appl Phys A 66 607-611

Steinbrecht RA (1989) Fine structure of thermo-/hygroreceptive sensilla in the silkmoth *Bombyx mori*: receptor membrane substructure and sensory cell contacts. Cell Tissue Res 255:49-57

Steinbrecht RA (1999) Bimodal thermo- and hygrosensitive sensilla. In: Harrison FW, Locke M (eds) Microscopic Anatomy of Invertebrates, vol 11B. Wiley-Liss, New York, pp 405-422

Steinbrecht RA, Lee J-K, Altner H, Zimmermann B (1989) Volume and surface of receptor and auxiliary cells in hygro-/thermoreceptive sensilla of moths (*Bombyx mori*, *Antheraea pernyi*, and *A. polyphemus*). Cell Tissue Res 255:59-67

Steinbrecht RA, Müller B (1991) The thermo-/hygrosensitive sensilla of the silkmoth, *Bombyx mori*: morphological changes after dry- and moist-adaptation. Cell Tissue Res 266:441-456

Tichy H (1987) Hygroreceptor identification and response characteristics in the stick insect *Carausius morosus*. J Comp Physiol A 160:43-53

Tichy H, Loftus R (1990) Response of moist-air receptor on antenna of the stick insect, *Carausius morosus*, to step changes in temperature. J Comp Physiol A 166:507-516

Tichy H, Loftus R (1996) Hygroreceptors in insects and a spider: Humidity transduction models. Naturwissenschaften 83:255-263

Tominaga Y, Yokohari F (1982) External structure of the sensillum capitulum, a hygro- and thermoreceptive sensillum of the cockroach, *Periplaneta americana*. Cell Tissue Res 226:309-318

Waldow U (1970) Elektrophysiologische Untersuchungen an Feuchte-, Trocken- und Kälterezeptoren auf der Antenne der Wanderheuschrecke *Locusta*. Z vergl Physiol 69:249-283

Yokohari F (1978) Hygroreceptor mechanism in the antenna of the cockroach *Periplaneta*. J Comp Physiol 124:53-60

Yokohari F (1981) The sensillum capitulum, an antennal hygro- and thermoreceptive sensillum of the cockroach, *Periplaneta americana* L. Cell Tissue Res 216:525-543

Yokohari F (1983) The coelocapitular sensillum, an antennal hygro- and thermoreceptive sensillum of the honeybee, *Apis mellifica* L. Cell Tissue Res 233:355-365

Yokohari F (1999) Hygro- and thermoreceptors. In: Eguchi E, Tominaga Y (eds) Atlas of Arthropod Sensory Receptors. Springer, Berlin Heidelberg New York Tokio, pp 191-210

Yokohari F, Tateda H (1976) Moist and dry hygroreceptors for relative humidity of the cockroach, *Periplaneta americana*. J Comp Physiol 106:137-152

Yokohari F, Tominaga Y, Tateda H (1982) Antennal hygroreceptors of the honeybee, *Apis mellifera* L. Cell Tissue Res 226:63-73

Zimmermann B (1991) Differentiation of the thermo-/hygrosensitive (no-pore) sensilla on the antenna of *Antheraea pernyi* (Lepidoptera, Saturniidae): a study of cryofixed material. Cell Tissue Res 266:427-440

VII.1 The Geomagnetic Field and its Role in Directional Orientation

Wolfgang Wiltschko and Roswitha Wiltschko

Fachbereich Biologie der J.W.Goethe-Universität, Zoologie, Siesmayerstrasse 70, 60054 Frankfurt a.M., Germany

Abstract
A magnetic compass has been demonstrated in vertebrates, arthropods and a nudibranch. Two different functional modes have been described: birds and marine turtles use an inclination compass based on the inclination of the field lines, while mammals, fishes and arthropods use a polarity compass based on their polarity. For amphibians, both mechanisms have been described. In birds and amphibians, magnetic compass orientation proved light-dependent, with normal responses being observed only at the short-wavelength end of the spectrum. This suggests an involvement of photopigments, which is in agreement with the hypothesis assuming magnetoreception by excited-state macromolecules. The nature of the non-light dependent primary processes of magnetoreception in turtles, mammals and arthropods are unclear; magnetite-based mechanisms are discussed. Animals use magnetic compass information in a wide variety of behaviors. Birds orient their homing flights and their migration with the help of the magnetic field, newly-hatched salmon and marine turtles use a magnetic compass to reach suitable habitats, amphibians and amphipods use it for orientation between land and water, insects for building activities. In many of these behaviors, the animals can also use celestial cues for locating directions. Ecological requirements determine how the various cues interact in a given situation.

Key words Geomagnetic field, magnetic compass, magnetoreception, orientation behavior, directional reference

1 Compass Orientation with the Help of the Geomagnetic Field

A magnetic sense is not among the classic five senses vision, hearing, smell, touch and taste. Man provided himself by technical means with the possibility to obtain magnetic information; we have been making use of a magnetic compass for navigation in exploration and seafaring for centuries. Yet the idea that animals might use the geomagnetic field in essentially the same way for such purposes appeared alien to most scientists. As a consequence, sensitivity to magnetic stimuli is one of the last sensory modalities discovered in animals. Even today, the mechanisms of magnetoreception are not known in detail, and we have only a vague idea about the multitude of uses of the magnetic compass by animals in their natural environment. This chapter attempts to give a brief overview of the present state of knowledge and to outline some of the ongoing reseach.

1.1 The Magnetic Field of the Earth

To understand magnetic orientation and magnetoreception, one must be familiar with the type of stimulus the geomagnetic field provides. A detailed description of the geomagnetic field has been published by Skiles (1985). Here we briefly summarize the properties of the geomagnetic field that provide directional information.

The geomagnetic field is a *dipole field.* The two poles lie in the vicinity of the geographic (rotational) poles. The field lines leave the ground at the antarctic pole,

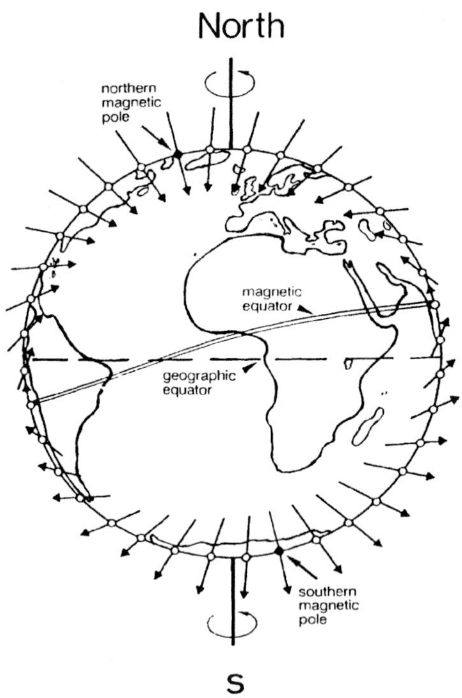

Fig. 1. Schematic view of the earth and its magnetic field. (R. Wiltschko and W. Wiltschko 1995)

curve around the earth and reenter its surface at the arctic pole. This means that the magnetic vector points upwards in the Southern and downwards in the Northern Hemisphere, being parallel to the earth's surface at the magnetic equator (Fig. 1). The angle between the magnetic vector and the horizon is called *inclination* or dip.

In most parts of the world, the field lines run roughly south-north. Yet magnetic north and geographic north do not always coincide. The angular difference, termed declination or variation, is considerable near the magnetic poles, but it decreases rapidly towards lower latitudes. The *total intensity* of the field decreases gradually from maximum values of about 60 000 nT at the poles to 30 000 nT near the magnetic equator. A number of spatial and temporal irregularities are superimposed on the general pattern described above. Yet these variations are generally too small to interfere with the animals' magnetic compass.

1.2 Demonstrating Magnetic Compass Orientation

To show that a species uses the magnetic field as a compass for determining directions, mere disorientation in the absence of suitable magnetic information is not sufficient. It must be demonstrated that the directional tendencies of the animals in question truly depend on the direction of the ambient magnetic field. Magnetic compass orientation can only be accepted when a deflection of magnetic north results in a shift of the animals' headings that qualitatively as well as quantitatively corresponds to the change in the magnetic direction, i.e. when the relationship between the deflection of magnetic north and the directional shift in the

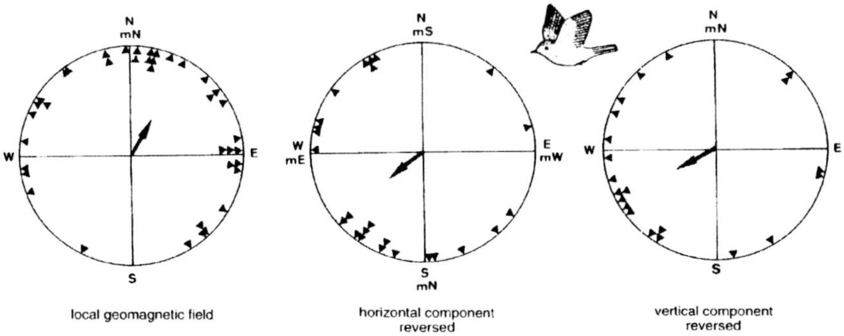

Fig. 2. Orientation of European robins in various magnetic fields. *mN* magnetic north; *mS* magnetic south. The *symbols* at the periphery of the circle indicate the headings recorded in single test nights, the *arrows* represent the mean vectors with their length being proportional to the radius of the circle = 1. (Data from W. Wiltschko and R.Wiltschko 1972)

animal's response is about 1:1. This means that testing for a magnetic compass is only possible when the test animals show a suitable orientation response that is reliably found and, in control tests, provides the reference against which the response in the experimentally altered magnetic field can be compared. This basic requirement has not always been easy to meet. In most behavioral situations, the various directions are equivalent, and there is no reason why an animal should prefer one direction over the others. However, certain behaviors, like the migration of birds, lead to strong spontaneous preferences of specific headings and thus provide a useful starting point for analysis.

Magnetic compass orientation was first demonstrated in European robins, *Erithacus rubecula* (Turdidae*)*, small passerine birds that migrate at night. Captive individuals become restless in their cages at the times of the year when their free-living conspecifics migrate. They prefer to stay at the side of their cage that points towards their migratory direction. When magnetic north was deflected with the help of coil systems, birds altered the direction of their preferences accordingly (Fig. 2 a,b), indicating that they oriented with the help of the magnetic field (W. Wiltschko 1968).

Meanwhile, magnetic compass orientation has been demonstrated in a number of other passerine migrants, in carrier pigeons, *Columba livia* f. *domestica*, in several species of fish, amphibians, marine turtles and mammals (see Fig. 4). A species of nudibranch sea slugs and, among the arthropods, various species of crustaceans and insects have also been shown to use a magnetic compass (see R. Wiltschko and W. Wiltschko 1995 for species lists). The list in Table 1 suggests that a magnetic compass may be rather widespread among free-moving animals. Many species, however, have escaped and still escape testing for magnetic orientation, because experimentalists could not yet find a reliable directional tendency for experimental manipulation.

Table 1. Magnetic compass orientation among animals

MOLLUSCA			
Gastropoda	1 order	1 family	1 species
ARTHROPODA			
Crustacea	3 orders	3 families	5 species
Insecta	5 orders	6 families	8 species
VERTEBRATA			
Chondrichtyes	1 order	1 family	1 species
Osteichthyes	2 orders	2 families	4 species
Amphibia	1 order	2 families	2 species
Reptilia	1 order	2 families	2 species
Aves	2 orders	10 families	18 species
Mammalia	2 orders	2 families	2 species

2 Functional Characteristics of Magnetic Compass Mechanisms

The functional mode of the magnetic compass has been analyzed in very few species only, with birds being by far the best-studied group, but this limited data base already shows that a common mechanism of magnetic compass does not exist among animals. Even among vertebrates, two types of mechanisms based on different features of the magnetic field have been described.

2.1 Functional Modes

According to the way in which directional information is derived from the field lines, two types of mechanisms can be distinguished: an *inclination compass* and a *polarity compass*.

2.1.1 The Inclination Compass

The inclination compass was first described for migratory birds. As already mentioned, the birds reversed their directional preference when the horizontal component of the ambient magnetic field was reversed (Fig. 2b). However, they reversed their preferences also when the vertical component was inverted (Fig. 2c). When both components were reversed, i.e. when the axis of the field lines remained constant, just changing its polarity, the test birds responded in the same way as they did in the geomagnetic field (W. Wiltschko and R. Wiltschko 1972). Birds obviously ignore the polarity of the magnetic field and derive directional information from the inclination of the axis of the field lines. This compass is fundamentally different from our technical compass: instead of magnetic North and South, it distinguishes *poleward,* the direction where the axis of the field lines forms the acute angle with gravity, and *equatorward*, where this angle is larger than 90° (Fig. 3). At the magnetic equator, where the field lines run horizontally, the two angles are equal and the inclination compass provides bimodal information. In cage tests, this leads to disoriented behavior (W. Wiltschko and R.Wiltschko 1972).

An inclination compass as described above was found in all bird species tested so far, in passerine migrants as well as in homing pigeons. It might thus be a common feature within the avian class (see R. Wiltschko and W. Wiltschko 1995). The inclination compass is not restricted to birds, however. In the critical test situation with the vertical component of the ambient field inverted, aquatic salamanders, *Notophthalmus viridescens* (Salamandridae), and young loggerhead sea turtles, *Caretta caretta* (Cheloniidae), reversed their headings, indicating that their magnetic compass, too, is an inclination compass (Phillips 1986a, Light et al. 1993, Goff et al. 1998).

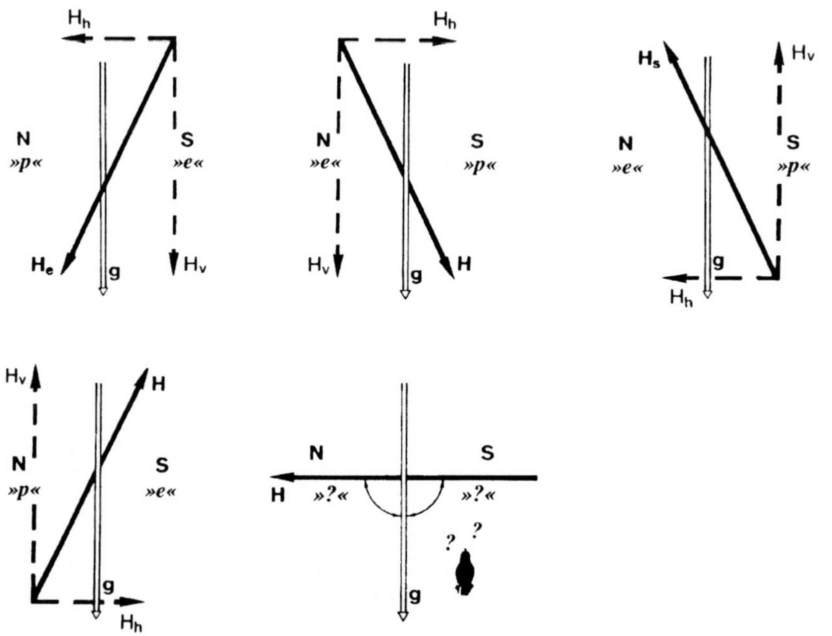

Fig. 3. Vertical cross-section through the magnetic field illustrating the inclination compass: H_e vector of the local geomagnetic field; H vector of an experimental magnetic field; H_s vector of an experimental field corresponding to that of the southern hemisphere; H_h, H_v horizontal and vertical components; g gravity vector; N north; S south. *»p«*, *»e«*, poleward, equatorward, the readings of the birds' magnetic compass. In a horizontal field, the directional information becomes bimodal. (R. Wiltschko and W. Wiltschko 1995)

2.1.2 The Polarity Compass

Experimental evidence suggests a second type of magnetic compass in salmonid fishes (Quinn et al. 1981) and rodents (Marhold et al. 1997a). These animals do not alter their headings when the vertical component of the magnetic field is inverted (Fig. 4c) so that their response suggests a compass that is based on the polarity of the magnetic field lines. Like our technical compass, it distinguishes between magnetic north and south. A polarity compass has also been described for amphibians: the same aquatic salamanders that rely on an inclination compass when orienting shoreward were reported to use a polarity compass when orienting towards home (Phillips 1986a).

The functional mode of the magnetic compass in invertebrates has been analyzed in only two species so far. Spiny lobsters, *Palinurus argus*, proved unaffected when the vertical component of the geomagnetic field was inverted (Lohmann et al. 1995). Flour beetles, *Tenebrio molitor*, were not tested under this condition; their unimodal orientation in a horizontal magnetic field, however, suggests that their compass is also based on the polarity of the field lines (Arendse and Vrins 1975). It is unknown whether this also applies to other crustacean and insect species and how widespread a polarity compass might be among arthropods.

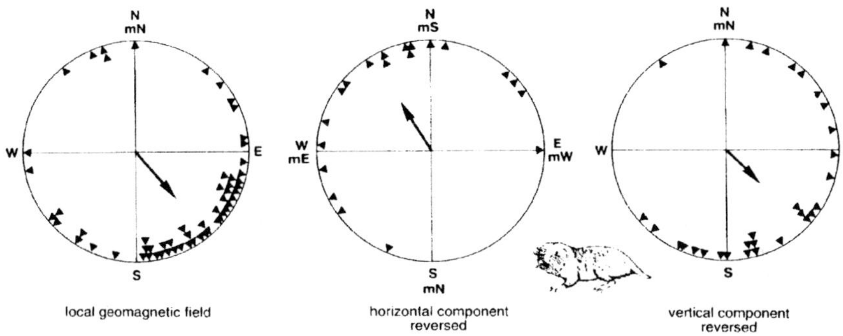

Fig. 4. Directions where a family of mole-rats *Cryptomys* sp. built their nests in a circular arena under various magnetic conditions. The *symbols* at the periphery of the circle indicate single nests, the *arrows* represent the mean vectors with the length proportional to the radius of the circle = 1. (Data from Marhold et al. 1997a)

2.2 The Intensity Range

Another point concerns the functional range of magnetic compass mechanisms: at what magnetic intensities can the magnetic field provide directional information? So far, this question has been studied only in birds by testing migrants in magnetic fields of various total intensities. European robins living in the local geomagnetic field with an ambient intensity of 46 000 nT were no longer oriented when tested in fields of 34 000 nT and below, or, more surprisingly, of 60 000 nT and above (Fig. 5). This means that a decrease or increase in total intensity by merely 25% led to a breakdown in orientation (W. Wiltschko 1968, 1978). The avian magnetic compass thus appears to be restricted to a narrow functional window that is closely tuned to the intensity of the ambient magnetic field (W. Wiltschko 1978).

This functional window is not fixed, however. Birds that had been exposed to fields of lower or higher intensity for at least 3 days adjusted themselves to those intensities and regained their ability to orient (W. Wiltschko 1968, 1978). Obviously, the potential functional range of the magnetic compass is much wider than the actual functional window. It even includes intensities beyond those found anywhere on earth today. After 3 days, orientation could be observed at inten sities of 16 000 or 150 000 nT. These adaptation processes represent neither a shift

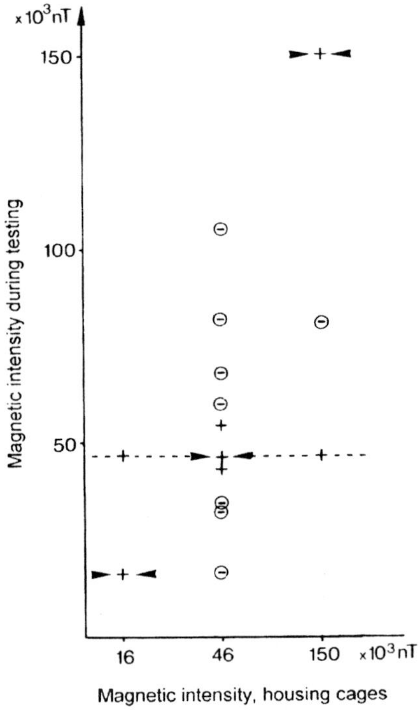

Fig. 5. Orientation behavior of European robins housed and tested under various magnetic conditions. The intensity of the local geomagnetic field is marked by a *dashed line*; test intensities that are identical with housing intensities are indicated by *two arrows*. + oriented behavior; - disoriented behavior. (Data from W. Wiltschko 1978)

nor a simple enlargement of the functional window. The birds adapted to lower or higher intensities did not lose the ability to orient in the local geomagnetic field, but they were disoriented at intermediate intensities (Fig. 5; W. Wiltschko 1978).

Similar studies from other animals are not yet available. Therefore it is unclear whether the magnetic compass mechanisms of other groups show similar restrictions in their functional range. In particular, it would be interesting to learn whether limited functional windows represent a peculiarity of an inclination compass or whether polarity compass mechanisms share this characteristic.

3 Mechanisms of Magnetoreception

One of the basic problems in magnetic orientation research concerns the question of how animals obtain the required directional information from the geomagnetic field. Where do the receptors lie, what physiological mechanisms are involved and what is the nature of the primary processes?

3.1 Possible Transducing Processes

A wide variety of mechanisms has been proposed as transducing processes (see R. Wiltschko and W. Wiltschko 1995). Today, the discussion focuses mainly on two hypotheses, namely (1) magnetoreception mediated by excited-state macromolecules, in particular photopigments, and (2) magnetoreception based on ferromagnetic material such as magnetite crystals. Other models are likewise considered, among them a hybrid model of the above-mentioned hypotheses (Edmonds 1996).

3.1.1 Magnetoreception by Photopigments

Hypotheses proposing an involvement of light-dependent processes in magnetoreception assume that, as a first step, certain macromolecules are elevated to singlet excited states by the absorption of a photon. A first hypothesis was suggested by Leask (1977); the presently discussed model was forwarded by Schulten and Windemuth (1986). These authors propose that the excited macromolecules may dissociate into radical pairs and, by hyperfine interactions, may be interconverted into triplet pairs. The yield of this step depends on the alignment of the molecules in the external field. Triplet products are chemically different from singlet pairs and might initiate processes that ultimately lead to magnetoreception (see Schulten and Windemuth 1986, Ritz et al. 2000).

To mediate magnetic information by this mechanism, an orderly array of macromolecules oriented in the various spatial directions would be required. This condition seems to be satisfied by the arrangement of photopigments in the vertebrate eyes. The light-dependent mechanisms proposed by Leask (1977) and Schulten and Windemuth (1986) would generate complex patterns of response in the retina. These patterns should be axial and symmetrical with respect to magnetic north and south and thus allow animals to detect the direction of the ambient magnetic field.

Leask (1977) and Schulten and Windemuth (1986) originally designed their models of magnetoreception with the functional characteristics of the avian magnetic compass in mind. The fact that light-dependent processes lead to axial rather than polar responses is in accordance with the inclination compass. Changes in intensity would alter the patterns on the retina without affecting their symmetry, so that birds might get used to the novel pattern and learn to interpret it after a while (Leask 1977, Ritz et al. 2000).

The ability to explain important features of the birds' responses under various magnetic conditions made this model rather attractive. A polarity compass as described for some fish, rodents and two arthropod species, on the other hand, appears incompatible with the axial response of the photoreceptor-based mechanisms described so far.

3.1.2 Magnetoreception based on Magnetite

The other prominent hypothesis assumes that particles of magnetite, Fe_3O_4, are involved in the processes mediating magnetic information. Obtaining directional information by ferromagnetic particles, which may act as miniature magnets, is an obvious possibility. Yet permanently magnetic material of biogenic origin was unknown until Lowenstam (1962) first described magnetite in the radula teeth of chitons. By remanence measurements and Mössbauer spectroscopy, magnetite crystals could be identified in the tissue of a wide variety of animals (see Kirschvink et al. 1985, R. Wiltschko and W. Wiltschko 1995 for summary), which promoted speculations about their possible role in magnetoreception.

The magnetic properties of magnetite particles depend on their size. Particles in the range of 0.05 to 1.00 μm are mostly single domains with a stable magnetic moment. The magnetic energy of orientation often just barely exceeds the background thermal energy, however. This problem could be overcome by an orderly array of particles in chains, clusters or lattices where adding their moments would increase their interactions with the geomagnetic field. Crystals of even smaller size are superparamagnetic; they have no stable moment, but may easily be aligned by an external field. Such particles might also be part of a receptor mechanism when arranged in a suitable array and properly aligned (see Kirschvink and Gould 1981, Shcherbakov and Winklhofer 1999).

Permanent magnetic material responds to polarity; hence mechanisms based on magnetite should be able to indicate the polar direction of the magnetic field. Yet magnetite-based mechanisms that produce axial responses are likewise possible, in particular when superparamagnetic particles are involved (e.g. Kirschvink and Gould 1981, Shcherbakov and Winklhofer 1999). Magnetoreception by magnetite could thus be involved in polarity compass as well as inclination compass mechanisms. Theoretical considerations also show that magnetite-based mechanisms would not only be able to mediate the direction of the magnetic field, but also be sensitive enough to detect small differences in intensity, provided that a sufficient number of magnetite particles were involved (e.g. Yorke 1979, Kirschvink and Gould 1981, Kirschvink 1989, Shcherbakov and Winklhofer 1999).

3.2 Experimental Evidence

The light-dependent initial processes of the photopigment model offer the possibility to test this hypothesis by a simple technique: If the assumptions of the model were correct, magnetoreception should be impossible in the absence of light.

Tests with homing pigeons indicated that magnetic compass information was disrupted in total darkness (W. Wiltschko and R. Wiltschko 1981). The same was true for the aquatic newt *Notophthalmus* (Salamandridae; Phillips and Borland

1992a). These findings suggest that the compass mechanisms of birds and amphibians might indeed be based on light-dependent processes. Salmon, marine turtles and rodents, in contrast, were able to orient in total darkness (see Table 2); their magnetic compass must thus be based on non-light dependent mechanisms. It is striking, however, that there is no correlation of light dependence and non-light dependence with the functional mode as polarity or inclination compass; all

Table 2. Functional properties of magnetic compass mechanisms

Group	Type of compass	Light required?	References
VERTEBRATA			
Osteichthyes	Polarity	No	Quinn and Brannon (1982)
Amphibia	Inclination	Yes	Phillips (1986a), Phillips and Borland (1992a)
	Polarity	Yes	Phillips (1986a), Phillips and Borland (1994)
Reptilia	Inclination	No	Light et al. (1993), Lohmann and Lohmann (1993)
Aves	Inclination	Yes	W. Wiltschko and R. Wiltschko (1972, 1981)
Mammalia	Polarity	No	Marhold et al. (1997a)
ARTHROPODA			
Crustacea	Polarity	No	Arendse (1978), Lohmann et al. (1995)
Insecta	Polarity	No	Arendse (1978), Schmitt and Esch (1993)

possible combinations are found among vertebrates. Among arthropods, amphipods, spiny lobsters, beetles and honey bees were found not to need light for orientation (Table 2), suggesting non-light-dependent magnetoreception. For fruitflies, however, wavelength-depedent responses have been reported (Phillips and Sayeed 1993).

3.2.1 Additional Evidence for Magnetoreception by Photopigments

In birds, an involvement of light-dependent mechanisms in magnetoreception is supported by electrophysiological data. Units of the accessory optic system in the nucleus of the basal optic root and in the tectum opticum of pigeons and passerines responded to changes in the direction of the ambient magnetic field. These responses were found to depend on the presence of light and an intact retina and optic nerve. Individual cells showed distinct peaks when magnetic north was aligned in a specific direction so that the input of a number of units, processed collectively and integrated, would represent all directions in space and thus provide a suitable basis for compass orientation (Semm et al. 1984; Semm and Demaine 1986; Beason and Semm 1987).

Behavioral studies aimed at analyzing a possible wavelength dependence of magnetoreception. Salamanders and birds were tested under nearly monochromatic light of various wavelengths, using oriented behavior as an indicator of whether or not the test animals could obtain directional information from the magnetic field in the test situation. Parallels, but also differences, between the two groups became apparent (Fig. 6).

Shoreward-oriented salamanders using their inclination compass showed normal orientation when tested under 400 nm violet and 450 nm blue light. A small range of disorientation followed, and from 500 nm blue-green to 600 nm orange, the salamanders shifted their orientation by about 90° counterclockwise with respect to the direction they preferred under white control light (Phillips and Borland 1992b). Salamanders that had been exposed for a certain period of time to only long wavelengths of 500 nm and beyond showed a corresponding clockwise shift when tested under white light. These data suggest that the directional deflection was caused by salamanders receiving a different type of directional information under long-wavelength light. Recent findings based on this response suggest the pineal as the site of magnetoreception (Deutschlander et al. 1999). For salamanders orienting homeward with their polarity compass, Phillips and Borland (1994) reported a somewhat different wavelength dependence. Orientation as with white light was observed only with wavelengths shorter than 400 nm violet; from 450 nm blue to longer wavelengths, the salamanders were no longer oriented. Discussing the latter finding, the authors point out that the homing behavior they used as criterion is a combination of navigational processes and compass orientation; disorientation could also have resulted from an interference with the navigational step.

A yet different wavelength dependence emerged in birds. Migrants showed normal migratory orientation from 425 nm to 565 nm, i.e. in the entire blue and green part of the visual spectrum. Below 588 nm yellow and beyond 630 nm red, they were disoriented (see Fig. 6). The wavelength range where the magnetic compass provides birds with correct information is thus much wider than the one shown in salamanders. Three species of passerine migrants were tested so far (W. Wiltschko et al. 1993, W. Wiltschko and R. Wiltschko 1995, 1999, Rappl et al. 2000).

Homing pigeons (R. Wiltschko and W. Wiltschko 1998) showed similar responses to the various wavelengths so that this wavelength-dependence may be typical for a wide variety of birds, if not for all. Magnetic compass orientation in birds does not depend on the pineal (e.g. Maffei et al . 1983, Schneider et al. 1994).

While the different responses of shoreward-oriented salamanders to short and longer wavelengths were indicating an involvement of two antagonistic receptors (Phillips and Borland 1992b), the wavelength dependence observed in birds appeared to resemble an all-or-none response. Recent experiments, however, indicate that the situation might be much more complex, as the intensity of the nearly monochromatic light was also found to affect the birds' behavior. Green light with a wavelength of 565 nm resulted in excellent orientation in the normal direction when it was presented at light levels of 0.0020 or 0.0030 Wm^{-2}; at 0.0150 Wm^{-2}, in contrast, a significant counterclockwise shift of about 70° was observed (W. Wiltschko and R. Wiltschko 2000). The interpretation of this finding is difficult. A change in response at higher intensities of white light was not observed. The crucial point appears to be that the relatively bright light consists only of a very narrow band of wavelengths; but how this disruption of the natural balance between the various wavelengths causes a change in behavior is unclear.

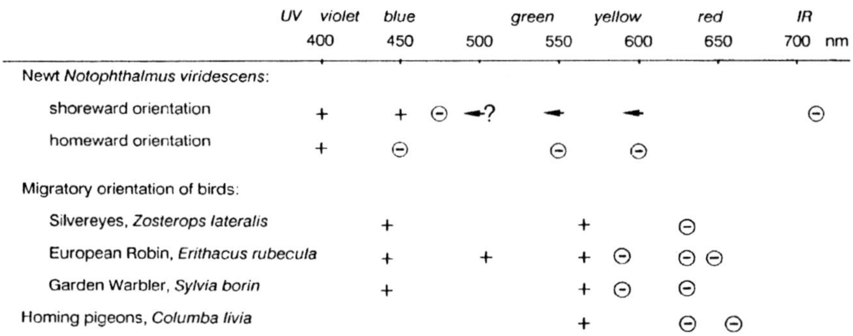

Fig. 6. Summary of vertebrate responses to light of various wavelengths. + indicates same response as under full spectrum light, - indicates disorientation; *arrows* indicate a shift in direction with respect to behavior under full spectrum light (Data from Phillips and Borland 1992b, 1994; W. Wiltschko et al. 1993; W. Wiltschko and R. Wiltschko 1995, 1999, in prep; R. Wiltschko and W. Wiltschko 1998; Rappl et al. 2000)

3.2.2 Non-Light Dependent Compass Mechanisms

Salmon, marine turtles, rodents and most arthropods do not require light for magnetic compass orientation (see Table 2). The nature of their non-light dependent compass mechanisms is still an open question. It is tempting to assume that it might be based on magnetite, but direct evidence is not yet available.

Magnetite has been identified by remanence measurements in the tissue of numerous animals, among them species known to use a magnetic compass (see Kirschvink et al. 1985, R. Wiltschko and W. Wiltschko 1995 for an overview). The crystals appeared to be single domains or superparamagnetic grains. In vertebrates, magnetite was repeatedly found in the ethmoid region above the upper mandible (e.g. Beason and Nichols 1984; Walker et al. 1997). Recent histological studies reported magnetite particles near thin terminal branches of nerve endings of simple mechanoreceptors in the upper mandible of pigeons (Holtkamp-Rötzler et al. 1997) and in the vicinity of nerves in the lamina of the olfactory lamellae of rainbow trouts, *Oncorhynchus mykiss* (Salmonidae; Walker et al. 1997). In both cases, the respective nerves were part of the *ramus ophthalmicus* of the *nervus trigeminus*. Electrophysiological recordings from this nerve produced responses to magnetic stimulation. However, the units seemed to respond to changes in the intensity rather than to changes in the direction of the magnetic field (Semm and Beason 1990; Walker et al. 1997). This speaks against the associated receptors providing compass information.

Attempts to interfere with magnetite-based receptors involved treating animals with a short, strong magnetic pulse designed to change the magnetization of single domain magnetite. The treatment had marked effects on the orientation behavior in migrants as well as homing pigeons (e.g. W. Wiltschko et al. 1994, 1998a; Beason et al. 1995, 1997; Munro et al. 1997). The manifestation of these effects, however, suggests that magnetic components of the navigational system rather than the magnetic compass were affected (see Beason and Semm 1996; Munro et al. 1997; W. Wiltschko et al. 1998a for discussion), an interpretation that is in agreement with the electrophysiological findings.

A case where a magnetic pulse might affect a compass response is reported from mole-rats, *Cryptomys* sp. (Bathyergidae). The pulse caused a roughly 90° counterclockwise shift in the nest position, which persisted for at least 3 months after the treatment (Marhold et al. 1997b). In arthropods, some histological, electrophysiological and behavioral data on magnetite are available from honeybees, yet the findings do not add up to a consistent picture (see R. Wiltschko and W. Wiltschko 1995 for discussion).

Taken together, our knowledge on possible magnetite-based receptors and what type of information they may provide is still rather limited. The source of magnetic compass information in animals that use light-independent mechanisms remains unclear. Magnetite-based receptors are still among the prime candidates. However, a modified radical pair mechanism, where the light-dependent initial step is replaced by another process producing excited state molecules (Ritz et al. 2000) seems also possible when an inclination compass is involved.

4 Behavioral Relevance of Magnetic Compass Orientation

In order to understand the biological significance of the magnetic compass as orientation mechanism, one must analyze its role in the various orientation tasks. Two points are to be considered: (1) Normally, the magnetic field is only one of a multitude of orientation cues available. It is not even the only source of directional information, as most animals can derive the same type of compass information from celestial cues. (2) A compass is only helpful if animals know what courses to take, i.e. whether their goal lies in the north, east etc. Therefore, the question of the origin of the respective compass courses cannot be separated from considerations on compass mechanisms. The final part of the present chapter is a brief overview on the widespread use of the magnetic compass and its interaction with other orientation cues.

4.1 Interrelations Between Magnetic and Celestial Compass Information

Day-active animals can use the sun as a compass. Sun-derived factors such as the pattern of polarized light also provide directional information, in particular in insects (see Wehner 1984). These celestial features vary with time; their exact position depends on the time of day, season and also on geographic latitude. This makes their use far more demanding than that of the magnetic field, where directions are immediately derived from sensory input. Most animals use celestial compass mechanisms that are established by learning processes, which allows a fine tuning of these mechanisms to the local situation.

The relative importance of magnetic and celestial cues has been analyzed in few species only, many of them birds. The results indicate a general tendency of day-active animals to prefer the sun compass.

In the case of conflict, e.g. when their internal clock was shifted, fish, anurans, birds and amphipods mostly followed their manipulated sun compass in spite of correct information from the magnetic field. A prominent case are homing pigeons that respond to clock-shifting with typical deflections in spite of their magnetic compass, indicating the preferred use of the sun compass. Yet under solid overcast, they are equally well oriented, relying on their magnetic compass (Fig. 7).

In certain ecological situations, the magnetic compass is of crucial importance. Near the equator, where the changes in sun azimuth are rather rapid and large seasonal changes occur, most litoral amphipods were found to rely on their magnetic compass; orientation deteriorated in the absence of magnetic information (e.g. Pardi et al. 1988, Ugolini and Pardi 1992).

Bird migration is another case where the magnetic compass plays a dominant role. Day migrants preferably rely on the magnetic field, possibly, because the magnetic compass, in contrast to the sun compass, is not affected by the latitudinal changes birds inevitably experience during migration (see Munro and Wiltschko 1993 for discussion).

The preferred use of visual cues often observed seems to suggest a hierarchy of cues, with the sun compass dominating over the magnetic compass. The two mechanisms are not independent from each other, however. In young homing pigeons, the magnetic compass provides the reference system for the learning processes which establish the sun compass, i.e. the sun compass is derived from the magnetic compass (W. Wiltschko et al. 1983, R. Wiltschko and W. Wiltschko 1990). The adjustment of the sun compass to seasonal changes of the sun's arc might also be controlled by the magnetic compass.

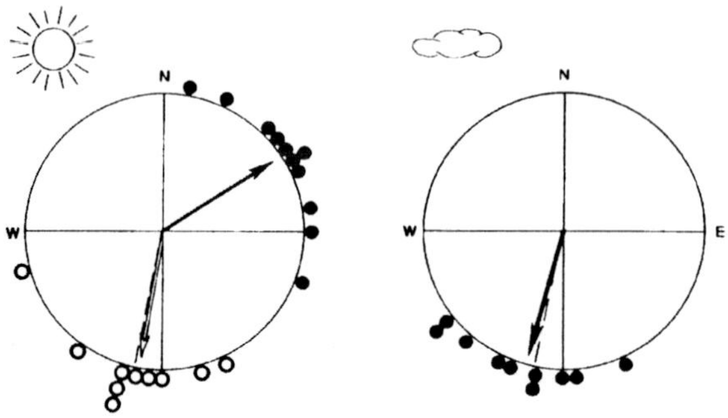

Fig. 7. Orientation of homing pigeons at a site 40 km north of their home loft. *Left:* Under sun; pigeons whose internal clock was shifted 6 h slow *(solid symbols)* show a characteristic deflection from untreated controls (*open symbols*). *Right:* Under solid overcast. The *symbols* at the periphery of the circle indicate vanishing bearings of individual pigeons; the *arrows* give the mean vectors proportional to the radius of the circle = 1

A similar situation with respect to celestial cues occurs during migration in night-migrating birds: they use the pattern of polarized light at sunset and a star compass at night for locating their migratory direction. Exposing birds to the natural sky in a magnetic field with north deflected resulted in a recalibration of these celestial cues, indicating that the directional significance of stars and sunset cues is controlled by the magnetic compass (Fig. 8; for summary, see R. Wiltschko et al. 1997).

In birds, the magnetic compass thus provides the directional reference for celestial compass mechanisms. It is unknown whether this is also true for other animal groups, because interactions between the various cue systems have not been systematically analyzed.

4.2 The Origin of Magnetic Compass Courses

In cases where animals have no direct contact with their goal, i.e where they cannot see or hear it, navigation consists of a two-step process (see Kramer 1959): in a first step, the animals determine the course to the goal as compass course in terms equivalent to 'go 225° SW'; in a second step, this course is located with a compass and transformed into a direction of movement, equivalent to a specification like 'go there' or 'this way'. The nature of the first step is very different in the various behaviors.

4.2.1 Innate Courses of Migration

Where a stable geographic relationship exists between the starting area and a distant goal, animals may rely on innate information on the compass course to be pursued. For young, naive animals, this is the only possible way to reach distant goals where they have never been before. The magnetic field serves as reference for the innate information.

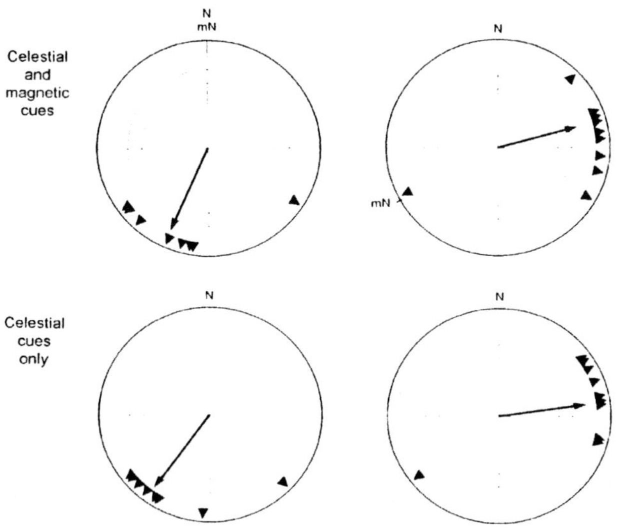

Fig. 8 Recalibration of sunset cues by the magnetic field: tests with Australian Silvereyes under clear sky at sunset. *Upper diagrams*: migrants tested in a magnetic field with North deflected to 240° WSW (*right*) showed a corresponding shift of their headings with respect to control birds tested in the local geomagnetic field (*left*), indicating the dominance of magnetic over celestial cues. *Lower diagrams*: when both groups were afterwards tested in the absence of magnetic cues, they continued in the same directions with the help of sunset cues, indicating recalibration of these cues according to the experimental magnetic field. The *symbols* at the periphery mark the mean headings of individual birds; the *arrows* represent the mean vectors with the length proportional to the radius of the circle = 1. (R. Wiltschko et al. 1999)

For their first migration to the still unknown wintering area, young migratory birds follow a route that is genetically coded as a sequence of compass courses (Helbig 1991). Experiments with hand-raised migrants indicate that the magnetic field provides a reference for this innate directional information, playing a most important role in the processes converting it into an actual course (e.g. Beck and Wiltschko 1982). In nature, celestial cues, in particular celestial rotation, are also involved. However, hand-raised birds develop normal migratory behavior when the magnetic field is the only cue available. The magnetic compass that also controls the directional significance of stars and sunset cues during migration thus represents a mechanism of primary importance for migratory orientation in birds (for details, see W. Wiltschko et al. 1998b).

A similar role of the magnetic field as reference for innate directions is indicated in salmon fry. Newly hatched sockeye salmon, *Oncorhynchus nerka*, must find a suitable nursing lake where they feed before they start their seaward migration. Fry of different origin showed spontaneous preferences of population-specific directions that were in agreement with the courses leading to nearby nursery lakes; the directional tendencies changed accordingly when magnetic north was deflected (e.g. Quinn 1980). Additionally, salmon fry show spontaneous rheotactic responses and are affected by various visual and possibly chemical cues so that the spontaneous tendency to swim in an innate magnetic direction is just one component in a multifactorial system leading the young fry to their first feeding grounds (for summary, see Groot 1982).

There is also support for innate magnetic courses in some arthropod species. Migrating monarch butterflies, *Danaus plexippus,* changed their preference accordingly when tested in a magnetic field with a deflected magnetic north (Etheredge et al. 1999). Laboratory-raised specimen of equatorial sandhoppers *Talorchestia martensii* (Talitridae) showed a spontaneous preference of the axis land-sea; this tendency could be altered by shifting magnetic north (Pardi et al. 1988). In talitrids, such innate tendencies can be modified and adjusted to the specific directional alignment of the home coastline (e.g. Ugolini et al. 1991), so that the directional tendencies of adult animals reflect a combination of innate and acquired information.

4.2.2 Acquired Magnetic Courses

Fixed directional relationships between start and goal are exceptions. In most cases, the direction to the goal varies according to the given situation. The animals use various means to determine the respective courses, with the specific processes depending on the behavioral context.

The orientation of newly hatched marine turtles is one of the best known examples for animals relying on innate preferences to determine the correct course. After emerging at their natal beaches, young turtles must cover a distance of several metres to reach the water and then swim a course roughly perpendicular to the shoreline to gain the open sea. Spontaneous responses to light stimuli from brighter parts of the horizon as well horizon elevation and beach slope guide the hatchlings towards the shoreline; once in the water, they head into the incoming waves (for details, see Salmon and Wyneken 1994). The direction of these movements is memorized as a magnetic course and then pursued for hours (Lohmann and Lohmann 1994, Goff et al. 1998), bringing the hatchling away from the shore into the open ocean.

Amphibians that orient along the axis perpendicular to the shore, the so-called Y-axis (see Ferguson 1967) also follow acquired courses. These courses are determined by the lay of the land in the home area; they are modified when the animals move or are placed in an aquarium with a different distribution of land and water. Salamanders memorize the Y-axis with respect to the magnetic field, using their magnetic compass to move shoreward or waterward (Phillips 1986b).

The most frequent cases of variable courses, however, concern homing, i.e. animals returning from foraging excursions or exploratory trips to their hiding places, burrows or nests. Homing animals determine the current home course by navigational processes. Homing behavior has been extensively studied in birds by displacement experiments with carrier pigeons, so that mechanisms of navigation in birds are far better known than in any other animal group.

Avian navigation is based on a multitude of factors that provide information on direction and position. These factors are integrated by learning processes to form complex mechanisms. The magnetic compass plays a crucial role in avian navigation, in particular in very young, inexperienced pigeons before the learned mechanisms become functional. Young birds depend almost exclusively on their magnetic compass, because it is the only mechanism available right from the beginning. They navigate by recording the net direction of the outward journey with the help of their magnetic compass; reversing this direction yields the home course (R. Wiltschko and W. Wiltschko 1978). Similar route-based mechanisms based on magnetic information have been indicated by displacement experiments with woodmice, *Apodemus sylvaticus* (Mather and Baker 1981), and domestic horses, *Equus caballus* (Baker 1989). It is unclear, however, how widespread this use of the magnetic compass might be among animals.

Birds use the simple route-based strategy described above only during a short transient stage. As soon as young pigeons become more experienced, they begin to integrate various factors and, by learning processes, establish the navigational 'map', which is assumed to be a directionally oriented mental representation of the distribution of various environmental gradients in the home region (for details, see Wallraff 1974; R. Wiltschko and W. Wiltschko 1999). The map is then preferentially used. Similarly, experienced migratory birds no longer rely on innate information, but take advantage of their experience from previous migrations, establishing a navigational map of their migration route. The nature of the respective learning processes escapes experimental analysis. It seems not unlikely, however, that the magnetic compass here, too, serves as directional reference for integrating the various factors forming the map.

4.3 The Magnetic Compass as Reference System

Section 4 of this chapter gave a few examples for the various uses of the geomagnetic field as a compass by animals. The magnetic compass is involved in a wide variety of other behaviors, like in the subterranean orientation of mole-rats

(Marhold et al. 1997a), the movements of spiny lobsters on the sea floor (Lohmann et al. 1995) or the building activity of insects like honeybees (e.g. De Jong 1982) and compass termites, *Amitermes* sp. (Termitidae; Jacklyn 1990). Other behaviors, like the directional tendencies of the marine slug *Tritonia* that change in the course of the lunar cycle (Lohmann and Willows 1987) have been shown to depend on the direction of the magnetic field, but are still poorly understood. The list of animals must remain rather incomplete, however. Many species have not yet been studied with regard to their possible use of magnetic compass information.

Even in animals that prefer visual cues for the orientation of their movements when available, the magnetic compass may be a mechanism of crucial importance. Birds use their magnetic compass to calibrate celestial and other visual cues. This may also be true for some other groups of animals. The magnetic field represents an omnipresent, reliable directional reference for all kinds of innate or acquired courses, and it may be this function that constitutes the most important biological role of the magnetic field. This role seems to be independent of the different functional modes of the magnetic compass and the possibly different primary processes of magnetoreception: all these mechanisms provide the same type of useful compass information for their bearers.

References

Arendse MC (1978) Magnetic field detection is distinct from light detection in the invertebrates *Tenebrio* and *Talitrus*. Nature 274:358-362

Arendse MC, Vrins JCM (1975) Magnetic orientation and its relation to photic orientation in *Tenebrio molitor* L (Coleoptera, Tenebrionidae). Neth J Zool 25:407-437

Baker RR (1989) Navigation and magnetoreception by horses and other non-human land mammals. In: Orientation and Navigation - Birds, Humans and other Animals. Royal Institute of Navigation, Cardiff, Paper 12

Beason RC, Nichols JE (1984) Magnetic orientation and magnetically sensitive material in a transequatorial migratory bird. Nature 309:151-153

Beason RC, Semm P (1987) Magnetic responses of the trigeminal nerve system of the Bobolink (*Dolichonyx oryzivorus*). Neurosci Lett 80:229-234

Beason RC, Semm P (1996) Does the avian ophthalmic nerve carry magnetic navigational information? J Exp Biol 199:1241-1244

Beason RC, Dussourd N, Deutschlander M (1995) Behavioral evidence for the use of magnetic material in magnetoreception by a migratory bird. J Exp Biol 198:141-146

Beason RC, Wiltschko R, Wiltschko W (1997) Pigeon homing: effects of magnetic pulses on initial orientation. Auk 114:405-415

Beck W, Wiltschko W (1982) The magnetic field as reference system for the genetically encoded migratory direction in Pied Flycatchers (*Ficedula hypoleuca* Pallas). Z Tierpsychol 60:41-46

De Jong D (1982) Orientation of comb building by honeybees. J Comp Physiol 147:495-501

Deutschlander ME, Borland SC, Phillips JB (1999) Extraocular magnetic compass in newts. Nature 400:324-325

Edmonds DT (1996) A sensitive optically detected magnetic compass for animals. Proc R Soc Lond B 263:295-298

Etheredge JA, Perez SM, Taylor OR, Jander R (1999) Monarch butterflies (*Danaus plexippus* L.) use a magnetic compass for navigation. Proc Nat Acad Sci USA 96:13845-13846

Ferguson DE (1967) Sun-compass orientation in anurans. In: Storm RM (ed) Animal Orientation and Navigation. Oregon State University Press, Corvallis, pp 21-32

Goff M, Salmon M, Lohmann KJ (1998) Hatchling sea turtles use surface waves to establish a magnetic compass direction. Anim Behav 55:69-77

Groot C (1982) Modification on a theme - a perspective on migratory behavior of Pacific salmon. In: Brannon EL, Salo EO (eds) Proc Salmon Trout Migratory Behav Symp. University of Washington, Seattle, pp 1-21

Helbig AJ (1991) Inheritance of migratory direction in a bird species: a cross-breeding experiment with SE- and SW-migrating Blackcaps (*Sylvia atricapilla*). Behav Ecol Sociobiol 28:9-12

Holtkamp-Rötzler E, Fleissner G, Hanzlik M, Wiltschko W, Petersen N (1997) Mechanoreceptors in the upper beak of homing pigeons (*Columba livia*) as putative structural candidates for magnetoreception. Verh Dtsch Zool Ges 90:290

Jacklyn, PM (1990) Orientation in meridional mounds of *Amitermes meridionalis* (Froggatt) and *Amitermes laurensis* (Mjöberg). PhD Thesis, University of Sydney, Sydney

Kirschvink JL (1989) Magnetite biomineralization and geomagnetic sensitivity in higher animals: an update and recommendations for future study. Bioelectromagnetics 10:239-259

Kirschvink JL, Gould JL (1981) Biogenic magnetite as a basis for magnetic field detection in animals. BioSystems 13:181-201

Kirschvink JL, Jones DS, MacFadden BJ (eds) (1985) Magnetite Biomineralization and Magnetoreception in Organisms. Plenum Press, New York

Kramer G (1959) Recent experiments on bird orientation. Ibis 101:399-416

Leask MJM (1977) A physico-chemical mechanism for magnetic field detection by migratory birds and homing pigeons. Nature 267:144-145

Light P, Salmon M, Lohmann KJ (1993) Geomagnetic orientation of Loggerhead Sea Turtles: evidence for an inclination compass. J Exp Biol 182:1-10

Lohmann KJ, Lohmann CMF (1993) A light-independent magnetic compass in the Leatherback Sea Turtle. Biol Bull 185:149-151

Lohmann KJ, Lohmann CMF (1994) Acquisition of magnetic directional preferences in hatchling Loggerhead Sea Turtles. J Exp Biol 190:1-8

Lohmann KJ, Willows AOD (1987) Lunar-modulated geomagnetic orientation by a marine mollusk. Science 235:331-334

Lohmann KJ, Pentcheff ND, Nevitt GA, Stetten GD, Zimmer-Faust RK, Jarrard HE, Boles LC (1995) Magnetic orientation of spiny lobsters in the ocean: experiments with undersea coil systems. J Exp Biol 198:2041-2048

Lowenstam HA (1962) Magnetite in denticle capping in recent chitons (Polyplacophora). Geol Soc Am Bull 73:435-438

Maffei L, Meschini E, Papi F (1983) Pineal body and magnetic sensitivity: homing in pineal-ectomized pigeons under overcast. Z Tierpsychol 62:151-156

Marhold S, Wiltschko W, Burda H (1997a): A magnetic polarity compass for direction finding in a subterranean mammal. Naturwissenschaften 84:421-423

Marhold S, Burda H, Kreilos I, Wiltschko W (1997b) Magnetic orientation in common mole-rats from Zambia. In: Orientation and Navigation - Birds, Humans and other Animals. Royal Institute of Navigation, Oxford, pp 5/1 – 5/9

Mather JG, Baker RR (1981) Magnetic sense of direction in woodmice for route-based navigation. Nature 291:152-155

Munro U, Wiltschko W (1993) Magnetic compass orientation in the yellow-faced honeyeater, *Lichenostomus chrysops*, a day-migrating bird from Australia. Behav Ecol Sociobiol 32:141-145

Munro U, Munro JA, Phillips JB, Wiltschko W (1997) Wavelength of light and pulse magnetization affect different magnetoreception systems in a migratory bird. Austr J Zool 45:189-198

Pardi L, Ugolini A, Faqi AS, Scapini F, Ercolini A (1988) Zonal recovering in equatorial sandhoppers: interaction between magnetic and solar orientation. In: Chelazzi G, Vannini M (eds) Behavioral Adaptation to Intertidal Life. Plenum Press, New York, pp 79-92

Phillips JB (1986a) Two magnetoreception pathways in a migratory salamander. Science 233:765-767

Phillips JB (1986b) Magnetic compass orientation in the Eastern Red-spotted Newt (*Notophthalmus viridescens*). J Comp Physiol A 158:103-109

Phillips JB, Borland SC (1992a) Magnetic compass orientation is eliminated under near-infrared light in the Eastern Red-spotted Newt Anim Behav 44:796-797

Phillips JB, Borland SC (1992b) Behavioral evidence for use of a light-dependent magnetoreception mechanism by a vertebrate. Nature 359:142-144

Phillips JB, Borland SC (1994) Use of a specialized magnetoreception system for homing by the Eastern Red-spotted Newt, *Notophthalmus viridescens*. J Exp Biol 188:275-291

Phillips JB, Sayeed O (1993) Wavelength-dependent effects of light on magnetic compass orientation in *Drosophila melanogaster*. J Comp Physiol A 172:303-308

Quinn TP (1980) Evidence for celestial and magnetic compass orientation in lake migrating Sockeye Salmon fry. J Comp Physiol 137:243-248

Quinn TP, Brannon EL (1982) The use of celestial and magnetic cues by orienting Sockeye Salmon smolts. J Comp Physiol 147:547-552

Quinn TP, Merrill RT, Brannon, EL (1981) Magnetic field detection in Sockeye Salmon. J Exp Zool 217:137-142

Rappl R, Wiltschko R, Weindler P, Berthold P, Wiltschko W (2000) Orientation behavior of Garden Warblers, *Sylvia borin*, under monochromatic light of various wavelengths. Auk 117:256-260

Ritz T, Adem S, Schulten K (2000) A model for vision-based magnetoreception in birds. Biophysic J 78:707-718

Salmon M, Wynecken J (1994) Orientation by hatchling sea turtles: mechanisms and implications. Herpetol Nat Hist 2:13-24

Schmitt DE, Esch HE (1993) Magnetic orientation of honeybees in the laboratory. Naturwissenschaften 80:41-43

Schneider T, Thalau HP, Semm P, Wiltschko W (1994) Melatonin is crucial for the migratory orientation of Pied Flycatchers (*Ficedula hypoleuca* Pallas). J Exp Biol 194:255-262

Schulten K, Windemuth A (1986) Model for a physiological magnetic compass. In: Maret G, Boccara N, Kiepenheuer J (eds) Biophysical Effects of Steady Magnetic Fields. Springer, Berlin Heidelberg New York, pp 99-106

Semm P, Beason RC (1990) Responses to small magnetic variations by the trigeminal system of the Bobolink. Brain Res Bull 25:735-740

Semm P, Demaine C (1986) Neurophysiological properties of magnetic cells in the pigeon's visual system. J Comp Physiol A 159:619-625

Semm P, Nohr D, Demaine C, Wiltschko W (1984) Neural basis of the magnetic compass: interaction of visual, magnetic and vestibular inputs in the pigeon's brain. J Comp Physiol A 155:283-288

Shcherbakov VP, Winklhofer M (1999) The osmotic magnetometer: a new model for magnetite-based magnetoreceptors in animals. Eur Biophys J 28:380-392

Skiles DD (1985) The geomagnetic field: its nature, history, and biological relevance. In: Kirschvink JL, Jones DS, MacFadden BJ (eds) Magnetite Biomineralization and Magnetoreception in Organisms. Plenum, New York, pp 43-102

Ugolini A, Pardi L (1992) Equatorial sandhoppers do not have a good clock. Naturwissenschaften 79:279-281

Ugolini A, Felicioni S, Macchi T (1991) Orientation in the water and learning in *Talitrus saltator* Montagu. J Exp Mar Biol Ecol 151:113-119

Walker MM, Diebel CE, Haugh CV, Pankhurst PM, Montgomery JC, Green CR (1997) Structure and function of the vertebrate magnetic sense. Nature 390:371-376

Wallraff HG (1974) Das Navigationssystem der Vögel. Schriftenreihe "Kybernetik", R Oldenbourg, München Wien

Wehner R (1984) Astronavigation in insects. Ann Rev Entomol 29:277-298

Wiltschko R, Wiltschko W (1978) Evidence for the use of magnetic outward-journey information in homing pigeons. Naturwissenschaften 65:112

Wiltschko R, Wiltschko W (1990) Zur Entwicklung des Sonnenkompaß bei jungen Brieftauben. J Ornithol 131:1-20

Wiltschko R, Wiltschko W (1995) Magnetic Orientation in Animals. Springer, Berlin Heidelberg New York

Wiltschko R, Wiltschko W (1998) Pigeon homing: effect of various wavelengths of light during displacement. Naturwissenschaften 85:164-167

Wiltschko R, Wiltschko W (1999) Compass orientation as a basic element in avian orientation and navigation. In: Golledge RG (ed) Wayfinding Behavior. Johns Hopkins University Press, London Baltimore, pp 259-294

Wiltschko R, Wiltschko W, Munro U (1997) Migratory orientation in birds: the effects and after-effects of exposure to conflicting celestial and magnetic cues. In: Orientation and Navigation - Birds, Humans and other Animals. Royal Institute of Navigation, Oxford, pp 6/1 – 6/14

Wiltschko R, Munro U, Ford H, Wiltschko W (1999) After-effects of exposure to conflicting celestial and magnetic cues at sunset in migratory Silvereyes *Zosterops l. lateralis*. J Avian Biol 30:56-62

Wiltschko W (1968) Über den Einfluß statischer Magnetfelder auf die Zugorientierung der Rotkehlchen (*Erithacus rubecula*). Z Tierpsychol 25:537-558

Wiltschko W (1978) Further analysis of the magnetic compass of migratory birds. In: Schmidt-Koenig K, Keeton WT (eds) Animal Migration, Navigation, and Homing. Springer, Berlin Heidelberg New York, pp 302-310

Wiltschko W, Wiltschko R (1972) Magnetic compass of European Robins. Science 176:62-64

Wiltschko W, Wiltschko R (1981) Disorientation of inexperienced young pigeons after transportation in total darkness. Nature 291:433-434

Wiltschko W, Wiltschko R (1995) Migratory orientation of European Robins is affected by the wavelength of light as well as by a magnetic pulse. J Comp Physiol A 177:363-369

Wiltschko W, Wiltschko R (1999) The effect of yellow and blue light on magnetic compass orientation in European Robins, *Erithacus rubecula*. J Comp Physiol A 184:295-299

Wiltschko W, Wiltschko R (2000): Light-dependent magnetoreception in birds: does directional information change with light intensity? Naturwissenschaften 86:36-40

Wiltschko W, Wiltschko R, Keeton WT, Madden R (1983) Growing up in an altered magnetic field affects the initial orientation of young homing pigeons. Behav Ecol Sociobiol 12:135-142

Wiltschko W, Munro U, Ford H, Wiltschko R (1993) Red light disrupts magnetic orientation of migratory birds. Nature 364:525-527

Wiltschko W, Munro U, Beason RC, Ford H, Wiltschko R (1994) A magnetic pulse leads to a temporary deflection in the orientation of migratory birds. Experientia 50:697-700

Wiltschko W, Munro U, Ford H, Wiltschko R (1998a) Effect of a magnetic pulse on the orientation of Silvereyes, *Zosterops l. lateralis*, during spring migration. J Exp Biol 201:3257-3261

Wiltschko W, Weindler P, Wiltschko R (1998b) Interaction of magnetic and celestial cues in the migratory orientation of passerines. J Avian Biol 29:606-617

Yorke ED (1979) A possible magnetic transducer in birds. J Theor Biol 77:101-105

VII.2 Electric Fields and Electroreception: How Electrosensory Fish Perceive Their Environment

Gerhard von der Emde

Institut für Zoologie, Universität Bonn, Poppelsdorfer Schloss, 53115 Bonn, Germany

Abstract

The electrosensory system of fishes is well matched to the conditions of the aquatic habitat, where electrical fields of biotic or abiotic sources are abundant. During passive electrolocation environmental low-frequency electrical fields are detected for orientation, navigation and prey detection. Weakly electric fish have perfected electro-orientation by developing active electrolocation. They actively emit high-frequency electrical signals with specialized electric organs. Objects are located by sensing the specific distortions they evoke on the electrical field which builds up during each electric organ discharge. The production of their own electric signals makes weakly electric fish independent of existing electrical fields and enables them to actively probe their environment. Several unique adaptations of the electrosensory system provide the means to detect and localize objects three-dimensionally in space and to analyze their electrical properties. The advantage of perceiving an electrically "colorful" picture of their surroundings led South American and African electric fishes to evolve capacitance detection, i.e., the ability to perceive capacitive object properties. Even though the fishes' sensory capabilities, which are the results of a convergent evolutionary process, look almost identical, South American and African fishes employ different receptor types and different neural algorithms to achieve them.

Key words Orientation, object detection, object properties, depth perception, weakly electric fish

1 Electric Fields in the Aquatic Environment

Animals use sensory signals of various modalities when gathering information about their environment. Besides visual signals, which play a dominant role in the sensory life of many species (including humans), signals of other physical qualities, such as acoustical, mechanical, chemical, and magnetic are also perceived. Furthermore, several fish species can sense electrical fields which are present in their natural aquatic habitats. When light conditions are poor, electric fields can still be used for orientation, navigation, and communication.

Aquatic electrical fields are of either abiotic or biotic origin. Abiotic fields are usually DC or low frequency fields with the exception of the high frequency fields caused by lightning (Hopkins 1999). Low-frequency abiotic fields can originate from geochemical sources, seismic activity, or the movement of water masses through the earth's magnetic field. Certain geological formations can cause electrical fields if they are connected through water, recalling DC batteries and forming a stationary pattern of electrical field lines. An electrosensitive organism swimming through this pattern will experience it as a low-frequency electrical field, depending on swimming speed and direction of movement (Kalmijn 1974).

Organisms living in the water are sources of biotic electrical fields. Muscle activity is accompanied by an electrical event at the muscle cell membrane, as well as by an electrical nerve potential. However, because of the multiple orientations of the many membranes over which electrical potentials are produced, the net electrical field outside an entire muscle or even a whole animal is extremely weak. Electrical fields produced by membrane-separated fluid compartments containing different ion compositions are probably more important than fields produced by muscles. An example are the gills of aquatic animals, which exchange oxygen, CO_2, and various ions between blood and water. Concentration differences of ions will lead to a net ionic flux across the thin gill membranes. Movements of gill covers, such as the operculum or the mouth, modulate these electrical fields in time at low frequencies. All other thin and therefore leaky barriers between internal body fluids and external media form additional sources of biotic electrical fields. When an animal is injured, the resulting ion movements may increase greatly and thus intensify the electrical field (Kalmijn 1974).

Most of the biotic electrical fields are DC or low frequency fields (0 to 30 Hz) similar to the abiotic fields. In many African and South American rivers, however, biotic electrical fields containing much higher frequencies are found as well. They originate from electric fishes, which have evolved specialized electric organs for the production of high frequency electrical signals, called electric organ discharges (EODs; Fig. 1). Electric organs have evolved, with one exception, from electrically excitable muscle tissue. The modified muscle cells are called electrocytes. They form regularly orientated stacks and are activated in synchrony. Thus, summation of the weak electrical potentials generated across the membranes of each single electrocyte results in a significant net electrical potential produced by the whole organ. Using their large electric organs, strongly electric fish living in freshwater habitats such as the South American electric eel can generate electrical potentials larger than 600 V. Strongly electric marine fish such as the electric ray *Torpedo* achieve voltages of only 60 V which are, however, accompanied by large electrical currents. In both fresh and marine waters, strongly electric fishes use their electric signals to stun or kill their prey and to defend themselves against predators (Moller 1995).

Electric organs of weakly electric fish produce EODs of only a few volts, which are undetectable for animals lacking specialized electroreceptors. These fish use them for orientation in their habitat and for electrocommunication. Many species of weakly electric fish have evolved in Africa and South America, where they live in tropical freshwater streams, rivers, and lakes. These aquatic habitats have several features in common. (1) They all contain low concentrations of ions resulting in a low electrical conductivity (between 10 and 150 μScm^{-1}), which favors the spatial spreading of electrical fields. (2) Water temperature often is rather high (between 25 and 35 °C). (3) The pH is low. Besides these similarities, different habitats vary considerably. The water may be fast-flowing or stagnant, the bottom may be sandy, rocky, or covered by decaying plant material. The visibility may be high or low because of turbid, muddy water. Water depth may be very shallow or rather deep. Great depths together with turbid waters result in habitats such as the Amazon River at depths below ca. 7 m with complete darkness all the time. In this diversity of habitats many different species of weakly electric fish have evolved in both Africa (more than 200 species) and South America (more than 100 species).

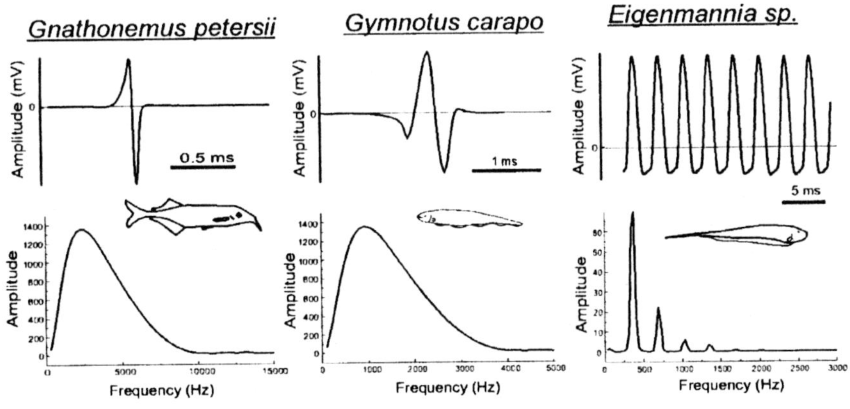

Fig. 1. Electric organ discharge wave forms (*upper traces*) of three species of weakly electric fish. *G. petersii* and *G. carapo* are pulse-fish from Africa and South America, respectively. *Eigenmannia* is a wave-species from South America. *Below each signal*, the FFT amplitude spectra of single EODs are shown

The variety of species and habitats is accompanied by a variety of electrical signals produced by different fish species (Fig. 1). EODs are species-specific and very stereotyped, but sometimes sexually dimorphic (Hopkins 1999). There exist two basic types of signals: (1) continuous, wave-type signals, which more or less resemble sinewaves, and (2) transient, pulse-type signals (Fig. 1). Wave-type signals are produced continuously. Interruptions are rare and serve certain purposes during electrocommunication. In contrast, each pulse-type EODs is followed by a pause which lasts longer than the signal itself.

The variety of EODs in different species of electric fish reflects the variety in habitat and life-style of these animals. Presumably, different types of signals are the result of an adaptation to different physical habitats and social structures. However, even within a certain habitat, demands on signal design for different purposes can be conflicting. For example, for electrocommunication it might be advantageous to produce high-amplitude and low-frequency electric signals. For active electrolocation of small objects (see below), signals containing higher frequencies are better suited. However, in order not to attract predators such as large catfishes or the electric eel (which use EODs for detecting their electrogenic prey), an electric fish should produce a low amplitude and high-frequency signal (Westby 1988; Stoddard 1999). Because of such conflicting constraints, signal evolution in weakly electric fish has been the result of a compromise satisfying the demands of electrolocation, electrocommunication, and predator avoidance. Interestingly, both wave-type and pulse-type EODs developed independently in Africa and South America. Presumably, each type of signal has certain advantages in a given ecological situation.

2 Sensing of Aquatic Electrical Fields

The ability to sense electrical fields appears to be a primitive vertebrate capacity (Bullock et al. 1983). Besides fishes, very few vertebrates, and no invertebrates are known to possess an electrical sense, which is always associated with an aquatic habitat. Among nonteleost fishes all taxa except two (the Myxiniformes and the Holostei) possess electroreceptor organs and related brain structures. The electroreceptor organs used by these species are called ampullary organs (Szabo and Fessard 1974) and they can be found in both freshwater and marine species. In marine elasmobranchs, the ampullary organs are called Ampullae of Lorenzini.

The common ancestor of all teleosts was not electroreceptive. For unknown reasons (New 1997), it lost its ability to detect weak electric fields. However, during teleost evolution, electroreception was "rediscovered" independently at least twice (maybe four times). In Africa, the mormyriforms, which produce high-frequency electric signals, and the nonelectrogenic African knifefish (Notopteriforms) developed an electric sense. Whereas the African notopteriforms possess only ampullary electroreceptor organs, the mormyriforms have, in addition, electroreceptors located in so-called tuberous organs. These receptors are sensitive to high-stimulus frequencies and thus can be used to detect the fish's own EODs as well as those of other fishes. In South America, gymnotiform weakly electric fish have developed electroreception independently of their African counterparts. Like mormyrids they possess both ampullary and tuberous electroreceptor organs. Catfishes (siluriforms), which are closely related to the gymnotiforms, are electroreceptive, but possess only low-frequency ampullary electroreceptor organs.

2.1 Passive Electrolocation: Sensing of Low-Frequency Electrical Fields

Because of their high sensitivity to low-frequency electric fields, ampullary electroreceptor organs are well suited for the detection of low-frequency environmental fields of abiotic or biotic origin. This passive electrolocation involves the detection of electrical fields which are not self-produced. Even though these natural fields are very low in amplitude, the ampullary electroreceptor organs of many fishes are sensitive enough to detect them. Skates, for example, respond reliably to voltage gradients of less than 0.01 μVcm^{-1}, which corresponds to a voltage of 1 V over 1000 km. The high sensitivity to low voltage gradients generally found in elasmobranchs enables these animals to detect electrical fields emanating from the gill epithelia of prey, such as flatfishes buried in the sand. Sharks have been found to be attracted by electrical dipole sources emitting prey like signals. Thresholds for feeding responses in the absence of any food odors are in the order of just 5 $nVcm^{-1}$ (Kalmijn 1974).

Elasmobranchs can also use electrical fields for orientation and navigation (Pals et al. 1982), either by detecting electrical fields of geochemical origin or by detecting magnetic fields indirectly through electrical induction. A moving water mass or a shark swimming through the earth's magnetic field induces an electrical current large enough to excite an elasmobranch's electroreceptors (Kalmijn 1974). Even though the exact mechanisms are debated (Paulin 1995), elasmobranchs use the information thus obtained to determine their magnetic compass heading. Likewise, freshwater fish use their ampullary electroreceptors for orientation. They detect voltage gradients in their habitat, which serve as electrical landmarks (Peters and Meek 1973).In addition to using tuberous receptors and active electrolocation (see below), weakly electric fish also employ their ampullary receptors for locating low frequency electric fields that originate, for example, from prey items. *Gnathonemus petersii* can find insect larvae (Chironomidae), its natural prey, by passive electrolocation, i.e., by detecting and evaluating electric gradients produced by the prey animal. Interestingly, passive electrolocation is just one of many senses that are used simultaneously by *G. petersii* in search for prey (von der Emde and Bleckmann 1998).

A hunting method based on electroreception is used by several South American predators of weakly electric fish. Catfishes and electric eels (Westby 1988; Moller 1995) prey on gymnotiforms by eavesdropping on their electrical discharges. Predator pressure by hunting catfishes, electric eels, and others might have caused several species of gymnotiforms, over evolutionary times, to shift the spectra of their EODs to higher frequencies in order to avoid detection (Stoddard 1999). Interestingly, throughout Peru (this is not true of other areas), wave fish dominate the large rivers while pulse fish are mainly found in smaller rivers. One of the reasons why pulse fish are rarely found in the large rivers may be the presence of catfishes, which detect the low frequencies present in the spectra of pulse-type EODs but not in those of wavetype signals (Mary Hagedorn, Kip Keller, pers. comm.).

2.2 Active Electrolocation: Sensing of Self-Produced Electrical Fields

The process of actively probing the environment by emitting an electrical signal and perceiving it with one's own tuberous electroreceptors was called active electrolocation by Lissmann and Machin (1958). The evolution of this ability constituted a qualitative step forward from the use of passive electrolocation. Passive electrolocation depends on existing electrical fields. In contrast, during active electrolocation, the fish provides its own source of electricity, using it to actively probe the environment and to detect and analyze objects even if these do not emit electrical energy.

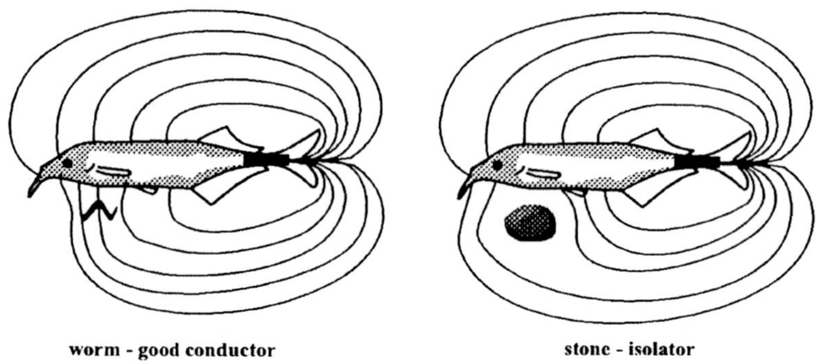

Fig. 2. Schematic drawings of the electric field generated by a *G. petersii* and distorted by a worm (good conductor, *left*) or by a stone (isolator, *right*). The fish is viewed from the side. Electrical field lines are depicted as *thin lines*. Even though they surround the fish three-dimensionally, only the vertical plane is shown here. The electric organ in the caudal peduncle is drawn in *black*, and the electroreceptive body surface of the fish in *gray*

African mormyrids emit brief electric pulses at a variable rate. The electrical current associated with each EOD flows through the water and builds up a three-dimensional electrical field around the fish's body (Fig. 2; Caputi et al. 1998). This electrical field lasts as long as an EOD, which may be only a fraction of a millisecond in some species. The current flowing through the water causes a voltage drop across the fish's skin that excites the tuberous electroreceptor organs, which are distributed over almost the entire body surface of the fish. Each discharge of the electric organ thus causes a locally occurring electric signal at each electroreceptor organ which serves as stimulus for the electroreceptors. The afferent nerve fibers innervating the receptor cells report the amplitude or the timing of the local signal to the brain. If an object with electrical properties different from those of the water comes close to a fish, it will distort the electrical field lines (Fig. 2). As a result, the voltage pattern at skin regions opposite the object will change and the affected electroreceptors will experience an altered local signal. The area on the skin surface thus affected by an object is called its electric image (Rasnow 1996; Caputi et al. 1998; von der Emde et al. 1998).

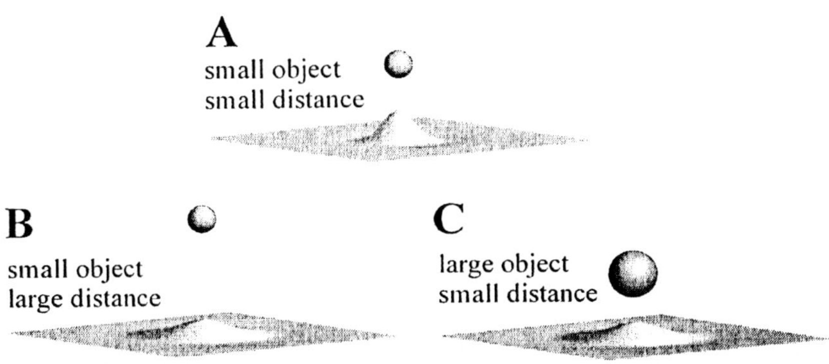

Fig. 3 A-C. Visual representation of electric images of spheres of different sizes at different distances from a surface, which represents the fish's skin. An amplitude change is represented on the vertical axis. An amplitude increase is shown as a evaluation from the plane, while an amplitude decrease is depicted as a dip. The images have a 'Mexican hat' profile (Caputi et al. 1998). In **A**, a small metal sphere is located close to the plane. The resulting amplitude increase is rather large. **B** The same small object located further away leads to an increase in image width and to a smaller amplitude increase in the image center. However, a very similar image as in **C** is produced by a close-by, larger sphere, which has a higher resistance than the small object. Despite the similarities of the images shown in B and C, the fish has no problem in measuring distances even of an unknown object

Different types of objects produce different types of images. A good conductor such as a rock containing metal inclusions or a water plant will raise the stimulus amplitude in the center of its image and diminish it in a small area at the image's edge. When plotted three-dimensionally, the image has a Mexican hat profile (Fig. 3). An insulator, like most stones, evokes an opposite type of image: throughout the image center the amplitude decreases while it increases in the image's rim area. In addition to amplitude changes, a change in timing of the local signals can also be evoked by an object. Objects with complex impedances due to a capacitive component cause phase shifts of the local signal within the electric image (Caputi et al. 1998).

3 Detection of Object Properties by Active Electrolocation

Lissmann and Machin (1958) proved that an electric fish can detect and discriminate between objects of different electrical properties like a glass rod and a metal rod. Glass is an electrical insulator and thus, if positioned close to an electrolocating fish, will lead to a decrease in voltage at electroreceptors in the fish's skin compared to the situation without the object present. In contrast, a metal rod will increase the local voltage perceived by the electroreceptors. According to Lissmann and Machin (1958), fish measure the local amplitude of their self-produced signals and discriminate between objects on the basis of amplitude changes. Each object is thought to project a black-and-white image onto the fish's skin, like an electrical shadow of varying intensity.

When investigating objects found in the natural habitat of African weakly electric fish, it turned out that inanimate objects, such as stones or dead plant material, mainly have Ohmic resistances. The impedance of these objects is not complex, i.e., its electrical resistance is identical for a DC signal and an AC signal such as an EOD. For an electrolocating fish this means that all frequencies of its EOD are similarly affected and no phase shifts occur. It follows that a resistive object changes only the local EOD amplitude but not its wave form. In contrast, all living matter possesses a complex impedance consisting of considerable capacitive and a resistive (Ohmic) components (Schwan 1963; Heiligenberg 1973; von der Emde 1990). Capacitive properties are particularly prominent in living objects such as plants, other fishes, or insect larvae (which constitute the main food items of African electric fish) presumably because of the many thin membranes separated by electrolytes. Capacitive objects affect the locally perceived EOD in several ways (von der Emde 1990): (1) The peak-to-peak amplitude of the EOD changes depending on the value of the impedance of the object. (2) The spectral composition of the EOD is altered because the lower frequencies within the EOD are attenuated more strongly by a capacitive load than the higher frequencies. (3) The wave form of the local EOD is distorted due to a frequency-dependent phase shift of the EOD. In the wave form of a *G. petersii* EOD, the relative amplitudes of the two phases change, and a third phase may occur (Fig. 5). Because capacitive properties occur mainly in living objects, capacitance detection might be particularly useful to electric fish in identifying food items or other fishes. In order to better understand how weakly electric fish perceive these capacitive object properties, a line of research was initiated in my laboratory.

First, we wanted to find out whether weakly electric fish can perceive capacitive object properties. We used a behavioral paradigm in which fish were trained to discriminate between a capacitive and a resistive object. We utilized so-called dipole-objects for these experiments that were made out of plastic tubing into which carbon electrodes were inserted (von der Emde 1990). The advantage of these objects is that their electrical properties can be changed quickly from outside the aquarium and they can thus be bestowed with any desired capacitive or resistive value. With their help we determined which range of capacitive values an electric fish can detect. It turned out that the range of detectable capacitances is species-specific: it depends on the frequency composition of the electric signal which the fish emits (see also Meyer 1982). Small fishes, which emit very short pulses that contain high frequencies (e.g., *Pollimyrus adspersus*), detect small but not large capacitances. Larger fishes that emit EODs of longer duration (e.g., *Mormyrus rume*) are superior in detecting larger capacitance. Fish of intermediate size, like *G. petersii,* which emit a signal of 400 μs duration, can detect an intermediate range of capacitances. In all cases, the detectable range of capacitive values included the values typical of many natural capacitive (living) objects. Natural capacitances range from a few nF, e.g., in small worms, to several tens of nF in larger living objects, such as leaves of water plants or other fishes. The smaller the object, the smaller its capacitance. This may explain why small fish, that can eat only small prey, produce short pulses that allow them to detect small capacitances.

3.1 Capacitance Detection in African Mormyrid Fish

After having determined the detectable range of capacitance of different fish species, it remained to be shown that mormyrids can perceive capacitive object properties separate from resistive properties. The problem is that both capacitive and resistive objects change the peak-to-peak amplitude of the local signal (see above). If the fish just measure the EOD amplitude, as suggested by Lissmann and Machin (1958), they should not be able to discriminate between a capacitive and a resistive object with the same impedance value.

Additional training experiments clarified this point. None of our trained fish ever had problems discriminating between a fixed capacitive object and a resistive object, whose value was changed from one testing session to the next (Fig. 4A). Even if the impedances of the two objects were numerically identical (which lead to an identical EOD amplitude change as perceived by the electroreceptors), each fish discriminated between them as quickly and accurately as it discriminated between objects of completely different impedance values. Mormyrids obviously perceive capacitive objects as objects having a quality different from resistive objects. Psychophysical multidimensional scaling experiments further showed that *G. petersii* can measure capacitive and resistive properties of natural objects quantitatively and independently from one another (von der Emde 1990; von der Emde and Ringer 1992).

The ability to detect capacitive object properties through active electrolocation was termed capacitance detection. In analogy to color vision, capacitive objects are perceived as having an electrical color. Capacitance detection thus colors the electrical shadow that is projected onto the fish's skin, according to Lissmann (1958). The sensory images an electric fish gathers from its surroundings at night are not like black-and-white pictures but they contain much more information about object properties than originally thought. Imagine, for example, a *G. petersii* hunting for insect larvae on the ground of a river at night. Small living prey items have electrical color, while the resistive background appears gray. As a consequence, prey stands out much more clearly and can be detected more easily by electrolocating fish (von der Emde and Bleckmann 1998).

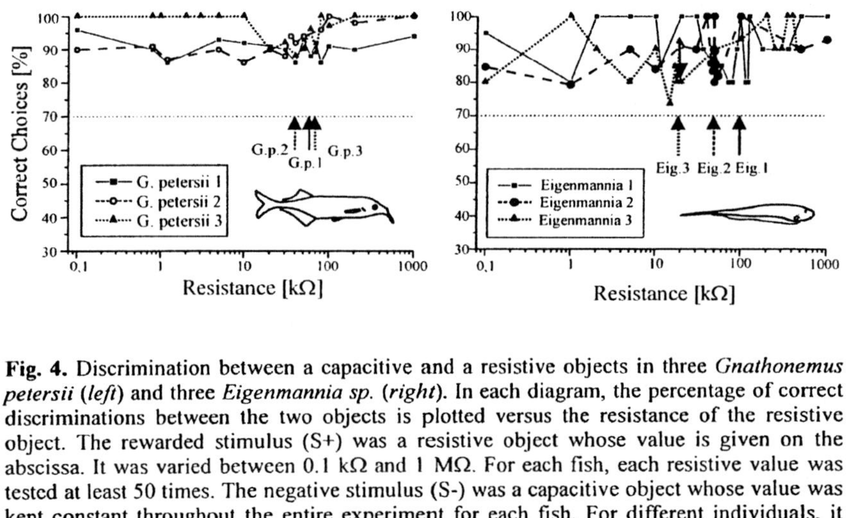

Fig. 4. Discrimination between a capacitive and a resistive objects in three *Gnathonemus petersii* (*left*) and three *Eigenmannia sp.* (*right*). In each diagram, the percentage of correct discriminations between the two objects is plotted versus the resistance of the resistive object. The rewarded stimulus (S+) was a resistive object whose value is given on the abscissa. It was varied between 0.1 kΩ and 1 MΩ. For each fish, each resistive value was tested at least 50 times. The negative stimulus (S-) was a capacitive object whose value was kept constant throughout the entire experiment for each fish. For different individuals, it had values between 1 and 20 nF. Each single *point* in the diagrams gives the percentage of correct discriminations between the capacitive object and a resistive object whose resistance is given on the *abscissa*. *Arrows* mark the calculated impedance values of the capacitive objects for each fish. The *dotted lines* mark the 70% correct level (threshold). Note that each fish's discrimination performances was above threshold for all object combinations. That means that each fish could discriminate between the capacitive object and every resistive object

How does capacitance detection work? Obviously, it is not enough for an electric fish to measure only the local EOD amplitude, because this can be identical for several different resistive and capacitive objects. We found that mormyrids measure an additional parameter, which is the local EOD *wave form* that is affected only by capacitive but not by resistive objects (Fig. 5A).

In order to find out whether wave form changes do influence electroreceptor responses, we recorded electrophysiologically from the afferent nerve fibers which innervate single electroreceptor organs. The tuberous organs of mormyrids that are used for active electrolocation are called mormyromasts (Szabo and Fessard 1974). They contain two morphologically different types of electroreceptor cells (A-type and B-type receptors), which project to two different brain areas. Both cell types respond to EOD amplitude changes in a similar way: A burst of action potentials is evoked in the afferent nerve fiber by each EOD. Increasing stimulus amplitude causes bursts with more spikes which are fired at a shorter latency. For more than 30 years only artificial squarewave or sinewave stimuli were used in electrophysiological experiments. It remained unclear, therefore, why each mormyromast organ contains two different types of receptor cells and how they differ physiologically (Bell 1990).

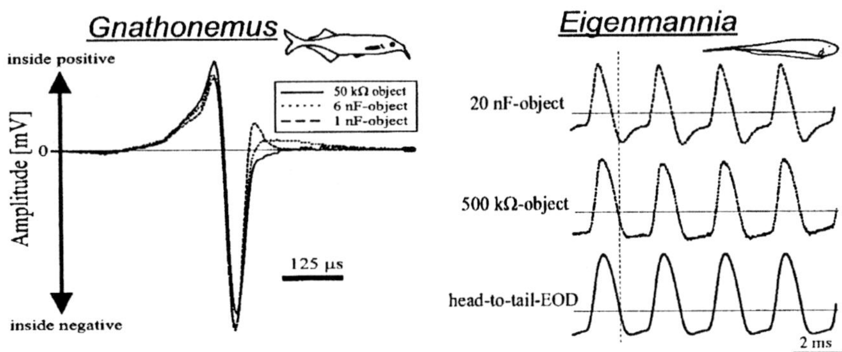

Fig. 5. Effect of different types of objects on the local EODs of *G. petersii* (*left*) and *Eigenmannia* (*right*). The objects were placed at a distance of 1 cm from the lateral side of the fish. Local EODs were recorded differentially with two small electrodes placed close to the pore of an electroreceptor organ between the object and the fish's skin. *Left* The solid line represents the local EOD recorded in the presence of a 50 kΩ object. Its wave form is identical to the simultaneously recorded head-to-tail EOD. *Dotted and dashed lines* show EODs in the presence of a 6 nF and a 1 nF capacitive object. The peak-to-peak amplitudes of all EODs shown were normalized to the same values. *Right Upper trace* shows the local EOD recorded in the presence of a 20 nF capacitive object, middle trace in the presence of a 50 kΩ resistive object. The *lower trace* shows the simultaneously recorded head-to-tail EOD. EOD amplitudes were normalized. *Dotted lines* indicate 0 mV. The *vertical dashed line* marks the timing of the zero crossing of the positive-negative transient of the head-to-tail EOD. Note that *G. petersii's* EODs are wave form distorted by capacitive objects while timing cues are affected only minimally. In contrast, large time shifts are induced in the EOD of *Eigenmannia* by the 20-nF object

Our experiments finally resolved this puzzle. We used natural EOD stimuli, which were prerecorded in the presence of several types of capacitive and resistive objects. The objects caused predictable distortions of the natural signal wave form. When these stimuli were played back to single mormyromast receptor organs, A- and B-cells responded differently. Responses of A-cells were not much affected by wave form distortions. B-cells, on the other hand, were extremely sensitive to even the smallest wave form changes. They responded as if the signal amplitude had increased strongly, even though it had been kept constant throughout the whole experiment. In summary, both receptor cell types responded to changes in signal amplitude but only the B-cells additionally responded to natural changes in signal wave form (von der Emde and Bleckmann 1997).

These results led us to develop a hypothesis about how the central nervous system of a mormyrid performs capacitance detection. In the presence of a purely resistive object only amplitude changes and no wave form distortions occur. When the brain compares A- and B-receptor responses, e.g., by subtracting the information they provide, it will detect no difference between the two. Now consider the result with a capacitive object. Such an object will change stimulus amplitude, but in addition it will distort the wave form of the stimulus. The A-receptors still respond only to stimulus amplitude. B-receptors, on the other hand, are strongly excited by the wave form distortions and therefore respond much stronger than the A-cells. A subtraction of A- and B-receptor responses by the brain will reveal these differences, which are proportional to the amount of wave form distortion and thus proportional to the capacitive value of the object. This mechanism would allow mormyrids to measure capacitive object properties quantitatively and therefore could explain our behavioral results. Recently obtained electrophysiological results suggest that our proposed subtraction mechanism might indeed exist in the lateral nucleus of the torus semicircularis of the *G. petersii* brain (von der Emde and Mohr, in prep.).

3.2 Capacitance Detection in South American Electric Fish

Apart from the fact that electric organs, electroreceptors and the electroreceptive brain structures of African mormyriforms and South American gymnotiforms developed separately, they are remarkably similar. This may relate to two main factors: (1) Both groups used preadaptations when developing electric organs (from muscle tissues), electroreceptors (primitive ampullary receptor organs and mechanoreceptive neuromasts), and corresponding brain structures (the octavo-lateralis system). The properties of these existing older structures set the tracks along which electrogenic and electrosensory structures of both fish groups developed during evolution. (2) Electric signals work best for orientation and communication in tropical, freshwater habitats with water of low electrical conductivity. The environmental conditions under which mormyriforms and gymnotiforms evolved their electric sense resembled each other, forming similar evolutionary pressures.

Can South American electric fish also detect capacitive object properties? In order to answer this question we trained several species of gymnotiforms, which emit either pulse-type signals (*Brachyhypopomus pinnicaudatus; Gymnotus carapo*) or continuous wave type EODs (*Eigenmannia sp.*), using the same method as we had used for the mormyrids. The results were clear: all individuals of each species tested discriminated between a capacitive object and any resistive objects (Fig. 4B). Even if the impedances of the two objects were identical, they had no difficulties in discriminating. Thus gymnotiforms, like mormyriforms, can perform capacitance detection (von der Emde 1998). The similarities in the behavior of the two groups of electric fish are, however, not reflected by the sensory mechanisms employed. As in mormyrids, the local signals stimulating an electrolocating gymnotiform fish are distorted when the fish approaches a capacitive object. However, the resulting changes in signal timing (the timing of the positive-negative zero crossing) are much larger than those occurring in mormyrid EODs. While in the latter the timing changes maximally by 3 μs, it changes by almost 200 μs in *Eigenmannia* (Fig. 5B).

In contrast to mormyrids, gymnotiforms do not possess mormyromasts with A- and B-type receptor cells, but instead have tuberous electroreceptor organs with a single population of receptor cells. In *Eigenmannia*, one type of such organ (P-type organs) responds only to EOD amplitude. The other type, called T-type receptor organs (t for timing), respond only to EOD timing, independent of stimulus amplitude (Scheich et al. 1973). When signal timing changes because of the proximity of a capacitive object, the t-type receptor organs report these changes to the brain. T-receptors in other skin areas, which lie outside the electric image of that object, do not experience a change in local signal timing. They continue to report the original EOD timing to the brain. Neurons in the torus semicircularis compare signal timing reported by t-receptors from different body regions (Heiligenberg 1991) and thus respond to the presence of a capacitive object.

In order to detect changes in signal wave form, the mormyrid brain compares A- and B-receptor responses from single mormyromast organs located at a single spot on the skin. In these fishes, changes in signal timing are too small and therefore of no importance. In contrast, *Eigenmannia* compares t-receptor responses originating from different skin locations. In this species, timing changes are the relevant cues while wave form changes most likely do not play a role. Thus, very similar-looking sensory capabilities in the two groups of weakly electric fishes are based on completely different sensory mechanisms (von der Emde 1998).

4 Three-Dimensional Location of Objects

For a long time it was thought impossible to localize an object in three-dimensional space using active electrolocation. Instead, it was assumed that the electric images provide ambiguous information when a fish has to determine object distance without knowing object size, impedance or shape. Figure 3 illustrates these ambiguities by showing the electric images of two small metal spheres located at different distances from the skin of a *G. petersii*.

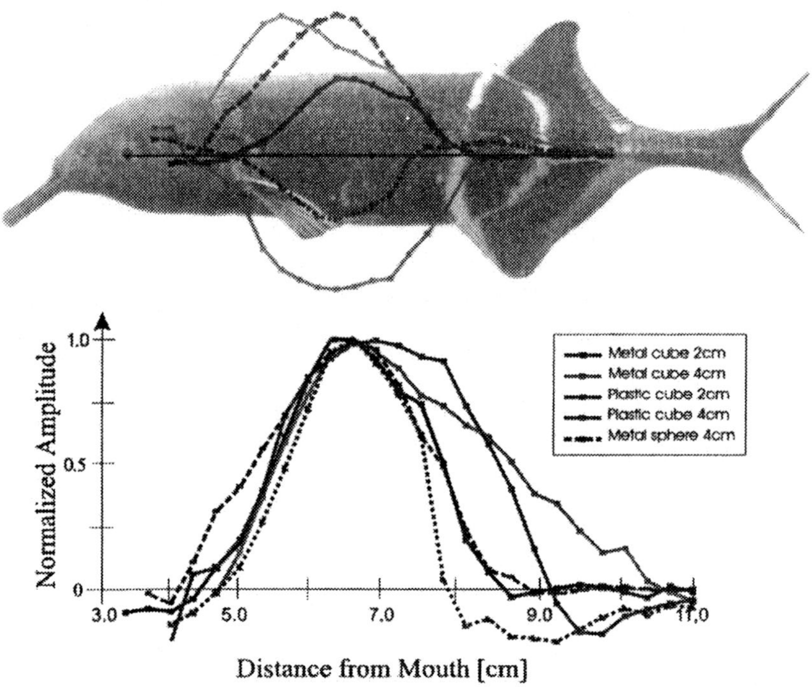

Fig. 6. *Above* Lateral view of a *G. petersii* with electric images superimposed at the position above which the objects were placed. *Below* Normalized electric images obtained from data shown above. For electric image measurements, five different objects were placed 12 mm from the fish's skin surface. Electric images were measured along the midline of the fish by moving a differentially amplified pair of electrodes from the fish's mouth towards its tail. In the *upper diagrams*, the modulation of the EOD peak amplitude (*ordinate*) caused by the object's presence is plotted versus position along the fish's body (0 cm = mouth opening). The *lower diagrams* were normalized by setting the maximal amplitude to 1 and shifting the curves horizontally to approximately align rostral (*left*) slopes. Note that rostral slopes of all cubeimages are similar, while the slope of the sphere is less steep

When the sphere is close, the electric image is small and the amplitude in the center region is high (Fig. 3A). If the same sphere is located at a larger distance, image width increases and image amplitude decreases (Fig. 3B). However, an image very similar to the one evoked by the faraway small sphere can be produced by a larger sphere of a higher resistance located at a closer distance (Fig. 3C). How can a fish distinguish between these two possibilities?

We considered it very unlikely that electric fish would not possess a means of determining object distance. Fish have to rely on active electrolocation during their nocturnal lives. They have to find food, locate obstacles, determine their distances to other fishes, and many similar tasks, all of which appear to be impaired without some means of distance determination. Therefore, we conducted behavioral training experiments to resolve this issue.

Individual *G. petersii* were trained to discriminate between two objects, each placed at a different distance behind a gate in the training tank. In order to obtain a food reward, the fish had to swim through that gate behind which the corresponding object was placed further away, compared with the object behind the other gate. First, we used two identical objects for training and fish learned in only 2 weeks to choose the correct object, no matter how large the absolute distance to the objects was. However, this performance did not demonstrate that fish could measure distance. With two identical objects, distance discrimination could have been based on pure amplitude or image width measurements.

In a second step, we presented two objects of different sizes or shapes or made of different materials. Now fish could no longer solve the distance discrimination task by measuring just image amplitude. Surprisingly, however, the fish needed no extra training to correctly measure the distances of these new objects. They immediately continued to choose the object located further away, as long as a certain threshold value of interobject distances was exceeded. This demonstrated that *G. petersii* can indeed measure object distance and that distance discrimination is independent of object size, shape or impedance (von der Emde et al. 1998).

To find the mechanism which the fish use for distance perception, we measured the electric images that our training objects projected onto the fish's skin (Fig. 6). We searched for image parameters that correlated only with object distance and were independent of other object properties. Ultimately, we found such a parameter combination: the ratio of the amplitude decline of the image amplitude at the image's edge (the slope of the image) and the maximal image amplitude in the center. The slope is a measure of the fuzziness of the image, or of how much the image is in focus. The slope/amplitude ratio decreases if an object moves away, and it does not depend (with one exception) on the type of object under investigation.

The electric image measurements also revealed that metal spheres yielded slope/amplitude ratios that were smaller than those of other objects placed at the same distance. Because smaller slope/amplitude ratios correspond to larger object distances, we hypothesized that if the fish indeed used this ratio to determine object distance, then they should judge spheres to be located further away than cubes, even if their actual distances were identical. Behavioral experiments exactly confirmed this prediction. The fishes made mistakes when discriminating between a sphere and a cube. They fell victim to an electrical illusion, which led them to misjudge the distance of the sphere (von der Emde et al. 1998).

The use of the slope/amplitude ratio constitutes a newly found mechanism of depth perception by animals. It enables electric fish to measure object distance with a single and stationary receptor surface. This mechanism is unique because all other mechanisms discovered so far employ either two receptor surfaces (e.g. the two retinae of our eyes) or movement of a single receptor surface. The mechanism is also unknown in human technology and could provide new solutions for distance determination by human-made sensors.

References

Bell CC (1990) Mormyromast electroreceptor organs and their afferent fibers in mormyrid fish. III. Physiological differences between two morphological types of fibers. J Neurophysiol 63: 319-332

Bullock TH, Bodznick DA, Northcutt RG (1983) The phylogenetic distribution of electroreception: Evidence for convergent evolution of a primitive vertebrate sense modality. Brain Res Rev 6: 25-46

Caputi AA, Budelli R, Grant K, Bell CC (1998) The electric image in weakly electric fish: Physical images of resistive objects in *Gnathonemus petersii*. J Exp Biol 201: 2115-2128

Heiligenberg W (1973) Electrolocation of objects in the electric fish *Eigenmannia* (Rhamphichthyidae, Gymnotoidei). J Comp Physiol 87: 137-164

Heiligenberg W (1991) Neural Nets in Electric Fish. MIT Press, Cambridge, London

Hopkins CD (1999) Design features for electric communication. J Exp Biol 202: 1217-1228

Kalmijn AJ (1974) The detection of electric fields from inanimate and animate sources other than electric organs. In: Fessard A (ed) Handbook of Sensory Physiology. Springer Verlag, Berlin, pp 148-200

Lissmann HW, Machin KE (1958) The mechanism of object location in *Gymnarchus niloticus* and similar fish. J Exp Biol 35: 451-486

Meyer JH (1982) Behavioral responses of weakly electric fish to complex impedances. J Comp Physiol 145: 459-470

Moller P (1995) Electric Fishes. History and Behavior. Chapman & Hall, London

New JG (1997) The evolution of vertebrate electrosensory systems. Brain Behav Evol 50: 244-252

Pals N, Peters C, Schoenhage AAC (1982) Local geo-electric fields at the bottom of the sea and their relevance for electrosensitive fish. Netherlands J Zool 32: 479-494

Paulin MG (1995) Electroreception and the compass sense of sharks. J theor Biol 174: 325-339

Peters RC, Meek J (1973) Catfish and electric fields. Experientia 29: 299-300

Rasnow B (1996) The effects of simple objects on the electric field of *Apteronotus*. J Comp Physiol A 178: 397-411

Scheich H, Bullock TH, Hamstra RHJ (1973) Coding properties of two classes of afferent nerve fibers: High-frequency receptors in the electric fish, *Eigenmannia*. J Neurophysiol 36: 39-60

Schwan HP (1963) Determination of biological impedances. In: Nastuk WL (ed) Physical Techniques in Biological Research. Academic Press, New York, pp 323-407

Stoddard PK (1999) Predation enhances complexity in the evolution of electric signals. Nature 400: 254-256

Szabo T, Fessard A (1974) Physiology of Electroreceptors. In: Fessard A (ed) Handbook of Sensory Physiology. Springer, Berlin, Heidelberg, New York, pp 59-124

von der Emde G (1990) Discrimination of objects through electrolocation in the weakly electric fish, *Gnathonemus petersii*. J Comp Physiol A 167: 413-421

von der Emde G (1998) Capacitance detection in the wave-type electric fish *Eigenmannia* during active electrolocation. J Comp Physiol A 182: 217-224

von der Emde G, Bleckmann H (1997) Wave form tuning of electroreceptor cells in the weakly electric fish, *Gnathonemus petersii*. J Comp Physiol A 181: 511-524

von der Emde G, Bleckmann H (1998) Finding food: Senses involved in foraging for insect larvae in the electric fish, *Gnathonemus petersii*. J Exp Biol 201: 969-980

von der Emde G, Ringer T (1992) Electrolocation of capacitive objects in four species of pulse-type weakly electric fish. I. Discrimination performance. Ethology 91: 326-338

von der Emde G, Schwarz S, Gomez L, Budelli R, Grant K (1998) Electric fish measure distance in the dark. Nature 395: 890-894

Westby GWM (1988) The ecology, discharge diversity and predatory behaviour of gymnotiform electric fish in the coastal streams of French Guiana. Behav Ecol Sociobiol 22: 341-354

Index

Bold numbers refer to figures or tables.

Printed in the United States
27928LVS00002B/6